LYNN E.H. TRAINOR is a member of the Department of Physics at the University of Toronto.

MARK B. WISE is a PhD candidate in physics at Stanford University.

The text takes an innovative approach to theoretical physics. It surveys the field in a way that emphasizes perspective rather than content per se, and identifies certain common threads, both conceptual and methodological, which run through the fabric of the subject today. Starting from recognized physical concepts, it demonstrates how these have led to the mathematical structures used extensively in physical theory and displays the unity of basic physics seen from the viewpoint of the symmetries of nature as manifested in the tensorial properties of physical quantities and physical laws.

The text focuses particularly on the linearity principle. A consideration of the description of physical events as viewed from different reference frames leads on to a discussion of the concepts of tensors and tensor fields as well as the theory of groups and group representations. As the concept of tensors is developed and broadened into multidimensional spaces, examples from elasticity theory, quantum mechanics, and the structure of elementary particles are considered. The nature of linear theories is illustrated repeatedly, but special attention is given to quantum mechanics and special relativity. The text also introduces non-linear geometry, in the sense of both Gauss and Riemann, and considers Riemannian geometry more extensively, as a prelude to a brief introduction to general relativity. The ten main chapters contain references, footnotes, and problems to assist the reader to pursue any topic further.

The volume can serve as a text or supplementary text in advanced undergraduate or graduate programs in theoretical physics and should also prove of interest to practising physicists, mathematicians, and theoretical chemists and biologists.

MATHEMATICAL EXPOSITIONS

Volumes Published

1 The Foundations of Geometry G. DE B. ROBINSON
2 Non-Euclidean Geometry H.S.M. COXETER
3 The Theory of Potential and Spherical Harmonics W.J. STERN-BERG and T.L. SMITH
4 The Variational Principles of Mechanics CORNELIUS LANCZOS
5 Tensor Calculus J.L. SYNGE and A.E. SCHILD
6 The Theory of Functions of a Real Variable R.L. JEFFERY
7 General Topology WACLAW SIERPINSKI (translated by C. CECILIA KRIEGER) (out of print)
8 Bernstein Polynomials G.G. LORENTZ (out of print)
9 Partial Differential Equations G.F.D. DUFF
10 Variational Methods for Eigenvalue Problems S.H. GOULD
11 Differential Geometry ERWIN KREYSZIG (out of print)
12 Representation Theory of the Symmetric Group G. DE B. ROBINSON
13 Geometry of Complex Numbers HANS SCHWERDTFEGER
14 Rings and Radicals N.J. DIVINSKY
15 Connectivity in Graphs W.T. TUTTE
16 Introduction to Differential Geometry and Riemannian Geometry ERWIN KREYSZIG
17 Mathematical Theory of Dislocations and Fracture R.W. LARDNER
18 n-gons FRIEDRICH BACHMANN and ECKART SCHMIDT (translated by CYRIL W.L. GARNER)
19 Weighing Evidence in Language and Literature: A Statistical Approach BARRON BRAINERD
20 Rudiments of Plane Affine Geometry P. SCHERK and R. LINGEN-BERG
21 The Collected Papers of Alfred Young, 1873-1940
22 From Physical Concept to Mathematical Structure: An Introduction to Theoretical Physics LYNN E.H. TRAINOR and MARK B. WISE
23 Lattice Path Combinatorics with Statistical Applications T.V. NARAYANA

LYNN E.H. TRAINOR and MARK B. WISE

From Physical Concept to Mathematical Structure: An Introduction to Theoretical Physics

MATHEMATICAL EXPOSITIONS NO. 22

UNIVERSITY OF TORONTO PRESS
Toronto Buffalo London

Toronto Buffalo London

Printed in Canada

Library of Congress Cataloging in Publication Data

Trainor, Lynn E.H., 1921-
From physical concept to mathematical structure.

(Mathematical expositions; no. 22 ISSN 0076-5333)
Includes bibliographical references and index.
1. Mathematical physics. I. Wise, Mark B., 1953- joint author.
II. Title. III. Series.
QC20.T68 530.1'5 78-11616
ISBN 0-8020-5432-3

To Rose and Helen

and in memory of Aline and Reg

Contents

Preface

For several years one of the authors (L.T.) has given a course on concepts and structures in physical theory to a select group of students in the senior years of specialist programs in mathematics and theoretical physics at the University of Toronto. The popularity of this course and the interest in the point of view taken provided the encouragement for expanding the lecture notes into a text that might be suitable for an introductory course in theoretical physics at the senior undergraduate level at other universities. Reaction from colleagues suggests that the book might also serve as a useful reference text even at the postgraduate level, both because the approach is different from that taken in other books and also because an attempt is made to make abstract concepts in several specialized fields of study more broadly accessible.

Although the present work treats in some detail concepts and structures embracing wide areas of theoretical and mathematical physics, it does not cover any single area comprehensively. The emphasis is, instead, on presenting the overall unity of the conceptual structures used. To some extent the objective is to whet the appetite of interest and curiosity rather than to satisfy or attenuate it. For the reader interested in pursuing specialized matters at greater depth, guidance is given in the form of footnotes and general references at the end of each chapter. Many of the problem sets have, indeed, been designed to motivate the student to pursue specialized topics further than is possible in the text itself.

Perhaps the classic work of Joos, Theoretical Physics,

represents the last serious attempt to present in one volume a reasonably comprehensive and unified treatment of all major areas of theoretical physics. The range and depth of new material that has emerged during the last half-century makes such an undertaking now difficult if not totally impossible. Another classic work, that of Morse and Feshbach, still serves a useful purpose, but one concerned more with covering the kinds of mathematical methods used than with exploring the various fields of theoretical physics themselves.

The present work shares neither the ambitions of Joos nor the objectives of Morse and Feshbach. It does not treat theoretical physics as a wide range of subject areas, nor is it a comprehensive account of modern mathematical techniques and methodology. Rather the emphasis is on a way of thinking about theoretical physics and the development of a perspective for young minds oriented toward a life of scientific enquiry. It is incidental to some degree that in the process the student does enrich his or her knowledge concerning modern discoveries and some of the details of their application. From a less philosophic point of view, the book is intended to fulfil two objectives. First, it attempts to introduce advanced concepts in theoretical physics at an early stage in the students' development and in a context which emphasizes perspective rather than content per se. Its second objective is to identify certain common threads, both conceptual and methodological, which run through much of the fabric of modern theoretical enquiry.

One of these threads is what we might loosely call the linearity principle. Whether by conscious design or by subconscious orientation, theoretical physics in the present era is characterized by attempts to see everything, in so far as possible, in a linear framework. Perhaps this orientation has been the inevitable outcome of using large computers. In any event the linearity principle is fulfilled in the choice made of physical constructs, usually the elements of a linear vector space; it also shows up in attempts to linearize the operations performed on such constructs, whether these operations involve transformation equations or the actual dynamics of systems of particles. To the extent they are successful, linear theories are comprehensible and aesthetically appealing; examples are afforded by special relativity and quantum mechanics. When linearization fails, mechanical and conceptual difficulties abound, as in the hydrodynamical equations and the equations of general relativity.

The technique used in the text is to raise very simple and basic questions, such as how observers in Toronto and Warsaw can compare results obtained from more or less identical experiments when they are using different coordinate frames for the description of physical events. Although such questions seem simple, they have far-reaching consequences and lead quickly and directly to the concepts of tensors and tensor fields on the one hand, and to the theory of groups and group representations on the other. Eventually a point is reached in the development when these conceptual structures merge in the sense that the symmetry classes of tensors characterize the irreducible representations of the transformation groups of interest in many areas of theoretical physics.

As the concept of tensors is developed and broadened into multidimensional spaces and eventually into non-linear geometries, examples from well-developed areas of theoretical physics are used, such as elasticity theory, quantum mechanics, and the structure of elementary particles. The nature of linear theories is illustrated time and again, but special attention is given to quantum mechanics and special relativity as the two most basic conceptual frameworks underlying modern developments in the theory of natural phenomena. The third basic concept, associated as much with gravitation as with relativity, namely Einstein's attempt to geometrize dynamics in the theory of general relativity, is presented as a counterexample, in principle, to the linearization process. Nonetheless, even here one has local linearization in tangent spaces, and global linearization where the density of matter is small.

The format is briefly as follows. Cartesian tensors and

transformation groups are introduced in Chapter 2 in the framework of coordinate transformations in two-dimensional space. In Chapter 3, extension is made to three-dimensions and the reader is brought to a fuller appreciation of the complexity of three-dimensional space in terms of both the existence of spinors and the double connectivity of the rotation group parameter space. In Chapter 4 some earthy examples of the tensor formulation of physical laws are given, principally the application of tensors to the theory of linear deformations, of elastic waves, and of hydrodynamic flow.

Chapter 5 develops the theme begun in Chapter 2, that tensors provide a fundamental approach to group representation theory, the symmetry classes of tensors forming the bases of irreducible representations of the matrix groups.

In Chapter 6 the schema of quantum mechanics is displayed as an example of a highly developed linear theory. The intention is not to teach the subject of quantum mechanics per se, but rather to integrate its structural form into the general flow of ideas established in the text. Students who have not been introduced to the history and phenomenology of quantum mechanics have a choice of doing supplementary reading or of omitting Chapter 6 in its entirety.

In Chapter 7 the horizons are expanded by introducing non-linear geometry, first in the sense of Gauss and then of Riemann, and the new geometry is illustrated by the example of curvilinear coordinates. A retreat is then made in Chapter 8 to another linear theory, the special theory of relativity, but the perspective is now broadened to enable one to make use of the non-Euclidean Lorentz metric in four-dimensional space-time. Quantum theory and special relativity, which provide the foundations for our present understanding of elementary particles and fields, are treated on a semi-postulational basis in an effort to achieve a desirable economy of space in the book. Chapter 9 considers the classification of elementary particles as an illustration of the significance of symmetry principles in the nature and organization of the physical universe.

In Chapter 10 we return to a more extensive study of Riemannian geometry as a prelude to a brief introduction to general relativity in Chapter 11. General relativity provides a beautiful example of a highly non-linear theory with important consequences for the large-scale structure of the universe.

Finally in Chapter 12, 'Horizons,' we take a brief look at modern conceptual developments in theoretical physics, particularly the new perspectives associated with innovative and frontier fields, such as molecular biology, elementary particle physics, and cosmology.

ACKNOWLEDGMENTS

We are indebted to many people who have expressed an interest in this manuscript or who have contributed to its final preparation. In particular we wish to thank J.R. Vanstone of the Mathematics Department, who gave valuable assistance in regard to certain aspects of Riemannian geometry, and C.J. Lumsden of the Physics Department, whose interest and enthusiasm gave us special inspiration along the way. Several secretaries contributed to the preparation of a difficult manuscript. We would especially like to thank Wendy Ross, who typed most of the first version, and Amy Parry, who prepared the final manuscript for publication.

Section 3 of Chapter 4 is based largely on an unpublished manuscript of C.J. Lumsden drawing attention to the close connection between hydrodynamical theory and the theory of small deformations.

Finally, we wish to express thanks for assistance and support from a number of organizations and agencies: the National Research Council of Canada, the Physics Department at the University of Toronto, Imperial Oil, and the Samuel Beatty Foundation. The kind cooperation of Professor Gilbert Robinson, secretary of the Editorial Board of the Mathematical Expositions Series, and of the Science Editor at the University of Toronto Press, Lorraine Ourom, is gratefully acknowledged.

Publication of the book has been made possible by a grant from the Publications Fund of the University of Toronto Press.

L.E.H.T.
M.B.W.

FROM PHYSICAL CONCEPT TO MATHEMATICAL STRUCTURE

An Introduction to Theoretical Physics

1 Introduction

1.1. GENERAL REMARKS

Physics is concerned with the observation and description of natural laws. As a result of our experience in growing up in and interacting with our environment, we develop certain more or less intuitive concepts, such as extension in space, passage of time, inertia of matter, temperature, heat flow. None of these concepts has meaning in isolation from the others; their definitions are interdependent and, to some extent, collective. Physical laws may be regarded as more or less precise and quantitative expressions of this interdependence. For example, consider Newton's second law in the form

$$\vec{F} = m\vec{a}. \tag{1}$$

Prior to our study of the explicit form of this 'law,' we have an intuitive feeling for or understanding of the three concepts involved: force, mass, and acceleration. The 'law' states that the three concepts are interrelated, and we are inclined to marvel at this apparent 'design.' However, closer examination reveals that mass and force cannot be defined separately from each other, so we are left to ponder which came first, the 'law' or the 'concepts.'

Since the concepts which are used depend upon the physical laws, their definitions are no more perfect than is the knowledge of the physical laws upon which they depend. We have become accustomed to think that physical laws are 'reality' and that the truth, once known, is inviolate. Clearly, however, a truth which relates to physical laws is only a relative truth; it is based on experience. As our experience widens, our concepts widen or change, and the physical laws must be modified to accommodate the changes. In this sense, there is no absolute truth revealed in physical sciences. If absolute truth exists, it cannot be known since experience is always limited. Truth, by this argument, is open-ended and the search for it amounts to an attempt to broaden experience. In his book on relativity, Bohm† likens the scientists' pursuit of truth to the learning experience of young children. What the concept of mass means to a layman is not the same as it means to an honours undergraduate, and neither is the same as the concept possessed by a field theorist or an elementary particle physicist: each of these develops a concept of mass which relates to his experience of it.

Naturally, objective scientists with differing experiences of mass have, nonetheless, overlapping concepts of it. In part, Newton and Einstein shared the same concept of mass, but Einstein's experience was greater and led him to a much more enriched perception. Nevertheless, even Einstein's concept was limited by his experience and was, by this token, imperfect. When we speak of mass, we have to some degree a common experience upon which to draw and so it is possible to talk to each other about it, but our experiences are not identical.

Despite some general beliefs to the contrary, physical science is not an exact science. It is a general attempt to describe and 'explain' natural phenomena in terms we think we understand; since these terms are relative to our experience, the description is always subjective to a considerable degree. Two scientists in different laboratories may perform the 'same experiment' and agree on the measurements, but they will never agree precisely upon the interpretation, since each depends upon his own particular experience when it comes to ascribing

---

†David Bohm, The Special Theory of Relativity (W.A. Benjamin Inc., New York, 1965).

significance to the results. In fact, the trouble is even deeper; since varying circumstances, including the personality of the experimenter and the skills of his technicians, enter into the design of actual experiments, no two experiments are ever quite the same.

Nonetheless, physical science is relatively exact since many experimenters share rather common experiences, and simple kinds of experiments can be almost exactly duplicated. We should always bear in mind, however, that our analysis of the results of experiments reflects not only the true meaning of nature, but the colour and modification arising from our particular way of looking at things.

At the outset we are troubled by our lack of understanding of the basic concepts. We possess intuitive ideas about the existence of <u>matter</u> moving about in <u>time</u> in a <u>space</u> continuum; but each of these concepts is an elusive one: Einstein's relativity shows that what space and time are is to some degree relative to the observer and that neither space nor time is independent of the existence of matter. To complicate things from the conceptual point of view, quantum physics teaches us that observers influence observation so that certain kinds of questions that seem a priori independent are in fact contradictory. Objectivity becomes a more difficult goal to achieve - physics is a land abounding in conceptual pitfalls.

Facing 'facts' is to some extent a counsel of despair. To the perfectionist it is disappointing to learn that physics is not an exact science, both because of the conceptual relativism discussed above, but also because even the simplest problems are mathematically so difficult that resort must be had to approximations, often crude and rarely exciting. From another point of view, it is impressive to survey what has been accomplished by painstaking measurements and dogged, if imperfect, attempts at interpretation. Occasionally knowledge thrusts forward with the force of genius, as in the discovery of special relativity by Einstein, but more and more the advance is slow and dependent upon the work of many people. It has been said that discovery is 1 per cent inspiration and 99 per cent perspiration and resolve. A brilliant idea is nothing if not pursued: moments of inspiration are followed by months or even years of determined effort. Even Einstein spent the latter half of his life pursuing, to limited avail, the Holy Grail, a unified theory of relativity, electromagnetism, and gravitation.

However imperfect our starting point, we must start somewhere. Leaving arguments about existence to the pure philosophers, we take it for granted that natural phenomena exist and that at least a partial and useful description is possible. Most of our intuitive concepts concern quantities like pressure, force, or temperature, which appear to be distributed in a space continuum in a manner which changes in time - indeed, the regular sequence of such changes provides a meaningful definition of time. It turns out that these simple, intuitive, and obvious concepts are not, in fact, obvious at all and are even in some degree incorrect. Nonetheless, imperfect as they are, they are still useful, so we shall temporarily adopt the view that space exists, that matter is distributed throughout space, and that Newton's laws form a valid description of the evolution of our universe.

We further avoid such embarrassing questions as: where does space end, or does it? and is it curved or not and if so by how much? Physics is largely the creation of models and the comparison of these models with the world as we see it. Our starting model envisages distributions of matter moving in a flat (Euclidean) 3-dimensional space, extending indefinitely in every direction. By an 'observer' we shall understand a set of recording instruments, distributed throughout space wherever needed, synchronized to any desirable accuracy, and stationary with respect to one another. Such a set of recording instruments enables us to monitor any distribution of matter, either directly or through its properties, and to express the results as a function of position and time coordinates. For example, an observer can record a temperature distribution $T(\vec{r},t)$ throughout some region of interest and over some appropriate time interval.

The process of recording one or more such distributions under controlled conditions (controlled conditions means, in effect, keeping other distributions constant or varying them in a specified manner, for example keeping the pressure constant when measuring a specific heat) is called an experiment. Understanding an experiment, in the last analysis, means setting up mathematical equations relating the distributions to each other, i.e. it is the formulation of a physical law. Theory has to do with discovering such relationships through either inductive or deductive processes and applying mathematical methods to work out the consequences. In short, the theorist sets up models, the experimenter tests their validity.

1.2. LINEAR THEORIES

Our first objective is to try to stand back and somewhat removed from the microcosmic world of physics in order to develop a perspective to guide us in our studies. We have already set out in 1.1 the primary 'world view' that we shall, at least temporarily, accept; that is, we use a model in which the existence of 3-dimensional Euclidean space is taken for granted, as well as the existence of Newtonian time and of inertial matter, even though we anticipate eventually modifying or abandoning this conceptual framework. Our task is now to survey and describe the kinds of phenomena which are in harmony with this 'world view.'

Of immediate concern is what kinds of matter inhabit space-time and how the various material entities affect one another. From a unified point of view one might pursue the possibility that only one kind of matter exists and that radically different appearances are only specialized manifestations of this one kind; such a point of view has been pursued by Heisenberg and is suggested by the Einstein principle of total interconvertibility of all forms of matter and energy. An alternative point of view is the one usually adopted in elementary particle physics, that a few specialized forms of matter exist from which all other forms are built up.

Having adopted a particular point of view it is then necessary to make a choice, based on intuition and collective experience, of the kinds of mathematical abstraction (models) which might be usefully employed to describe the physical phenomenon of immediate concern. For that is what theoretical physics is all about. In looking for suitable mathematical descriptors, we are guided both by preconditions and by predilections. One precondition, for example, is the assumption of an infinite, 3-dimensional space continuum. A predilection might be a special inclination toward the beauties of classical analysis, or toward the power of modern point set topology. Preconditions are objective limitations based on axioms; predilections can be variously experiential limitations or creative prejudices.

Our point of view in the present volume is necessarily pedagogical; we are not out to create new theories, but to survey and interrelate what has already been done. In doing so, however, we shall try to be both philosophical and research-oriented. Although we know beforehand the kinds of conceptual and mathematical structures which we shall inevitably use, we shall try to provide motivation for their use to demonstrate either intrinsic utility or axiomatic inevitability or both.

At the outset we are faced with several conceptually difficult propositions. Do we, for example, regard that matter distributions are continuous or discrete distributions, or neither, or both? Are the fundamental and universal excitations of matter wave or particle phenomena, or neither, or both? To some extent, the answers to such questions depend subjectively on what are our predilections and what seems simple or convenient. As a general rule the principle of simplicity is overpowering. The process of understanding is the process of simplification and familiarity, at least to a considerable degree. We have a greater ability, it seems, to conceptualize and invent mathematical equations than we have to solve them. For this reason, and possibly for others, the history of mathematical physics shows a predominant inclination toward linear theories and the use of linear equations.

Just as the straight line seems a simpler structure than the curve, so a linear theory is simpler than any that is non-linear. In the course of our further developments, what is meant by a linear theory will be codified and made transparent. Suffice it to say here that a dual linearity is a structural component of much of mathematical physics. The description of a 'state' of a system is usually achieved in terms of linear entities, such as vectors and functions which can be algebraically added in a linear manner to describe new possible 'states' of the system. Fundamentally, this type of linearity is inherent in the notion of a linear (vector) space to which we shall return in a moment. The second aspect of linearity is an operational linearity. In a linear theory states of a system are altered or transformed into new states by linear operators. Again, we shall leave detailed codification until later, but two illustrations at this point may help to motivate our further considerations.

Our first illustration is a familiar one of interest in both classical and modern physics. Consider the position vector $\vec{r}$ of a point P in 3-dimensional Euclidean space with respect to a fixed origin O, as illustrated in Figure 1.1. If we associate with O three mutually perpendicular axes, the vector $\vec{r}$ can be described and the point P located relative to O by three algebraic quantities x,y,z, the so-called Cartesian components of the vector $\vec{r}$. The collection of all such position vectors with respect to O forms a linear vector space. A linear process, the well-known vector addition, permits one to compose any two such vectors to form a third.

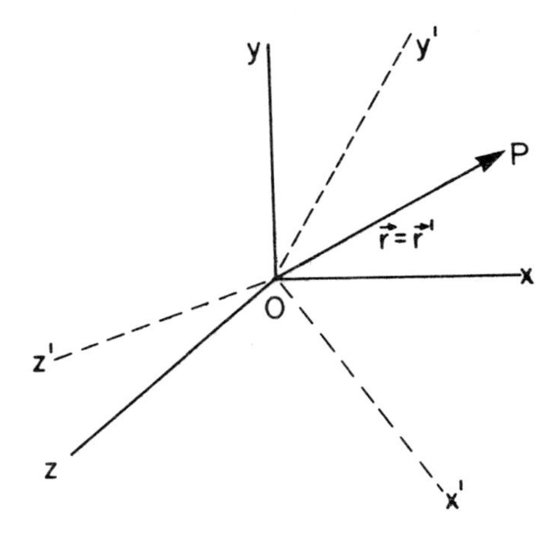

Figure 1.1. Primed (dotted) and unprimed (solid) axes as alternative coordinate axes for the fixed point P.

One can also consider, however, linear operations on a linear vector space. Keeping points O and P fixed, let us make a new choice of coordinate axes, the dotted line axes of Figure 1.1, again a set of three mutually perpendicular axes. The vector $\vec{r}$ is in nowise altered by this choice, but its description is, for we can now specify the point P with respect to O by three new algebraic quantities x',y',z' referred to the dotted axes. As we shall see, the primed coordinates can be obtained from the unprimed coordinates by means of a linear matrix operator whose description depends upon the relative orientation (direction cosines) of the primed axes with respect to the unprimed axes. The operator is linear in the sense that if $\vec{r}_3 = \vec{r}_1 + \vec{r}_2$ is the vector addition of two position vectors $\vec{r}_1$ and $\vec{r}_2$, then the components of $\vec{r}_3$ referred to the primed axes may be obtained directly from the unprimed components by means of a linear matrix operator, or by linear addition of the primed components of $\vec{r}_1$ and $\vec{r}_2$. These statements will be made quantitatively precise in Chapter 2.

A less familiar example is the structure of modern quantum theory. For simplicity we limit ourselves to a one-particle system without internal structure and in the context of non-relativistic quantum theory. Moreover, our statements will be made without detailed justification and by way of illustration only. According to quantum mechanics (see Chapter 6), the possible 'states' of such a system are the normalized, square-integrable, complex-valued functions in 3-dimensions. A function $\phi(\vec{r})$ representing a possible state of the system has the interpretation of a probability amplitude such that the square modulus $|\phi|^2$ times a volume element d^3r is the probability of finding the particle within d^3r in a 'position measurement.' The functions ϕ form a linear vector space, the linearity having the physical connotation that the usual addition of two such functions, each representing a possible state of the system, is again such a function representing still another possible state of the system. As a physical statement, this circumstance is referred to as the principle of linear superposition.

Again, linearity within the vector space is supplemented by an operational linearity which transforms one state into another. In quantum mechanics, all information about the system, or at least all information it is possible to know, is contained in the wave function (probability amplitude). Information may be extracted about the system by a series of measurements, each measurement by its nature changing the state of the system. In the mathematical formulation, measurements are represented by linear differential operators or transition operators which alter the state of the system or transform it from one state into another.

Our two examples are far-reaching in their significance. The simple ideas contained in the linear mapping of a vector space onto itself can be generalized to products of linear vector spaces and hence to the theory of generalized tensors. In quantum mechanics, such product spaces can be associated with the description of many-particle systems, or with the description of particles possessing 'internal structure.'

1.3. LINEAR OPERATORS AND TRANSFORMATION GROUPS

Let us for the moment accept the general structure of linear theories without worrying over detailed interpretation, viz. that the theory has 'objects' or 'states' describable by vectors (or functions) in a linear vector space and that the theory corresponds to a statement about what kinds of states can occur and what kinds of linear operators are involved. For example, the dynamics of a quantum system are given by the specification of a linear operator H (the Hamiltonian) which, acting on the state (wave function) of the system, transforms it into a 'state' describing the rate of change of the state of the system in time:

$$H\phi = i\hbar \frac{\partial \phi}{\partial t} . \tag{2}$$

This is the famous Schroedinger equation, of which more will be said later.

In anticipation, we now assert that linear operators in physics often form groups, or are the structure operators for continuous groups of transformations. Without digressing at this point into the theory of groups and group representations, we can already anticipate the importance of this statement since group theory is a highly developed field of mathematics. The appearance of transformation groups in physical theory means that a multitude of organizational detail associated with the mathematical structure of groups has immediate significance within the interpretation of the physical theory. Appeal to our two examples will illustrate this point.

Consider again an origin O and the linear vector space of all position vectors $\vec{r}$. As we have already observed, $\vec{r}$ can be specified in terms of components x,y,z corresponding to the choice of axes shown with full lines in Figure 1.1 or in terms of x',y',z' corresponding to the dotted axes. Clearly infinitely many such choices of axes can be made, each giving a different component description of the same vector $\vec{r}$. Each different choice of axes can be regarded as obtained from a standard set (say the x,y,z axes of Figure 1.1) by a suitable rigid rotation of axes about the origin O. It is well known that two successive rotations are equivalent to a single rotation, and in fact the set of all operations of rotation forms a group, the so-called rotation group in three dimensions R_3. It is a transformation group, viz. the group of rotations transforming a system of orthogonal coordinates into all possible orientations about a common centre O.

As a second example, we consider the Schroedinger equation, which can be re-expressed using the Taylor's expansion of $\phi(\vec{r},t)$ in the time variable:

$$(3) \qquad \phi(\vec{r}, t + dt) = \phi(\vec{r},t) + (\partial\phi/\partial t)\, dt + \ldots$$
$$= (1 - \frac{i\, dt}{\hbar} H)\, \phi(\vec{r},t) + \ldots,$$

a result that follows immediately from equation (2). Let T be a finite time interval and associate with it the infinitesimal dt = T/n, where n is a large positive integer. Taking the limit $n \to \infty$ one obtains from (3)

$$(4) \qquad \phi(\vec{r}, t + T) = \underset{n\to\infty}{\text{Limit}}[(1 - iTH/n\hbar)^n\, \phi(r,t)]$$
$$= e^{-iTH/\hbar}\, \phi(\vec{r},t).$$

Thus, the operator $e^{-iTH/\hbar}$ is the time translation operator† that replaces the state of the system at time t by the state of the system at time t + T. It is easily seen that successive time translations T_1 and T_2 are equivalent to a single time translation $T_3 = T_1 + T_2$, viz.

$$e^{-iT_1H/\hbar} \cdot e^{-iT_2H/\hbar} = e^{-i(T_1+T_2)H/\hbar},$$

†Exponential operators are defined by their series expansions.

so that the time translation operators form a group.

Transformation groups arise in mathematical physics basically because the system possesses some fundamental symmetries with which are associated important invariance properties. In our first example above, we can say that the length of the vector $\vec{r}$ joining O to P is invariant under rotations of the coordinate system used to describe it. Alternatively, one can say that this invariance reflects the isotropy of space. In the second example, the symmetry principle is that all time intervals are equivalent: the system is invariant under time translations. Because of the particular structure of quantum mechanics and the special role played by the Hamiltonian as a time development operator, time translation invariance is equivalent to energy conservation.

In succeeding chapters the theme introduced here will be developed and illustrated by examples encompassing many aspects of mathematical physics. Essentially we shall be concerned with the generality of application of the notion of groups of linear operators on a linear vector space. The generalization to products of such spaces leads on the one hand to the mathematics of multilinear mapping and tensor analysis, and on the other to the representation theory for transformation groups. Although we shall not be concerned with mathematical existence theorems and proofs per se, nonetheless our pursuit of physical theory will give us insight into the deep connections which exist between the irreducible matrix representations of transformation groups and the symmetry classes of tensors.

Finally, let us note some important branches of physics where linear theories are inappropriate, but where the constructs so natural and basic to linear theory still have utility. One such branch is hydrodynamical theory, where the fluid velocity, itself a linear construct, obeys non-linear equations. Chemical kinetics and the description of chemical reactions are but other examples. Non-linear thermodynamics, which plays a fundamental role in the description of biological phenomena, is still another example where linear theory is inadequate and where new approaches seem essential to further progress. We return to discuss a few of these examples in a later chapter.

2
Cartesian tensors and transformation groups

2.1. THE POSITION VECTOR AND ITS GENERALIZATION: THE ROTATION GROUP IN 2-DIMENSIONS

We begin our studies by using the familiar, but not entirely transparent, example of the covariance of physical laws under coordinate rotation, an example already introduced briefly in Chapter 1. The underlying idea is to recognize the apparent isotropy of space by expressing physical laws in terms of natural constructs, viz. tensors and spinors. Once the idea has been established in this particular case, its extension to more varied circumstances will be straightforward.

We begin by asking what may seem like silly or trivial questions, such as: if one observer using a particular coordinate system describes a physical situation in a specific fashion, what equations will he use in a coordinate system which has been rotated from the original position? In other words, we shall examine the covariance of physical laws under simple changes of coordinate systems. Far from trivial, such considerations will lead us to a useful classification of most physical constructs and will give us a deeper insight into the hidden assumptions in 'ordinary classical physics.'

As noted previously we shall largely be concerned with distributions in space and time. A 2-dimensional coordinate system is, in effect, a 2-dimensional net for identifying points in space, much as the streets and avenues on a city map identify locations throughout a city. (A more complicated example is the use of longitude and latitude on a sphere, but this is a curved rather than a flat space.) It is evident that before talking about how our distributions behave under coordinate transformations, such as rotations and translations, we shall first have to examine how our description of space points in the net itself changes under these transformations.

Consider first the 2-dimensional Cartesian coordinate system used to provide a 'net' in 2-dimensional flat spaces. (We shall discuss later what is meant by non-flat or non-Euclidean spaces.) Let the point P have coordinates (x,y), or in an alternative notation convenient for summation (x_1,x_2), with respect to system O (O implies observer O and the coordinate system he uses), and let (x',y') be the coordinates of the same point P described in the system O' which is rotated with respect to O counterclockwise through an angle θ, as in Figure 2.1. From simple geometry

$$x' = x\cos\theta + y\sin\theta, \quad y' = -x\sin\theta + y\cos\theta, \tag{1}$$

which we can write in matrix form

$$\begin{bmatrix} x' \\ y' \end{bmatrix} = \begin{bmatrix} \cos\theta & \sin\theta \\ -\sin\theta & \cos\theta \end{bmatrix} \begin{bmatrix} x \\ y \end{bmatrix}, \tag{2}$$

or more simply

$$x'_k = a_{kj}x_j. \tag{3}$$

Here we have used the summation convention that repeated indices are summed (i.e. in this case $a_{kj}x_j = a_{k1}x_1 + a_{k2}x_2$). The a_{kj} are elements of the orthogonal matrix (see Appendix B)

$$A = \begin{bmatrix} a_{11} & a_{12} \\ a_{21} & a_{22} \end{bmatrix} = \begin{bmatrix} \cos\theta & \sin\theta \\ -\sin\theta & \cos\theta \end{bmatrix}$$

and are just the direction cosines $a_{jk} = \hat{e}'_j \cdot \hat{e}_k$ ($\hat{e}_k$ is the unit vector in the direction of the x_k axis and $\hat{e}'_k$ is the unit vector in the direction of the x'_k axis). In standard vector notation we can write for the same physical vector $\vec{r} = \vec{r}\,'$ either

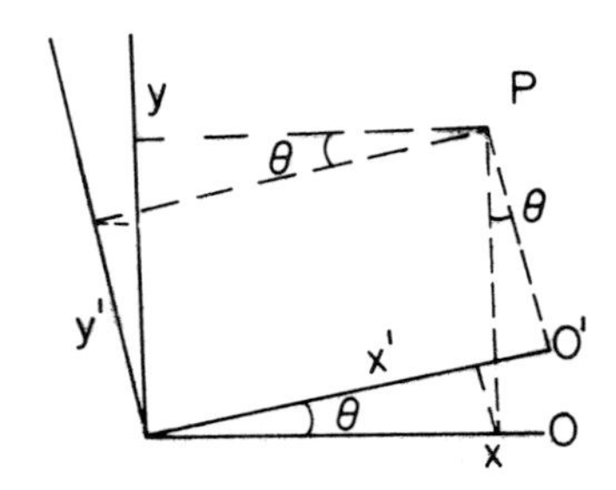

Figure 2.1. The relationship between two Cartesian coordinate systems rotated with respect to each other through an angle θ.

(4) $\vec{r} = x_1\hat{e}_1 + x_2\hat{e}_2$

in the unprimed system or, in the primed system,

(5) $$\begin{aligned}\vec{r}\,' &= x_1'\hat{e}_1' + x_2'\hat{e}_2' \\ &= (a_{11}x_1 + a_{12}x_2)\hat{e}_1' + (a_{21}x_1 + a_{22}x_2)\hat{e}_2' \\ &= x_1(a_{11}\hat{e}_1' + a_{21}\hat{e}_2') + x_2(a_{12}\hat{e}_1' + a_{22}\hat{e}_2').\end{aligned}$$

Comparing the results in equations (4) and (5) we get

(6) $\hat{e}_1 = a_{11}\hat{e}_1' + a_{21}\hat{e}_2', \quad \hat{e}_2 = a_{12}\hat{e}_1' + a_{22}\hat{e}_2',$

or, using the summation convention,

(7) $\hat{e}_k = a_{jk}\hat{e}_j'.$

Multiplying equation (7) by $a_{\ell k}$ and summing over k gives (see Appendix B)

(8) $\hat{e}_\ell' = a_{\ell k}\hat{e}_k$

and taking the dot product of both sides of equation (7) with $\hat{e}_\ell'$ yields

(9) $a_{\ell k} = \hat{e}_\ell'\cdot\hat{e}_k$

which confirms the definition of a_{jk} given above.

In Euclidean space, the unit vectors and the components of a position vector transform in the same way if the coordinate system is Cartesian (see Equations (3) and (8)). We shall see later that this is not true for an arbitrary coordinate system.

For a space of three dimensions or greater the number of distinct axes of rotation possible is much greater than for the simple 2-dimensional space considered here, but the same general considerations still apply; that is, two orthogonal coordinate systems rotated with respect to one another in such a manner as to have direction cosines $a_{jk} = \hat{e}'_j \cdot \hat{e}_k$ between the jth axis of one and the kth axis of the second have the property that the primed and unprimed components of a point P are related by $x'_j = a_{jk}x_k$ where again the a_{jk} form an orthogonal matrix. The orthogonality conditions $a_{jk}a_{\ell k} = \delta_{j\ell}$ express the fact that the two directions $\hat{e}'_j$ and $\hat{e}'_\ell$ are either the same or orthogonal when viewed from the unprimed coordinate system depending on whether $j = \ell$ or $j \neq \ell$.

We are now prepared to give an analytical definition for a vector which accords with the standard descriptive definition but which lends itself to generalizations and which is in any case more convenient for our purposes.

An ordered set of two algebraic quantities (n in n-dimensional space) which transform under rotations in the same manner as the components of a position 'vector' is said to be a <u>vector</u>. What kinds of physical quantities are vectors is a matter of experience and experiment, but we have the well-known examples of force $\vec{F}$, velocity $\vec{v}$, momentum $\vec{p}$, and acceleration $\vec{a}$. In other words we say that the force $\vec{F}$ is a vector since experiment shows that its components F'_k in a rotated coordinate system are related to its components F_j in the original coordinate system by

$$F'_k = a_{kj}F_j \tag{10}$$

where the a_{kj} are the same direction cosines (specifying the rotation) which were previously defined.

If $\vec{B}$ and $\vec{C}$ are two vectors, then their sum $\vec{E} = \vec{B} + \vec{C}$ is also a vector since

$$\text{(11)} \quad \begin{aligned} E'_k &= B'_k + C'_k = a_{kj}B_j + a_{kj}C_j = a_{kj}(B_j + C_j) \\ &= a_{kj}E_j. \end{aligned}$$

Also the product of a vector and a number α from the real number field R is again a vector since, if $C_k = \alpha B_k$, then

$$\text{(12)} \quad C'_k = \alpha B'_k = \alpha a_{kj}B_j = a_{kj}\alpha B_j = a_{kj}C_j.$$

In fact, it is easily shown that vectors defined in this way satisfy the formal requirements of the elements of a linear vector space (see Appendix A).

We turn now to a study of the operators A which express the components of a fixed vector, say $\vec{r}$, in a rotated coordinate system in terms of the components in the original or standard system. This 'rotation operator' is linear since it acts linearly, according to equations (11) and (12), on the vectors of the linear vector space. For two dimensions we have seen that A can be expressed as a matrix acting on the components of $\vec{r}$ arranged as a column matrix.

Clearly the matrix A is a function of the rotation angle θ and we can write $A(\theta)$. It is also clear that two successive rotations of angles θ_1 and θ_2 are equivalent to a single rotation of angle $\theta_1 + \theta_2$ and that this process is represented by matrix multiplication. Let

$$\text{(13a)} \quad \begin{bmatrix} x'_1 \\ x'_2 \end{bmatrix} = \begin{bmatrix} \cos\theta_1 & \sin\theta_1 \\ -\sin\theta_1 & \cos\theta_1 \end{bmatrix} \begin{bmatrix} x_1 \\ x_2 \end{bmatrix}$$

and

$$(13b)\quad \begin{bmatrix} x_1'' \\ x_2'' \end{bmatrix} = \begin{bmatrix} \cos\theta_2 & \sin\theta_2 \\ -\sin\theta_2 & \cos\theta_2 \end{bmatrix} \begin{bmatrix} x_1' \\ x_2' \end{bmatrix} ;$$

then

$$(14)\quad \begin{bmatrix} x_1'' \\ x_2'' \end{bmatrix} = \begin{bmatrix} \cos\theta_2 & \sin\theta_2 \\ -\sin\theta_2 & \cos\theta_2 \end{bmatrix} \begin{bmatrix} \cos\theta_1 & \sin\theta_1 \\ -\sin\theta_1 & \cos\theta_1 \end{bmatrix} \begin{bmatrix} x_1 \\ x_2 \end{bmatrix}$$

$$= \begin{bmatrix} \cos(\theta_1 + \theta_2) & \sin(\theta_1 + \theta_2) \\ -\sin(\theta_1 + \theta_2) & \cos(\theta_1 + \theta_2) \end{bmatrix} \begin{bmatrix} x_1 \\ x_2 \end{bmatrix}.$$

Hence matrix multiplication composes properly for successive rotations. Moreover, if we consider all possible rotation matrices, they form a group: closure is satisfied according to (14); an identity exists, viz.

$$(15)\quad A(0, \text{mod } 2\pi) = \begin{bmatrix} 1 & 0 \\ 0 & 1 \end{bmatrix} ;$$

an inverse exists, viz. $A(-\theta)$; and the associative law holds generally for matrix multiplication:

$$(16)\quad \big(A(\theta_1)A(\theta_2)\big)A(\theta_3) = A(\theta_1)\big(A(\theta_2)A(\theta_3)\big).$$

The group of all coordinate rotations in 2-dimensions is denoted by R_2. It is easily seen that the correspondence between coordinate rotations and orthogonal 2 x 2 matrices of unit determinant is one to one, and that this group of matrices also corresponds to the structure of R_2 (i.e., the two groups differ only by the labelling of their elements).

The theme developed in Chapter 1 is now partially exposed: we have seen manifest the dual linearity, on the one hand, of the vector space itself (position, velocity, force, etc.) and, on the other, of the transformation (rotation) operators linking observations made in rotated frames of reference. What is true of 2-dimensions is true of 3-dimensions (to which we shall return in Chapter 3) and, in fact, can be easily generalized to Cartesian coordinates in n-dimensions.

2.2. INNER AND OUTER PRODUCTS OF VECTORS: TENSORS

It is a matter of experience that certain physical quantities, such as mass m, time t, and temperature T, do not change under coordinate rotation (i.e. $m = m'$, $t = t'$, and $T = T'$). In addition to these 'obvious' scalars it is evident that the dot product or scalar product of any two vectors is also a scalar and hence the magnitude or length of any vector, as well as the angle between two vectors, is a scalar (since these quantities are defined in terms of dot products). For example, if $\vec{F}$ is a constant force which linearly displaces a body from O to $\vec{r}$, then W, the work done, is a scalar:

$$(17)\qquad W' = \vec{r}'\cdot\vec{F}' = x_k'F_k' = a_{kj}a_{k\ell}x_jF_\ell$$
$$= \delta_{j\ell}x_jF_\ell = x_jF_j = \vec{r}\cdot\vec{F} = W.$$

If the force is not constant, the work done is the line integral along the length L of the curve,

$$(18)\qquad W = \int_0^L \vec{F}(s)\cdot\hat{t}\ ds,$$

where ds is the displacement in the direction of the tangent $\hat{t}$ to the curve. Since $\vec{F}\cdot\hat{t}$ is a scalar and ds is a scalar, and since sums of scalars are scalars, the integral as a whole is a scalar.

It is evident that one can form mathematical constructs which are more complicated than vectors in so far as their transformation properties are concerned. For example, consider the nine products

$$(19)\qquad x_kF_j = S_{kj},\quad k,j = 1,2,3,$$

of the components of two vectors $\vec{r}$ and $\vec{F}$. Clearly, under rotation, these nine quantities transform bilinearly as

$$S'_{kj} = x'_k F'_j = a_{kn} a_{jm} x_n F_m = a_{kn} a_{jm} S_{nm}; \tag{20}$$

they are said to form the components of a <u>second-rank tensor</u> (here in 3-dimensional space); in this terminology a vector is a first-rank tensor and a scalar is a tensor of zero rank. The question now is: are there any physical quantities (presumably intuitively defined) which behave as the components of a second-rank (or higher) tensor? The answer is not obvious, but is 'yes' nonetheless.

A second-rank tensor is an element of the <u>outer product</u> of two linear vector spaces (see Appendix A) which are of the same dimension and whose elements are restricted to transform in the same way under coordinate rotation. However, not every second-rank tensor can be written as the outer product of two vectors (problem 8).

A third-rank tensor (in 3-dimensional space) is similarly an ordered set of 27 quantities transforming under rotation according to

$$S'_{jk\ell} = a_{jp} a_{kq} a_{\ell r} S_{pqr} \tag{21}$$

where the a_{jk} are the direction cosines defining the actual rotation. In general a tensor of rank r in D-dimensional space is an ordered set of D^r components which transform under rotation like

$$S'_{i_1 \ldots i_r} = a_{i_1 j_1} \ldots a_{i_r j_r} S_{j_1 \ldots j_r} \tag{22}$$

where the $a_{\ell m}$ are again the direction cosines defining the rotation in D-dimensional space.

If $T_{i_1 \ldots i_r}$ and $S_{i_1 \ldots i_r}$ are two tensors of rank r in D-dimensional space, their sum $(S \oplus T)_{i_1 \ldots i_r}$ is also a tensor of rank r in D-dimensional space with components defined by

(23) $$(S \oplus T)_{i_1 \ldots i_r} = S_{i_1 \ldots i_r} + T_{i_1 \ldots i_r}.$$

Note that tensor addition is only defined between tensors of the same rank in spaces of the same dimension.

The components of a tensor may or may not be all independent. For example, a second-rank tensor in 3-dimensional space T_{jk} can be written as

(24) $$T_{jk} = \tfrac{1}{2}(T_{jk} + T_{kj}) + \tfrac{1}{2}(T_{jk} - T_{kj}) = S_{jk} + A_{jk}$$

where the tensor S_{jk} is symmetric under permutation of indices

(25) $$S_{jk} = S_{kj}$$

and has six independent components, and A_{jk} is antisymmetric under permutation of indices

(26) $$A_{jk} = -A_{kj}$$

and has only three independent components. Note that

(27) $$\text{Trace } A = A_{kk} = 0.$$

Quite generally the trace of any second-rank tensor is a scalar quantity since

(28) $$T'_{kk} = a_{kj}a_{ki}T_{ji} = \delta_{ji}T_{ji} = T_{jj}.$$

The symmetry properties of tensors play an important role in the representation of certain groups. This subject will be discussed more fully in Chapter 5.

We have seen that the algebraic products of vector components transform as the components of a higher rank tensor. Similarly if $S_{i_1 \ldots i_p}$ and $T_{j_1 \ldots j_q}$ are respectively pth and qth rank tensors in D-dimensional space the D^{p+q} components composed of products of the components of S and T,

$$(29) \qquad (S \otimes T)_{i_1 \ldots i_p j_1 \ldots j_q} = S_{i_1 \ldots i_p} T_{j_1 \ldots j_q},$$

transform like those of a (p + q)th-rank tensor. We call $S \otimes T$ the outer product (or tensor product) of S with T.† The following properties of outer products are immediate consequences of the definition:

(i) $S \otimes (T \otimes P) = (S \otimes T) \otimes P$ associativity

(ii) $P \otimes (S \oplus F) = (P \otimes S) \oplus (P \otimes F)$ distributivity over addition

where S, F, P, and T are all tensors in the same space, S and F having the same rank.

By contrast the 'scalar product' of vectors corresponds to an inner product which reduces the rank by two. We can generalize the notion of the scalar product in the following way. Consider an arbitrary outer product of tensors of various ranks. Let the overall tensor rank be m (the sum of the ranks of the various tensors in the product). If we set two indices equal and sum over their values from 1 to D in D-dimensional space, the resulting quantities are tensors of rank m - 2. For example, consider the following fifth-rank tensor in 3-dimensions:

$$T_{jk\ell} S_{mn} = B_{jk\ell mn}.$$

Setting $\ell = m$ and summing, we obtain a set of 27 quantities

---

†We sometimes call a tensor and its components by the same name. In this paragraph, for example, we use $S_{i_1 \ldots i_p}$ to denote both the entire tensor (with each subscript taking on all the values 1,2,...D) and also the single component corresponding to specific values of $i_1, \ldots, i_p$. Sometimes the entire tensor is denoted simply by S. The latter notation has the advantage of making the distinction between tensor and component clear but fails to indicate the rank of the tensor.

$$(30) \qquad R_{jkn} = T_{jkm}S_{mn} = B_{jkmmn}$$

$$= T_{jk1}S_{1n} + T_{jk2}S_{2n} + T_{jk3}S_{3n}$$

which transform as the 27 components of a third-rank tensor in 3-dimensions. This process of setting $\ell = m$ and summing is called contraction and corresponds to taking an <u>inner product</u>. More than one contraction can take place. For example, if we set $j = k$ and sum from 1 to 3, we obtain another set of 27 quantities

$$(31) \qquad P_{\ell mn} = T_{kk\ell}S_{mn} = B_{kk\ell mn}$$

$$= T_{11\ell}S_{mn} + T_{22\ell}S_{mn} + T_{33\ell}S_{mn}$$

which also transform as the 27 components of a third-rank tensor in 3-dimensions.

Clearly, the ordinary scalar product is the special case of contraction of two first-rank tensors (vectors); for example,

$$(32) \qquad W = x_k F_k \quad \text{or} \quad r^2 = x_k x_k .$$

A second example is the trace of a second-rank tensor, which is a scalar (equation (28)).

2.3. MOMENT OF INERTIA AS A SECOND-RANK TENSOR

As our first detailed example consider the rotation of a rigid body with one point fixed. The instantaneous motion is characterized by an angular velocity $\vec{\omega}$ about an instantaneous axis of rotation. As the motion proceeds $\vec{\omega}$ may change in either direction or magnitude or both.

The angular momentum vector is given by the cross-product

(33) $$\vec{L} = \sum_{\lambda=1}^{N} m_\lambda \vec{r}_\lambda \times \vec{v}_\lambda$$

where λ numbers the atoms composing the rigid body and m_λ, $\vec{r}_\lambda$, $\vec{v}_\lambda$ are respectively the mass, position, and velocity of the λth atom. The components of the angular momentum depend on what coordinate system is used when calculating the sum given by equation (33) as regards both the orientation of the coordinate axes and the location of the origin. For convenience one usually picks a coordinate system whose origin coincides with the fixed point of the rigid body (sometimes, however, the centre of mass of the body is more convenient and the origin is placed there). If one can view the rigid body as a continuum of matter (as is often convenient) the sum in equation (33) goes over into the integral

(34) $$\vec{L} = \iiint \rho(\vec{r})\, \vec{r} \times \vec{v}\, dV$$

where $\rho(\vec{r})$ is the density of matter and dV is a volume element. The angular momentum vector $\vec{L}$ need not coincide with $\vec{\omega}$ in direction. If no external torques act (and we are using an inertial coordinate system) then

(35) $$\frac{d\vec{L}}{dt} = 0, \quad \vec{L} = \text{constant},$$

and the motion reduces to a spin of instantaneous magnitude ω about an instantaneous axis $\hat{\omega}$ which wanders about the fixed direction $\vec{L}$. In any case, one can easily show that the three components of $\vec{L}$ in any coordinate system are related to the three components of $\vec{\omega}$ in the same coordinate system by a set of linear equations

(36) $$L_k = I_{kj}\omega_j.$$

The instantaneous velocity $\vec{v}$ of an atom located at $\vec{r}$ is evidently given by $\vec{v} = \vec{\omega} \times \vec{r}$ in equation (34). Hence, we can write

(37) $$\vec{L} = \iiint \rho(\vec{r})[\vec{r} \times (\vec{\omega} \times \vec{r})]dV.$$

But for any three vectors $\vec{A}$, $\vec{B}$, and $\vec{C}$

(38) $$\vec{A} \times (\vec{B} \times \vec{C}) = \vec{B}(\vec{A}\cdot\vec{C}) - \vec{C}(\vec{A}\cdot\vec{B})$$

so that equation (37) becomes

(39) $$\vec{L} = \iiint \rho(\vec{r})[(\vec{r}\cdot\vec{r})\vec{\omega} - (\vec{r}\cdot\vec{\omega})\vec{r}]dV,$$

or in component form (using the identity $\omega_k = \omega_j\delta_{jk}$)

(40) $$L_k = \iiint \rho(\vec{r})(x_\ell x_\ell \delta_{jk}\omega_j - x_k x_j \omega_j)dV.$$

Since all points in the rigid body have the same instantaneous angular velocity $\vec{\omega}$, the components ω_j may be factored out from under the integral sign in equation (40), yielding

(41) $$L_k = I_{kj}\omega_j$$

where

(42) $$I_{jk} = \iiint \rho(\vec{r})(x_\ell x_\ell \delta_{jk} - x_k x_j)dV.$$

The nine quantities I_{jk} have the symmetric property

(43) $$I_{kj} = I_{jk}.$$

The three diagonal components are called moments of inertia and are given by

(44) $$I_{kk} = \iiint \rho(\vec{r})(x_i^2 + x_j^2)dV, \quad i,j,k \text{ cyclic},$$

and the three independent off-diagonal components

(45) $$I_{jk} = -\iiint \rho(\vec{r})x_j x_k dV$$

are called products of inertia. Equation (41) expresses the fact that the moment of inertia tensor I_{kj} can be regarded as a linear operator mapping the vector $\vec{\omega}$ into the vector $\vec{L}$.

It is a matter of experiment that the law $\vec{L} = I\vec{\omega}$ also holds for rotated observers. The question we now raise is: what kind of transformation properties do the I_{jk} have as a consequence? Consider a rotation of coordinates such that

$$L'_k = a_{kj}L_j \quad \text{and} \quad \omega'_k = a_{kj}\omega_j.$$

Upon inversion the latter becomes $\omega_j = a_{kj}\omega'_k$. Now

(46) $$L'_k = a_{kj}L_j = a_{kj}I_{jn}\omega_n = a_{kj}I_{jn}a_{mn}\omega'_m$$

$$= (a_{kj}a_{mn}I_{jn})\omega'_m = I'_{km}\omega'_m$$

where we have set $I'_{km} = a_{kj}a_{mn}I_{jn}$ in order to make the physical law covariant (i.e. have the same form in all rotated frames of reference). From the form of I_{jk} (see equation (42)) we can obtain the result $I'_{km} = a_{kj}a_{mn}I_{jn}$ directly, and so the law connecting $\vec{L}$ and $\vec{\omega}$ may be looked upon as covariant by virtue of the fact that the I_{jk} transform as the components of a second-rank tensor.

2.4. TENSORS AS MULTILINEAR MAPS

Contraction brings forward the possibility of interpreting tensors as multilinear maps. Consider for example a third-rank tensor $S_{jk\ell}$. Contracting this tensor with the vectors u_ℓ, v_m, and w_n on the first, second, and third indices respectively gives the scalar $S_{jk\ell}u_jv_kw_\ell$. Thus the tensor S is an operator with three slots or arguments, and when each argument is filled with a vector a scalar is produced:

(47) $$S(\underset{\substack{\text{slot}\\1}}{\vec{u}}, \underset{\substack{\text{slot}\\2}}{\vec{v}}, \underset{\substack{\text{slot}\\3}}{\vec{w}}) = S_{jk\ell}u_jv_kw_\ell, \text{ a scalar.}$$

S is called a multilinear operator or map since it is linear in each argument. Linearity in the second argument, for example, means that

$$S(\vec{u}, a\vec{v}^1 + b\vec{v}^2, \vec{w}) = aS(\vec{u}, \vec{v}^1, \vec{w}) + bS(\vec{u}, \vec{v}^2, \vec{w}) \tag{48}$$

where a, b are scalars and $\vec{v}^1$, $\vec{v}^2$ are vectors. If one of the slots is left empty the tensor operates on two vectors to produce another vector:

$$S(\ ,\vec{v},\vec{w}) = S_{jk\ell}v_k w_\ell, \tag{49}$$

which is a vector.

The moment of inertia tensor I may be interpreted as an operator with two arguments or slots. If the angular velocity vector $\vec{\omega}$ is placed in the second argument, the tensor operates on it to produce another vector, the angular momentum $\vec{L}$:

$$I(\ ,\vec{\omega}) = I(\ ,\omega(\ )) = L(\ ) \tag{50}$$

where the vectors $\vec{L}$ and $\vec{\omega}$ are likewise interpreted as operators with one slot. If $\vec{L}$ is inserted in the first argument and $\vec{\omega}$ in the second, the moment of inertia tensor operates on these two vectors, mapping them into a scalar:

$$I(\vec{L},\vec{\omega}) = L(\vec{L}) = \vec{L}\cdot\vec{L}, \tag{51}$$

which is a scalar.

REFERENCES

1 R.L. Halfman, Dynamics, Particles, Rigid Bodies and Systems. Addison-Wesley, Reading, Mass., 1962
2 H. Jeffreys, Cartesian Tensors. Cambridge at the University Press, 1963
3 G. Temple, Cartesian Tensors. Methuen & Co. Ltd., London, 1960

PROBLEMS

1 The wedge or exterior product of two vectors $\vec{u}$ and $\vec{v}$ is denoted $\vec{u} \wedge \vec{v}$ and defined by the difference between two tensor products:

$$\vec{u} \wedge \vec{v} = (\vec{u} \otimes \vec{v}) = (\vec{v} \otimes \vec{u})$$

or in component form

$$(\vec{u} \wedge \vec{v})_{ij} = \vec{u}_i \vec{v}_j - \vec{u}_j \vec{v}_i .$$

(i) Show that
(a) $\vec{u} \wedge \vec{v} = -\vec{v} \wedge \vec{u}$,
(b) $\vec{u} \wedge \vec{u} = 0$,
(c) $(a_1\vec{v} + a_2\vec{u}) \wedge \vec{w} = a_1(\vec{v} \wedge \vec{w}) + a_2(\vec{u} \wedge \vec{w})$, $\quad a_1, a_2 \in R$.

(Note that for 3-dimensional vectors $\vec{u} \wedge \vec{v} \equiv \vec{u} \times \vec{v}$ where by $\vec{u} \times \vec{v}$ we mean the vector cross-product, not the tensor product.)

2 Suppose some physical law has the form

$$B_{ijk\ell} = A_{ij} M_{k\ell}$$

where $B_{ijk\ell}$ is a fourth-rank tensor and $M_{k\ell}$ is a non-zero second-rank tensor. If the law is covariant under rotation show that A_{ij} is a rank 2 tensor.

3 An object consists of mass 'points' connected to each other by rigid weightless bars. Find the moment of inertia tensor for the following objects: 5 grams at (0,0,0); 5 grams at (2,3,-1); 10 grams at (0,2,1); 10 grams at (-1,-1,2).

4 In some coordinate system with base vectors $\hat{e}_1$, $\hat{e}_2$, $\hat{e}_3$ the moment of inertia tensor for a rigid body is

$$I_{jk} = \begin{bmatrix} 2/3 & -1/4 & -1/4 \\ -1/4 & 2/3 & -1/4 \\ -1/4 & -1/4 & 2/3 \end{bmatrix} .$$

Find the coordinate system in which all products of inertia vanish and evaluate the moment of inertia tensor in that frame.

5 It is frequently convenient to use a coordinate system fixed in the body when considering the dynamics of a rigid body. This is, of course, a non-inertial frame if the body is rotating, and one must be careful to adjust Newton's laws accordingly. $J_k = I_{kj}\omega_j$ remains a valid equation but the components J_k of the angular momentum are those as seen by an observer in the rotating frame.

Evaluate the moment of inertia tensor I_{jk} for an extremely thin rod of length ℓ and uniform density ρ using a body-fixed coordinate system with origin at

(i) one end of the rod,

(ii) the centre of mass of the rod.

6 Evaluate the moment of inertia tensor I_{jk} (using body-fixed axes, uniform density ρ, and origin of your choice) for each of the following:

(i) a spherical mass of radius R,

(ii) a cylindrical mass of height ℓ and of base R,

(iii) a cube with edges of length ℓ.

7 Let X_i and x_i be the coordinate axes of two body-fixed coordinate systems with origins Q and O respectively, such that the coordinate axes of both systems have the same orientation but the origins O and Q differ by a vector $\vec{a}$ in location (one reference frame translated with respect to the other), as illustrated in Figure 2.2. Show that

$$I_{ij} = J_{ij} - M[a_\ell a_\ell \delta_{ij} - a_i a_j]$$

where M is the total mass of the rigid body, J_{ij} is the moment of inertia tensor referred to the origin Q, and I_{ij} is the moment of inertia tensor referred to the origin O which is located at the centre of mass of the rigid body.

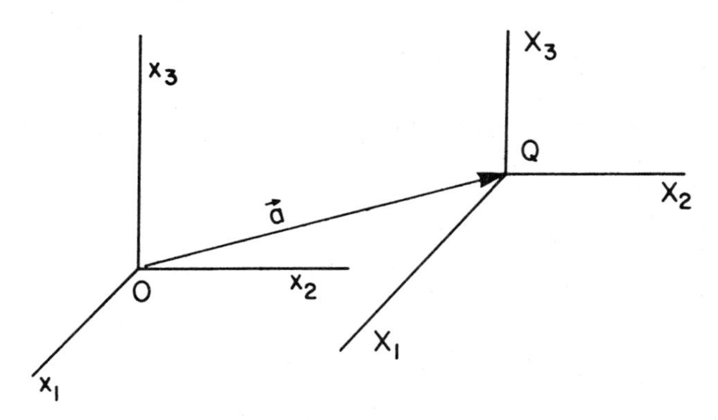

Figure 2.2. Body-fixed coordinate systems separated by a displacement $\vec{a}$.

8 Construct a second-rank tensor which is not simply the outer product of two vectors.

3
Rotations, reflections, and more about tensors

3.1. INTRODUCTION

In Chapter 2 we have seen how the notion of tensors in 3-dimensional Euclidean space arises in a natural way from our attempt to formulate mathematical relationships between physical quantities in such a way that a comparison between 'rotated observers' is facilitated. It was then also natural to seek a generalization of the tensor concept to multilinear transformations on an n-dimensional linear vector space. Because of the importance of the multilinear transformations associated with coordinate rotations in a wide class of physical problems, we return here to a more detailed examination of the rotation problem.

We restrict ourselves in this chapter to observers in 3-dimensions sharing a common origin but using coordinate axes with different orientations. One observer, say O, may be regarded as the 'standard' observer using coordinate axes lined up with the corners of a rectangular room (e.g. the 'lab'). The second (primed) observer O' may be considered to have his axes embedded in a rigid body with one point fixed, viz. the common origin of coordinates. Originally the two sets of axes coincide, but the primed observer then elects to rotate his axes (carried in the rigid body) to a new 'primed' position. According to a theorem in classical kinematics[†] any rotational displacement of a rigid body with one point fixed, however achieved, can be accomplished by a suitable single

---

[†]E.T. Whittaker, A Treatise on the Dynamics of a System of Particles and Rigid Bodies (Cambridge University Press, Cambridge, 1937).

rotation about an appropriate single axis. It follows that two successive rotations, of differing angles about differing axes, are equivalent to a single rotation so that the operations of rotation form a group. Such operations can be represented by matrices and successive rotations by matrix multiplications (group representations), as we shall now show. However, the general subject of groups and group representations is a very broad one, and we defer detailed consideration to Chapter 5.

Representations of the group of rotations by matrices in 3-dimensional space are of interest not only from a purely mathematical point of view but also as a valuable tool in physics. For example, attempts to represent the group of rotations in 3-dimensional space by 2-dimensional unitary matrices leads to a whole new classification of physical entities with significance in quantum physics.

3.2. REPRESENTATION OF THE GROUP OF ROTATIONS IN 3-DIMENSIONAL SPACE

An arbitrary rotation of a Cartesian coordinate system with axes x,y,z into new positions x',y',z' may be achieved by three successive rotations performed in a certain sequence (see Figure 3.1). First the coordinate axes x,y,z are rotated counterclockwise through an angle ϕ about the z axis. The resultant coordinate system has axes labelled x_1,y_1,z. Then the x_1,y_1,z axes are rotated through an angle θ counterclockwise about the x_1 axis to produce a second set of intermediate axes x_1,y_1',z'. Finally the x_1,y_1',z' axes are rotated counterclockwise through an angle ψ about the z' axis bringing them into coincidence with the x',y',z' axes. Each of the successive steps has a simple matrix formulation similar to the 2-dimensional rotation matrix given in Chapter 2. Let P be a point in 3-dimensional space with coordinates (x,y,z), (x_1,y_1,z), (x_1,y_1',z'), and (x',y',z') in the coordinate systems with axes x,y,z, x_1,y_1,z, x_1,y_1',z', and x',y',z' respectively. From our considerations in Chapter 2 we can write

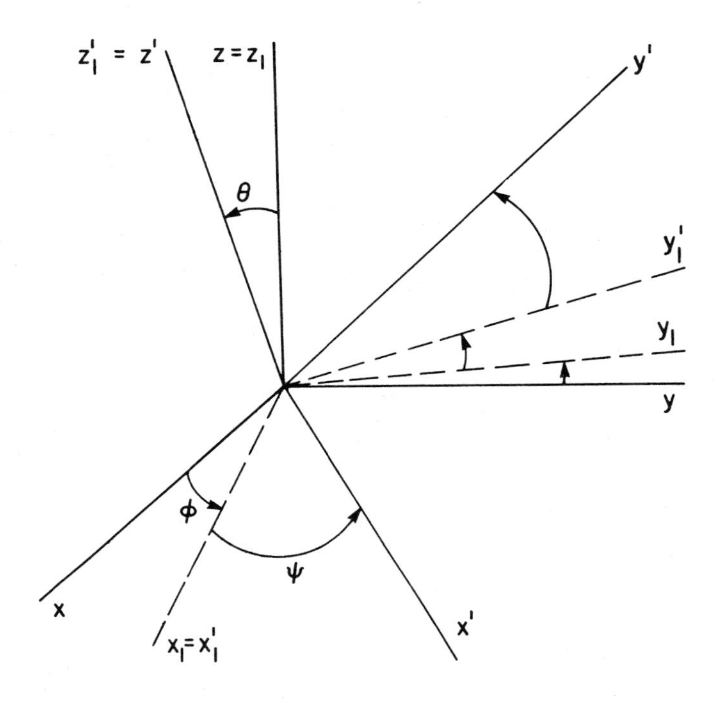

Figure 3.1. Euler angles which are defined by successive rotations through angles ϕ, θ, and ψ carry the 'body' coordinates through successive positions $x,y,z \rightarrow x_1,y_1,z \rightarrow x_1,y_1',z' \rightarrow x',y',z'$.

$$\text{(1a)} \quad \begin{bmatrix} x_1 \\ y_1 \\ z \end{bmatrix} = \begin{bmatrix} \cos\phi & \sin\phi & 0 \\ -\sin\phi & \cos\phi & 0 \\ 0 & 0 & 1 \end{bmatrix} \begin{bmatrix} x \\ y \\ z \end{bmatrix},$$

$$\text{(1b)} \quad \begin{bmatrix} x_1' \\ y_1' \\ z' \end{bmatrix} = \begin{bmatrix} 1 & 0 & 0 \\ 0 & \cos\theta & \sin\theta \\ 0 & -\sin\theta & \cos\theta \end{bmatrix} \begin{bmatrix} x_1 \\ y_1 \\ z \end{bmatrix},$$

and

$$(1c)\qquad \begin{bmatrix} x' \\ y' \\ z' \end{bmatrix} = \begin{bmatrix} \cos\psi & \sin\psi & 0 \\ -\sin\psi & \cos\psi & 0 \\ 0 & 0 & 1 \end{bmatrix} \begin{bmatrix} x_1 \\ y_1' \\ z' \end{bmatrix}$$

so that

$$(2)\qquad \begin{bmatrix} x' \\ y' \\ z' \end{bmatrix} = \begin{bmatrix} \cos\psi & \sin\psi & 0 \\ -\sin\psi & \cos\psi & 0 \\ 0 & 0 & 1 \end{bmatrix} \begin{bmatrix} 1 & 0 & 0 \\ 0 & \cos\theta & \sin\theta \\ 0 & -\sin\theta & \cos\theta \end{bmatrix} \begin{bmatrix} \cos\phi & \sin\phi & 0 \\ -\sin\phi & \cos\phi & 0 \\ 0 & 0 & 1 \end{bmatrix} \begin{bmatrix} x \\ y \\ z \end{bmatrix}$$

which, on performing the matrix multiplications, can be written

$$(3)\qquad \begin{bmatrix} x' \\ y' \\ z' \end{bmatrix} = A(\phi,\theta,\psi) \begin{bmatrix} x \\ y \\ z \end{bmatrix},$$

where the rotation matrix $A(\phi,\theta,\psi)$ is given by

$$(4)\qquad A(\phi,\theta,\psi) =$$

$$\begin{bmatrix} \cos\psi\cos\phi - \cos\theta\sin\phi\sin\psi & \cos\psi\sin\phi + \cos\theta\cos\phi\sin\psi & \sin\psi\sin\theta \\ -\sin\psi\cos\phi - \cos\theta\sin\phi\cos\psi & -\sin\psi\sin\phi + \cos\theta\cos\phi\cos\psi & \cos\psi\sin\theta \\ \sin\theta\sin\phi & -\sin\theta\cos\phi & \cos\theta \end{bmatrix}.$$

A is orthogonal and has determinant +1. Its nine components a_{ij} are just the direction cosines of the angles between new and old axes which specify the rotation. The matrices $A(\phi,\theta,\psi)$ are said to form a representation of the group of rotations in 3-dimensional space and the angles ϕ,θ,ψ are called Euler angles. Every possible rotation (group element) is determined by an appropriate choice of the angles ϕ,θ,ψ and hence the Euler angles are called 'group parameters.'

3.3. A DOUBLE-VALUED REPRESENTATION OF THE GROUP OF ROTATIONS IN 3-DIMENSIONAL SPACE

Let $Q = \begin{bmatrix} \alpha & \beta \\ \gamma & \delta \end{bmatrix}$ be a two-by-two unitary matrix with complex entries and determinant equal to +1. We regard Q as a linear transformation on a linear vector space over the field of complex numbers. Although eight real quantities are needed to specify the four entries in Q uniquely (two for each complex number a + ib), they are not independent, since the components of Q must satisfy the unitary conditions

$$\begin{aligned} \alpha^*\alpha + \beta^*\beta &= 1, \\ \gamma^*\gamma + \delta^*\delta &= 1, \\ \alpha^*\gamma + \beta^*\delta &= 0, \end{aligned} \tag{5}$$

as well as the determinantal condition

$$\alpha\delta - \beta\gamma = 1. \tag{6}$$

These conditions reduce the number of independent, real-valued quantities to three, the same number necessary to specify a rotation (see problem 2).

Let P be a self-adjoint or Hermitian (i.e. $P^+ = P$) traceless 2 x 2 matrix. Thus P has the form

$$P = \begin{bmatrix} c & a - bi \\ a + bi & -c \end{bmatrix} \tag{7}$$

where a,b,c are real and i is the imaginary unit. The Hermitian and traceless character of P is invariant under similarity transformation so that if

$$P' = QPQ^+ \qquad (Q^+ = Q^{-1}), \tag{8}$$

then

$$P' = \begin{bmatrix} c' & a' - b'i \\ a' + b'i & -c' \end{bmatrix}, \tag{9}$$

where c',a',b' are again real. Since the determinant of a matrix is invariant under similarity transformations, it follows that

(10) $$\det P = -(a^2 + b^2 + c^2) = -(a'^2 + b'^2 + c'^2) = \det P'$$

so that the quadratic form $(a^2 + b^2 + c^2)$ is an invariant. Under rotation the quantity $x^2 + y^2 + z^2$ remains invariant (where x,y,z are the coordinates of a point in 3-dimensional space). We have already seen that the unitary matrices with determinant 1 have the right number of independent quantities to specify a rotation. Thus if we put $a = x$, $b = y$, $c = z$ in P so that

(11) $$P = \begin{bmatrix} z & x - iy \\ x + iy & -z \end{bmatrix}$$

we should be able to construct a Q for each rotation R, such that

(12) $$\begin{bmatrix} z' & x' - iy' \\ x' + iy' & -z' \end{bmatrix} = P' = QPQ^+$$

where R maps (x,y,z) onto (x',y',z') (this is usually denoted by R: (x,y,z) → (x',y',z')) in the manner given by equation (3). Let Q_1, Q_2, Q_3 be unitary matrices with determinant 1 corresponding to rotations R_1, R_2, and R_3 respectively such that

(13) $$\begin{aligned} R_1&: (x,y,z) \rightarrow (x',y',z'), \\ R_2&: (x',y',z') \rightarrow (x'',y'',z''), \\ R_3&: (x,y,z) \rightarrow (x'',y'',z''). \end{aligned}$$

The last of equations (13) implies that $R_2 \cdot R_1 = R_3$. Now:

$$Q_3PQ_3^+ = P'' = Q_2P'Q_2^+ = Q_2Q_1PQ_1^+Q_2^+;$$

but it can be shown that $Q_1^+Q_2^+ = (Q_2Q_1)^+$ so that

(14) $$Q_3PQ_3^+ = (Q_2Q_1)\ P\ (Q_2Q_1)^+.$$

Thus

(15) $$Q_3 = \pm Q_2 Q_1.$$

The plus and minus signs in equation (15) reflect the fact that if Q corresponds to a rotation R, then so does -Q since the matrix P is multiplied by both -Q and $-Q^+$, thus cancelling the effect of the minus sign.

The matrix which corresponds to a rotation ϕ about the z axis $Q_{\hat{z}}\phi$ (or $-Q_{\hat{z}}\phi$) is given by

(16) $$Q_{\hat{z}}\phi = \begin{bmatrix} e^{i\phi/2} & 0 \\ 0 & e^{-i\phi/2} \end{bmatrix}$$

since

$$\begin{bmatrix} z & x_1-iy_1 \\ x_1+iy_1 & -z \end{bmatrix} = \begin{bmatrix} e^{i\phi/2} & 0 \\ 0 & e^{-i\phi/2} \end{bmatrix} \begin{bmatrix} z & x-iy \\ x+iy & -z \end{bmatrix} \begin{bmatrix} e^{-i\phi/2} & 0 \\ 0 & e^{i\phi/2} \end{bmatrix}$$

$$= \begin{bmatrix} z & (x\cos\phi+y\sin\phi)-i(-x\sin\phi+y\cos\phi) \\ (x\cos\phi+y\sin\phi)+i(-x\sin\phi+y\cos\phi) & -z \end{bmatrix}$$

which is in agreement with the usual transformation equation (1a). Similarly, the matrix which corresponds to a rotation θ about the x axis $Q_{\hat{x}}\theta$ (or $-Q_{\hat{x}}\theta$) is given by

(17) $$Q_{\hat{x}}\theta = \begin{bmatrix} \cos\theta/2 & i\sin\theta/2 \\ i\sin\theta/2 & \cos\theta/2 \end{bmatrix}$$

since

$$\begin{bmatrix} z' & x_1'-iy_1' \\ x_1'+iy_1' & -z' \end{bmatrix} = \begin{bmatrix} \cos\theta/2 & i\sin\theta/2 \\ i\sin\theta/2 & \cos\theta/2 \end{bmatrix} \begin{bmatrix} z & x_1-iy_1 \\ x_1+iy_1 & -z \end{bmatrix} \begin{bmatrix} \cos\theta/2 & -i\sin\theta/2 \\ -i\sin\theta/2 & \cos\theta/2 \end{bmatrix}$$

$$= \begin{bmatrix} z\cos\theta - y_1\sin\theta & x_1 - i(y_1\cos\theta + z\sin\theta) \\ x_1 + i(y_1\cos\theta + z\sin\theta) & -z\cos\theta + y_1\sin\theta \end{bmatrix}$$

which is in agreement with equation (1b). The matrix corresponding to a rotation through Euler angles ϕ,θ,ψ is thus given by a matrix Q (or -Q) where (see equation (15))

$$(18)\qquad Q = Q_{\hat{z}}\psi\ Q_{\hat{x}}\theta\ Q_{\hat{z}}\phi$$

$$= \begin{bmatrix} e^{i\psi/2} & 0 \\ 0 & e^{-i\psi/2} \end{bmatrix} \begin{bmatrix} \cos\theta/2 & i\sin\theta/2 \\ i\sin\theta/2 & \cos\theta/2 \end{bmatrix} \begin{bmatrix} e^{i\phi/2} & 0 \\ 0 & e^{-i\phi/2} \end{bmatrix}$$

$$= \begin{bmatrix} e^{i(\psi+\phi)/2}\cos\theta/2 & ie^{i(\psi-\phi)/2}\sin\theta/2 \\ ie^{-i(\psi-\phi)/2}\sin\theta/2 & e^{-i(\psi+\phi)/2}\cos\theta/2 \end{bmatrix}.$$

The appearance of half-angles in equation (18) results in the association of both Q and -Q with the same rotation (i.e. $Q_{\hat{z}}0 = I$ and $Q_{\hat{z}}2\pi = -I$ but both 2π and 0 are the same rotation). Thus we call the set of Q matrices defined by equation (18) a double-valued representation of the group of rotations in 3-dimensional space (i.e. $R_1 \cdot R_2 = R_3 \Rightarrow Q_1 \cdot Q_2 = \pm Q_3$ where Q_1, Q_2, and Q_3 are two-by-two unitary matrices with determinant 1 corresponding to rotations R_1, R_2, and R_3 respectively).

3.4. ROTATIONS IN QUATERNION FORM

The P matrix given in equation (11) can be written as the sum of three matrices

$$(19)\qquad P = x\sigma_1 + y\sigma_2 + z\sigma_3 \equiv \vec{x}\cdot\vec{\sigma}$$

where σ_1, σ_2, σ_3 are the Pauli spin matrices

$$(20)\qquad \sigma_1 = \begin{bmatrix} 0 & 1 \\ 1 & 0 \end{bmatrix},\quad \sigma_2 = \begin{bmatrix} 0 & -i \\ i & 0 \end{bmatrix},\quad \sigma_3 = \begin{bmatrix} 1 & 0 \\ 0 & -1 \end{bmatrix}$$

and

$$(21)\qquad \vec{x} = (x,y,z),\quad \vec{\sigma} = (\sigma_1,\sigma_2,\sigma_3).$$

The rotation matrices $Q_{\hat{x}}\theta$, $Q_{\hat{y}}\theta$, and $Q_{\hat{z}}\theta$ may also be expressed in terms of the Pauli spin matrices and the identity I, that is, in a quaternion[†] basis:

$$
\begin{aligned}
Q_{\hat{x}}\theta &= I\cos\theta/2 + i\sigma_1\sin\theta/2, \\
Q_{\hat{y}}\theta &= I\cos\theta/2 + i\sigma_2\sin\theta/2, \qquad (22)\\
Q_{\hat{z}}\theta &= I\cos\theta/2 + i\sigma_3\sin\theta/2.
\end{aligned}
$$

Each Pauli spin matrix is associated with a rotation about one of the coordinate axes and may be thought of as a unit rotator for that axis. The matrix for a rotation of a Cartesian coordinate system about an arbitrary axis $\hat{n} = (\cos\alpha, \cos\beta, \cos\gamma)$, $\cos^2\alpha + \cos^2\beta + \cos^2\gamma = 1$, counterclockwise[††] through an angle θ is given by

$$(23) \qquad Q_{\hat{n}}\theta = (\cos\theta/2)I + (i\sin\theta/2)(\sigma_1\cos\alpha + \sigma_2\cos\beta + \sigma_3\cos\gamma)$$

$$(24) \qquad = (\cos\theta/2)I + (i\sin\theta/2)(\vec{\sigma}\cdot\hat{n})$$

which is in agreement with the results of problem 2 of Chapter 5. If we interpret e^A, where A is a matrix, as the exponential power series

$$(25) \qquad e^A = \sum_{p=0}^{\infty} (1/p!)A^p$$

then equation (24) may be written

$$(26) \qquad Q_{\hat{n}}\theta = e^{i(\theta/2)\vec{\sigma}\cdot\hat{n}}$$

[†]Quaternions are defined in problem 2 of Chapter 5. They are essentially 4-component objects with specific rules of addition and multiplication; they play a fundamental role in the theory of the motion of rigid tops.

[††]Equivalently, one could say it is the matrix for a rotation of a vector clockwise through an angle θ in a fixed coordinate system. When physical vectors are fixed and the coordinate system is rotated, we refer to the passive case. The active case corresponds to rotating the physical vector in the opposite sense but in a fixed coordinate system.

since

$$e^{i(\theta/2)\vec{\sigma}\cdot\hat{n}} = \sum_{p=0}^{\infty} (1/p!)(i\theta\vec{\sigma}\cdot\hat{n}/2)^p$$

$$= \sum_{p=0}^{\infty} [1/(2p)!](i\theta/2)^{2p}(\vec{\sigma}\cdot\hat{n})^{2p}$$

$$+ \sum_{p=0}^{\infty} [1/(2p+1)!](i\theta/2)^{2p+1}(\vec{\sigma}\cdot\hat{n})^{2p+1}$$

$$= I\sum_{p=0}^{\infty} [(-1)^p/(2p)!](\theta/2)^{2p}$$

$$+ i(\vec{\sigma}\cdot\hat{n})\sum_{p=0}^{\infty} [(-1)^p/(2p+1)!](\theta/2)^{2p+1}$$

$$= (\cos\theta/2)I + (i\sin\theta/2)(\vec{\sigma}\cdot\hat{n}).$$

We have used the results

$$\sigma_1^2 = \sigma_2^2 = \sigma_3^2 = I, \tag{27}$$
$$\sigma_i\sigma_j = -\sigma_j\sigma_i = i\sigma_k, \quad i,j,k \text{ cyclic,}$$

which together imply that $(\vec{\sigma}\cdot\hat{n})^{2p} = I$ where p is a positive integer.

3.5. SPINORS

We define a first-rank spinor ξ_k as an ordered set of two components which transform under rotation in the following way:

$$\begin{bmatrix} \xi_1' \\ \xi_2' \end{bmatrix} = \begin{bmatrix} e^{i(\psi+\phi)/2}\cos\theta/2 & ie^{i(\psi-\phi)/2}\sin\theta/2 \\ ie^{-i(\psi-\phi)/2}\sin\theta/2 & e^{-i(\psi+\phi)/2}\cos\theta/2 \end{bmatrix} \begin{bmatrix} \xi_1 \\ \xi_2 \end{bmatrix} \tag{28}$$

or, in summation notation,

$$\xi_i' = \alpha_{ij}\xi_j, \quad i,j \in \{1,2\}, \tag{29}$$

where α_{ij} are the four components of the transformation matrix

$$(30) \qquad \alpha = \begin{bmatrix} e^{i(\psi+\phi)/2}\cos\theta/2 & ie^{i(\psi-\phi)/2}\sin\theta/2 \\ ie^{-i(\psi-\phi)/2}\sin\theta/2 & e^{-i(\psi+\phi)/2}\cos\theta/2 \end{bmatrix}$$

An ordered set of two components which transform by the complex conjugate of the transformation matrix given in equation (30) is also called a first-rank spinor but of the second kind[†] and is denoted by $\xi_{\dot{k}}$:

$$(31) \qquad \xi'_{\dot{k}} = \alpha^*_{\dot{k}\dot{\ell}}\xi_{\dot{\ell}}, \qquad \dot{k},\dot{\ell} \in \{\dot{1},\dot{2}\},$$

where the $\alpha^*_{\dot{k}\dot{\ell}}$ are the components of the transformation matrix

$$(32) \qquad \alpha^* = \begin{bmatrix} e^{-i(\psi+\phi)/2}\cos\theta/2 & -ie^{-i(\psi-\phi)/2}\sin\theta/2 \\ -ie^{i(\psi-\phi)/2}\sin\theta/2 & e^{i(\psi+\phi)/2}\cos\theta/2 \end{bmatrix}.$$

A second-rank spinor $\xi_{\ell m}$ is defined as an ordered set of four components which transform like the products $\xi_1\xi_1$, $\xi_1\xi_2$, $\xi_2\xi_1$, $\xi_2\xi_2$ where ξ_1 and ξ_2 are the components of a first-rank spinor. Thus

$$(33) \qquad \xi'_{\ell m} = \alpha_{\ell s}\alpha_{mr}\xi_{sr}, \qquad \ell,m,s,r \in \{1,2\}.$$

An ordered set of four components $\xi_{\dot{\ell}\dot{m}}$ which transform like

$$(34) \qquad \xi'_{\dot{\ell}\dot{m}} = \alpha^*_{\dot{\ell}\dot{s}}\alpha^*_{\dot{m}\dot{r}}\xi_{\dot{s}\dot{r}}$$

and an ordered set of four components $\xi_{\ell\dot{m}}$ which transform like

$$(35) \qquad \xi'_{\ell\dot{m}} = \alpha_{\ell s}\alpha^*_{\dot{m}\dot{r}}\xi_{s\dot{r}}$$

are also second-rank spinors. Higher-rank spinors are defined analogously.

[†]O. Laporte and G. Uhlenbeck, Phys. Rev. 37, 1380 (1931). Dotted indices are not different from undotted indices, but simply indicate that the spinor involved is of the second kind.

The four components $P_{1\dot{1}}$, $P_{1\dot{2}}$, $P_{2\dot{1}}$, $P_{2\dot{2}}$ defined by

$$(36)\qquad P = \begin{bmatrix} z & x - iy \\ x + iy & -z \end{bmatrix} = \begin{bmatrix} P_{1\dot{1}} & P_{1\dot{2}} \\ P_{2\dot{1}} & P_{2\dot{2}} \end{bmatrix}$$

transform under rotation like (see equations (12) and (18))

$$(37)\qquad P'_{s\dot{r}} = \alpha_{s\ell} P_{\ell\dot{m}} \alpha^{*}_{\dot{r}\dot{m}} = \alpha_{s\ell} \alpha^{*}_{\dot{r}\dot{m}} P_{\ell\dot{m}}$$

so that the $P_{\ell\dot{m}}$ are components of a mixed second-rank spinor (we say mixed because one index is dotted and one undotted). Thus a vector is associated with a mixed second-rank spinor and in general an rth-rank tensor is associated with a mixed spinor of rank 2r (with r dotted indices and r undotted indices).

Equation (28) defines a double-valued transformation. Thus for a physical quantity to transform as an odd-rank spinor it must be double-valued (i.e. we say ψ is double-valued if $\psi(\phi + 2\pi) = -\psi(\phi)$). Even-rank spinors, however, are not double-valued (the minus sign occurs an even number of times so the effect cancels) so that the single-valued tensor quantities appear as even-rank spinors. In this sense spinors represent a generalization of tensors, allowing for both single- and double-valued quantities.

Spinors are not just mathematical curiosities, but play a fundamental role in relativistic quantum theory, as we shall see in Chapter 9. In fact, it turns out that the wave-functions of all elementary particles of half-odd integer spin, the so-called fermions, are spinor quantities and transform under rotations according to the above rules.

3.6. REFLECTIONS AND INVERSIONS - PSEUDOTENSORS

In 2-dimensional space we can define two distinct sets of axes (as illustrated in Figure 3.2) which cannot be brought into coincidence with each other either by rotation or translation of axes or both. One set can be obtained from the other,

Figure 3.2. Right- and left-handed coordinate frames in 2-dimensions.

however, by inverting the direction of either one (but not both) of the axes. We call such a transformation a reflection. It is a different kind of transformation of coordinates from either rotations or translations in that only a finite number of distinct transformations occur (two in 2-dimensions) and the change is dramatic (i.e. $-\infty$ changes to $+\infty$). By contrast, rotations and translations are continuously infinite in number and can be arbitrarily small (so-called infinitesimal transformations). Reflections are called discrete transformations because they are not continuously parametrized. For example, in the 2-dimensional case, reflection of the x-axis changes the coordinates of a point P from (x,y) to (x',y') where

$$\begin{bmatrix} x' \\ y' \end{bmatrix} = \begin{bmatrix} -1 & 0 \\ 0 & 1 \end{bmatrix} \begin{bmatrix} x \\ y \end{bmatrix} \tag{38}$$

and similarly reflection of the y-axis changes the coordinates of P from (x,y) to (x',y') where

$$\begin{bmatrix} x' \\ y' \end{bmatrix} = \begin{bmatrix} 1 & 0 \\ 0 & -1 \end{bmatrix} \begin{bmatrix} x \\ y \end{bmatrix} \tag{39}$$

Reflections† do not form a group since the product of two reflections is a rotation (i.e. an orthogonal matrix with determinant +1) and thus they lack the closure property. In 2-dimensions, for example, simultaneous reflection along both axes, a process we shall refer to as inversion, is equivalent to a rotation through $\pm\pi$ modulo 2π. Thus inversion is already

†By including the identity, the set I,R, where R is a reflection in 1-dimension, forms a group isomorphic to the symmetric group of permutations on two things (see Chapter 5).

contained in the group of rotations and is a distinctly different process for reflection along a single axis. The interesting thing, however, is the fact that a lack of uniqueness exists in choosing a Cartesian coordinate system in 2-dimensional space. Two essentially different choices can be made - each is equally good, but the descriptions differ. In order that a set of measurements referred to a coordinate system be meaningful it is necessary not only to specify the position of the coordinate system (relative to some standard position) in terms of rotation and translation parameters, but also to specify whether it is of one type or the other.

In three dimensions (Figure 3.3) the same situation exists but it is more apparent and more interesting. A right-handed coordinate system (RHS) cannot be brought into coincidence with a left-handed coordinate system (LHS) by any combination of rotations or translations, however clever. In addition to rotations and translations, a discrete transformation of reflection or inversion (three simultaneous reflections of orthogonal axes) is required. Two simultaneous reflections will not do, being equivalent to a rotation only. (Note that in three dimensions inversion is not the same as a rotation.) In specifying the location of a coordinate system relative to a standard it is necessary to state the required rotation, the required translation, and the required inversion (i.e. one or none) to achieve coincidence. This duality of geometric description has a deep physical significance as we shall see in part here but will pursue in greater detail in Chapters 6 and 9.

Consider now a position vector $\vec{r}$ (with components r_j) in 3-dimensional space; we clearly have

(40) $$r'_j = -r_j$$

where $\vec{r}\,'$ is the same physical vector as $\vec{r}$ (i.e. $\vec{r}\,' = \vec{r}$) but described in the inverted (i.e. primed) coordinate system.[†] Any vector which behaves in a like manner under inversion we shall call a true vector, a polar vector, or, more simply, just a 'vector.'

[†]In the active case, where the inversion is performed on the vector and the coordinate system remains unchanged, we have $\vec{r}\,' = -\vec{r}$.

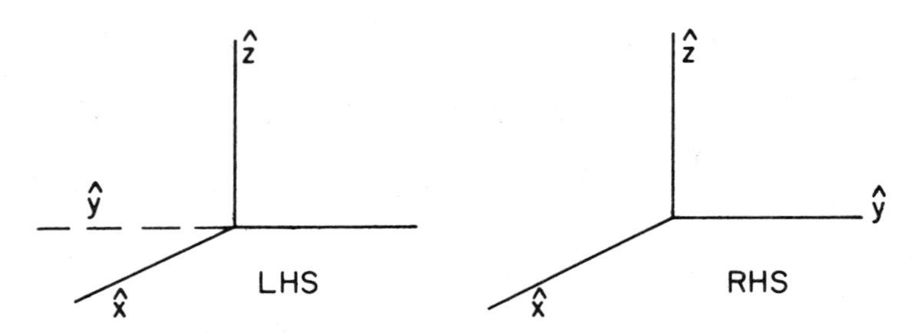

Figure 3.3. Right- and left-handed coordinate frames in 3-dimensions.

Some vector quantities, however, do not behave in this fashion under inversion, but rather satisfy

(41) $L'_j = L_j$.

Such vectors we shall call axial or pseudovectors. They possess the correct rotational properties but the wrong inversion property. Important examples are not hard to come by. Angular velocity vectors

(42) $\vec{\omega} = \vec{r} \times \vec{v}$

are clearly of this type, as are the angular momentum vectors

(43) $\vec{L} = \vec{r} \times m\vec{v}$.

In general the cross-product of two polar vectors yields an axial vector since if $\vec{v}$ and $\vec{u}$ are polar vectors then

(44) $$(\vec{v}' \times \vec{u}')_i = v'_j u'_k - v'_j u'_k = (-v_j)(-u_k) - (-v_k)(-u_j)$$
$$= v_j u_k - v_k u_j = (\vec{v} \times \vec{u})_i, \quad i,j,k \text{ cyclic.}$$

A particularly interesting example of a pseudovector is the magnetic field $\vec{B}$. If we assume the electric charge e is unaffected by reflection (e' = e) then from $\vec{F} = e\vec{E}$, where $\vec{E}$ is the electric field vector and $-F_j = F'_j$ (i.e. force is a true vector), it follows that $\vec{E}$ is a true vector. However, $\vec{B}$ is defined as a cross-product

(45) $$d\vec{B} = \frac{i d\vec{\ell} \times \vec{r}}{r^3}$$

in Ampere's law so that $\vec{B}$ is a pseudovector.

Once it is admitted that pseudovectors exist, other pseudotensors follow necessarily. Consider first the scalar quantity

(46) $$s = \vec{r}\cdot\vec{r} = (-r_j')\cdot(-r_j') = \vec{r}'\cdot\vec{r}' = s'$$

so that the length of a vector is unchanged under inversion. Clearly also quantities such as time and temperature are unaffected by inversion of the coordinate system. We call such quantities true scalars or simply scalars. By contrast, the 'scalar' $K = \vec{r}\cdot\vec{B}$, the component of $\vec{B}$ along a fixed direction $\vec{r}$, is a pseudoscalar since it is invariant under rotation but changes sign under inversion:

(47) $$K' = \vec{r}'\cdot\vec{B}' = -\vec{r}\cdot\vec{B} = -K.$$

Similar statements can be made about higher-order tensors; for example true second-rank tensors have components that do not change sign under inversion but second-rank pseudotensors do. This leads us to draw up the following table.

Tensor rank	Pseudonym	Sign change under inversion
zero	true scalar	+
	pseudoscalar	-
one	true vector	-
	pseudovector	+
two	true tensor	+
	pseudotensor	-
three	true tensor	-
	pseudotensor	+
etc.	etc.	etc.

3.7. INVARIANT TENSORS

We call a tensor invariant if its components are the same in all coordinate systems. We have already encountered one example of an invariant tensor, namely δ_{ij}. To show that the Kronecker delta is a second-rank tensor we must verify that its definition

$$\delta_{ij} = \begin{cases} 1 & \text{if } i = j, \\ 0 & \text{if } i \neq j, \end{cases} \tag{48}$$

is consistent with the transformation properties of a second-rank tensor (under rotation). This follows from the orthogonality of the rotation matrix, as follows:

$$\begin{aligned} \delta'_{ij} &= a_{ik}a_{j\ell}\delta_{k\ell} \\ &= a_{ik}a_{jk} = \delta_{ij}. \end{aligned} \tag{49}$$

Also under reflection $\delta_{ij} = \delta'_{ij}$ so that δ_{ij} is a true second-rank tensor rather than a pseudo second-rank tensor.

In 3-dimensional space one can define another invariant tensor by

$$\varepsilon_{ijk} = \begin{cases} +1 & \text{if } i \neq j \neq k,\ i,j,k \text{ cyclic}, \\ -1 & \text{if } i \neq j \neq k,\ i,j,k \text{ anticyclic}, \\ 0 & \text{otherwise}. \end{cases} \tag{50}$$

ε_{ijk} is called the Levi-Civita tensor. To prove that ε_{ijk} is a third-rank tensor we must show that its definition is consistent with the transformation properties of a third-rank tensor (under rotation). This follows at once from the relationship of ε_{ijk} to the structure of determinants:

$$\varepsilon'_{\ell mn} = a_{\ell i}a_{mj}a_{nk}\varepsilon_{ijk} \tag{51}$$

$$= \det \begin{bmatrix} a_{1\ell} & a_{2\ell} & a_{3\ell} \\ a_{1m} & a_{2m} & a_{3m} \\ a_{1n} & a_{2n} & a_{3n} \end{bmatrix}$$

The determinant of a matrix changes sign if two columns are interchanged and is zero if two columns are the same. Rotation matrices, however, have determinant one, so that when $\ell = 1$, $m = 2$, and $n = 3$

$$\varepsilon'_{123} = \det \begin{bmatrix} a_{11} & a_{12} & a_{13} \\ a_{21} & a_{22} & a_{23} \\ a_{31} & a_{32} & a_{33} \end{bmatrix} = 1. \tag{52}$$

Hence, one has

$$\varepsilon'_{\ell mn} = \begin{cases} +1 & \text{if } \ell \neq m \neq n, \quad \ell,m,n \text{ cyclic,} \\ -1 & \text{if } \ell \neq m \neq n, \quad \ell,m,n \text{ anticyclic,} \\ 0 & \text{otherwise,} \end{cases}$$

and $\varepsilon_{\ell mn}$ is a third-rank tensor. Under inversion $\varepsilon'_{ijk} = \varepsilon_{ijk}$ so that the Levi-Civita tensor is a third-rank pseudotensor.

The notion of a Levi-Civita tensor is easily generalized to a space of arbitrary dimension. In n-dimensional space the Levi-Civita tensor $\varepsilon_{i_1 \ldots i_n}$ is defined by

$$\varepsilon_{i_1 \ldots i_n} = \begin{cases} +1 & \text{if } \begin{pmatrix} 1,2,\ldots,n \\ i_1,i_2,\ldots,i_n \end{pmatrix} \text{is an even permutation,} \\ -1 & \text{if } \begin{pmatrix} 1,2,\ldots,n \\ i_1,i_2,\ldots i_n \end{pmatrix} \text{ is an odd permutation,} \\ 0 & \text{otherwise.} \end{cases} \tag{53}$$

(In equation (53), $i_1, i_2 \ldots i_n$ is a rearrangement of the integers $1,2,3\ldots n$.) The proof of the tensor character of $\varepsilon_{i_1 \ldots i_n}$ is similar to the 3-dimensional case.

3.8. AXIAL VECTORS IN 3-SPACE AS SECOND-RANK TENSORS

Consider the three components of a vector product of two vectors $\vec{c} = \vec{a} \times \vec{b}$, that is

$$c_i = a_j b_k - a_k b_j, \quad i,j,k \text{ cyclic.} \tag{54}$$

From the right-hand side of this equation we should expect this quantity to transform as an antisymmetric second-rank tensor, say T_{jk}, whose components are given by

$$T_{jk} = a_j b_k - a_k b_j, \tag{55}$$

and this is, indeed, the case as can easily be verified. How then do we justify writing cross-products in vector notation? To answer this consider the vector d_i formed from the double contraction

$$d_i = \varepsilon_{ijk} T_{jk}/2 = \varepsilon_{ijk}(a_j b_k - a_k b_j)/2 = \varepsilon_{ijk} a_j b_k. \tag{56}$$

Putting in the values of ε_{ijk}, we see that d_i is in fact equal to c_i. Thus in 3-dimensional space, antisymmetric second-rank tensors have three independent components which happen to transform as the components of a vector. However, d_i is not a true vector but a pseudovector since it does not change sign under inversion.

The pseudotensor ε_{ijk} can be used to construct pseudotensors of other ranks, for example

$$\varepsilon_{ijk} s_k = P_{ij} \tag{57}$$

(where s_k is a true vector) is a pseudotensor of rank two.

3.9. VECTOR AND TENSOR FIELDS

When we say that the electric field $\vec{E}$ is a vector, we mean it is a vector function $\vec{E}(\vec{r},t)$ defined at each point in space-time (i.e. it is a set of three functions of space-time coordinates, such that at each point of space-time $(\vec{r},t)$, say, the three numbers $E_x(\vec{r},t)$, $E_y(\vec{r},t)$, $E_z(\vec{r},t)$ transform under coordinate rotation as the components of a vector). Mass distributions are scalar fields, whereas moments of inertia and products of inertia for a rigid body with one point fixed are 2nd-rank tensor functions of the position of the fixed point.

It turns out that the wave-fields which define the properties of elementary particles usually have simple tensor properties (e.g. photons are described by a vector field, π-mesons by a pseudoscalar field, etc.). The number of independent tensor components bears a simple relation to the spin quantum number s, viz. it equals 2s + 1, for example:

Particle	Spin	Number of independent components
vector photon	one	3
pseudoscalar meson	zero	1
tensor graviton	two	5

Note that 2nd-rank, traceless symmetric tensors have five independent components in 3-dimensions.

However, the situation is complicated by the fact that the wave functions which describe particles are not directly measurable; only their square modulus is measurable as a probability function. This means that a wave function can be double-valued. For example, if we consider only dependence on azimuth ϕ,[†] we can have

$$\psi(\phi + 2\pi) = -\psi(\phi)$$

since this preserves the single-valuedness of probability

$$|\psi(\phi + 2\pi)|^2 = |\psi(\phi)|^2.$$

[†]Here ϕ is regarded as an azimuthal variable which specifies either the angular position of the centre of mass of the particle or the angular orientation of its internal structure, if any, with respect to the centre of mass. Note that the requirement of single-valuedness is distinct from the usual statement that $\psi(\phi)$ and $\psi(\phi)e^{i\alpha}$, where α is an arbitrary phase constant, represent the same state.

Naturally one has

$$\psi(\phi + 4\pi) = \psi(\phi)$$

for such double-valued functions.

Thus a more basic form than tensors exists, viz. spinors with a correspondence as follows:

Spinor rank	Tensor character	Spin
0	scalar	0
1	no equivalent	1/2
2	vector	1
3	no equivalent	3/2
4	2nd-rank tensor	2
	etc.	

The spin 1/2 particles such as the electron, proton, and neutron and the half-odd integer 'resonances' have a basically spinor character.

When composite particles are formed, the spinor character of the complex is determined by the laws of angular momentum coupling (and vice versa). Analysis of this leads to the vector-coupling rule for the possible J values†:

$$\vec{J}_1 + \vec{J}_2 = \vec{J}, \quad J = (J_1 + J_2), (J_1 + J_2 - 1), \ldots, |(J_1 - J_2)|$$

in integral steps. Thus nuclei, for example, have half-odd integer spins or integer spins depending upon whether the number of particles (nucleons) is odd or even.

Finally, let us remark that a very deep connection exists between 'spin' and 'statistics,' systems having half-odd integer spins obeying Fermi statistics and systems having integer spins obeying Bose statistics. The fermions (with spins 1/2, 3/2, 5/2, etc.) are anti-social particles in the

†The vector $\vec{J}$ is the total angular momentum. In the case where $\vec{J}$ is due to intrinsic (i.e. internal) angular momentum only, $\vec{J}$ is called the spin. $\vec{J}$ is related to the magnitude of the angular momentum vector $|\vec{J}|$, by $|\vec{J}|^2 = \hbar^2 J(J + 1)$ (see Chapter 6).

sense that no two identical fermions can exist in the same quantum state (Pauli exclusion principle) whereas the bosons prefer to be in the same state. The former behaviour leads to the stable structure of matter as we know it (electrons, protons, and neutrons are the 'building blocks'), while the latter property makes it possible to attain intense macroscopic fields, e.g. those associated with electromagnetic energy radiated from a broadcast antenna or a laser tube.

3.10. COVARIANCE OF PHYSICAL LAWS

Laws of physics are said to be covariant under a set of transformations if the form of the law is the same in every transformed coordinate system. For example, Newton's second law

(58) $$\vec{F} = m\vec{a}$$

is covariant under the group of rotations (actually it is covariant under the full orthogonal group, i.e. including reflections) in 3-dimensional space since $\vec{F}$ and $\vec{a}$ are vectors and m is a scalar. The vector and scalar nature of $\vec{a}$ and m respectively is evident; but how do we justify calling $\vec{F}$ a vector? The force on a particle is defined as the negative of the gradient of the potential energy of the particle V. The potential energy depends on the magnitude of the separation of particles and is a scalar:

(59) $$V' = V.$$

Now

(60) $$x'_k = a_{kj}x_j \Rightarrow \frac{\partial x'_k}{\partial x_\ell} = a_{kj}\,\frac{\partial x_j}{\partial x_\ell} = a_{kj}\delta_{j\ell} = a_{k\ell}.$$

Thus

(61) $$F_\ell = -\frac{\partial V}{\partial x_\ell} = -\frac{\partial V'}{\partial x'_k}\,\frac{\partial x'_k}{\partial x_\ell} = F'_k a_{k\ell}$$

so that force is indeed a vector.

Newton's law is covariant with respect to a larger group of transformations than rotations. In particular it preserves form under transformations of the Galilean type

(62) $$\vec{r}' = \vec{r} - \vec{v}t$$

or in component form

$$x'_k = x_k - v_k t$$

corresponding to a coordinate system moving uniformly with relative velocity $\vec{v}$ with respect to a standard one. One has

(63) $$\frac{d\vec{r}'}{dt} = \frac{d\vec{r}}{dt} - \vec{v}$$

but $\vec{a}' = d^2r'/dt^2 = d^2r/dt^2 = \vec{a}$ and $m = m'$.

The difference between two position vectors is invariant under Galilean transformations

(64) $$\vec{r}_1' - \vec{r}_2' = \vec{r}_1 - \vec{v}t - (\vec{r}_2 - \vec{v}t) = \vec{r}_1 - \vec{r}_2$$

so that the potential energy is also invariant under Galilean transformations

$$V = V'.$$

In this case, however, $\partial x'_k/\partial x_\ell = \delta_{k\ell}$ so that

(65) $$F_\ell = -\frac{\partial V}{\partial x_\ell} = -\frac{\partial V'}{\partial x'_k}\frac{\partial x'_k}{\partial x_\ell} = F'_k\delta_{k\ell} = F'_\ell$$

and Newton's law in the uniformly moving frame remains unaltered in the form

(66) $$\vec{F}' = m'\vec{a}'.$$

Coordinate systems in which the force law takes on its simplest form are said to be inertial. The earth frame (or laboratory frame) is not quite inertial if the solar frame is, since the earth rotates about its own axis and revolves around the sun. Experience shows that the sun is a better inertial frame than the earth, and the frame of the fixed stars is still a better inertial frame than the sun.

To see that a frame rotating with respect to an inertial frame is not inertial consider the following. Assume that the laboratory frame, with coordinate axes x,y,z, is inertial and consider a frame with coordinate axes x',y',z' rotating with constant velocity $\vec{\omega}$ rad/sec with respect to it. Let $\vec{A}$ be a vector with components A_x, A_y, A_z in the inertial laboratory frame and components A'_x, A'_y, A'_z in the rotating frame. Without loss of generality we may assume that $\vec{\omega} = \omega\hat{z}$. Then from Figure 3.4

$$(67) \quad \vec{A} = A'_x\, \hat{x}' + A'_y\, \hat{y}' + A'_z\, \hat{z}'$$

$$= A'_x(\hat{x}\cos\omega t + \hat{y}\sin\omega t) + A'_y(-\hat{x}\sin\omega t + \hat{y}\cos\omega t) + A'_z\hat{z}'.$$

Thus

$$(68) \quad \frac{d\vec{A}}{dt} = \omega A'_x(-\hat{x}\sin\omega t + \hat{y}\cos\omega t) + \omega A'_y(-\hat{x}\cos\omega t - \hat{y}\sin\omega t)$$

$$+ \frac{dA'_x}{dt}\,\hat{x}' + \frac{dA'_y}{dt}\,\hat{y}' + \frac{dA'_z}{dt}\,\hat{z}'$$

$$= \vec{\omega} \times \vec{A} + \left(\frac{d\vec{A}}{dt}\right)_R$$

where $(d\vec{A}/dt)_R$ denotes the rate of change of A as seen by an observer at rest in the rotating frame. Applying equation (68) to the vector $d\vec{A}/dt$ rather than $\vec{A}$ itself yields

$$(69) \quad \frac{d^2\vec{A}}{dt^2} = \vec{\omega} \times \left[\vec{\omega} \times \vec{A} + \left(\frac{d\vec{A}}{dt}\right)_R\right] + \frac{d}{dt}\left[\vec{\omega} \times \vec{A} + \left(\frac{d\vec{A}}{dt}\right)_R\right]_R$$

$$= \vec{\omega} \times (\vec{\omega} \times \vec{A}) + 2\vec{\omega} \times \left(\frac{d\vec{A}}{dt}\right)_R + \left(\frac{d^2\vec{A}}{dt^2}\right)_R .$$

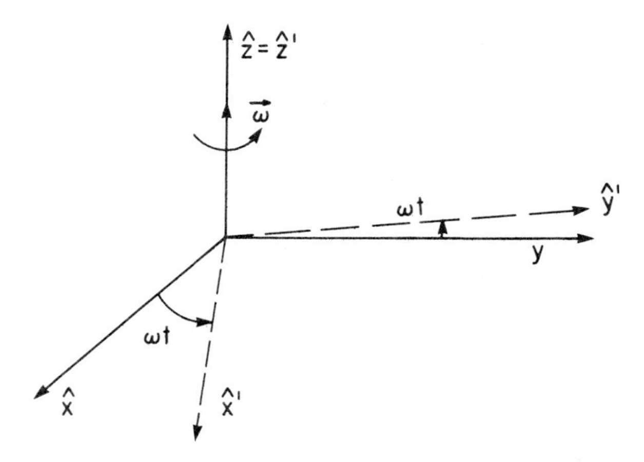

Figure 3.4. Laboratory and rotating frames.

If $\vec{A}$ is the position vector $\vec{r}$ of a particle of mass m

$$(70) \quad \frac{d^2\vec{r}}{dt^2} = \left(\frac{d^2\vec{r}}{dt^2}\right)_R + 2\vec{\omega} \times \left(\frac{d\vec{r}}{dt}\right)_R + \vec{\omega} \times (\vec{\omega} \times \vec{r})$$

or $\quad \vec{a} = \vec{a}' + 2\vec{\omega} \times \vec{v}' + \vec{\omega} \times (\vec{\omega} \times \vec{r})$

where $\vec{a}'$ and $\vec{v}'$ are respectively the acceleration and velocity of the particle as seen by an observer at rest in the rotating frame. Thus Newton's law $\vec{F} = m\vec{a}$ becomes

$$(71) \quad \vec{F}' = m'\vec{a}' + 2m'\vec{\omega} \times \vec{v}' + m'\vec{\omega} \times (\vec{\omega} \times \vec{r}')$$

in a frame rotating with constant angular velocity $\vec{\omega}$. (Note that we have used $\vec{F} = \vec{F}'$, $\vec{r} = \vec{r}'$, and $m = m'$.) From equation (71) we see that Newton's law is clearly not covariant under changes to rotating frames. The term $-2m\vec{\omega} \times \vec{v}'$ is called the Coriolis force and $-m\vec{\omega} \times (\vec{\omega} \times \vec{r})$ is the usual expression for the centripetal force.

For a rigid body rotating with angular velocity $\vec{\omega}$, the law $d\vec{L}/dt = \vec{N}$ becomes

$$(72) \quad \left(\frac{d\vec{L}}{dt}\right)_R + \vec{\omega} \times \vec{L} = \vec{N},$$

which, using coordinate axes fixed in the body, can be written

$$I\,\frac{d\vec{\omega}}{dt} + \vec{\omega} \times (I\vec{\omega}) = \vec{N}. \tag{73}$$

These are the famous Euler equations for the motion of a rigid body.

REFERENCES

1 Elie Cartan, The Theory of Spinors. MIT Press, Cambridge, Mass. 1966
2 H. Goldstein, Classical Mechanics. Addison Wesley, Cambridge, Mass., 1950
3 A. Kyrala, Theoretical Physics: Applications of Vectors, Matrices, Tensors and Quaternions. W.B. Saunders, Philadelphia, 1967
4 C.W. Misner, K.S. Thorne, and J.A. Wheeler, Gravitation. W.H. Freeman Co., San Francisco, 1973
5 Max Morand, Géométrie spinorielle. Masson et Cie, Paris, 1973

PROBLEMS

1 Prove that a rotation of coordinate axes labelled x,y,z through Euler angles ϕ,θ,ψ is equivalent to a rotation of coordinate axis x,y,z counterclockwise through an angle ψ about the z-axis followed by a rotation of the new coordinate axes counterclockwise through an angle θ about the original x-axis, followed by a rotation through an angle ϕ counterclockwise about the original z-axis.

2 Show that any two-by-two unitary matrix with determinant 1 may be written in the form

$$Q = \begin{bmatrix} \alpha & \beta \\ -\beta^* & \alpha^* \end{bmatrix}$$

with the condition that $\alpha\alpha^* + \beta\beta^* = 1$.

3 (i) Show that as the angle of rotation gets very small (i.e. infinitesimal rotation) the result of a rotation of a coordinate system has the form

$$x'_k = (\Omega_{kj} + \delta_{kj})x_j \quad \text{or} \quad x'_k - x_k = dx_k = \Omega_{kj}\, x_j$$

where Ω_{kj} is a 3 x 3 antisymmetric matrix whose components are infinitesimals.

(ii) Show that infinitesimal rotations commute.

(iii) Prove that infinitesimal rotations are pseudovector quantities.

4 If ξ_i and η_i are first-rank spinors show that

$$\xi_1\eta_2 - \xi_2\eta_1$$

is invariant under rotations.

5 Prove that $\varepsilon_{ijk}\varepsilon_{k\ell m} = \delta_{i\ell}\delta_{jm} - \delta_{im}\delta_{j\ell}$.

6 Express the following vector quantities or identities in tensor form:

(i) the triple scalar product $a\cdot(\vec{b} \times \vec{c})$.

(ii) curl curl $\vec{F}$ = grad(div $\vec{F}$) - (div grad)$\vec{F}$,

(iii) $\vec{F} = m\vec{\omega} \times (\vec{\omega} \times \vec{r}) + 2m\vec{\omega} \times \vec{v} + m\vec{a}$,

(iv) curl $(\vec{a} \times \vec{b}) = (\vec{b}\cdot\vec{\nabla})\vec{a} - (\vec{a}\cdot\vec{\nabla})\vec{b} + \vec{a}$ div $\vec{b} - \vec{b}$ div $\vec{a}$.

7 Express in spinor form the tensor $T_{jk} = x_j x_k$.

8 Prove that Maxwell's equations

$$\vec{\nabla}\cdot\vec{E} = 4\pi\rho,$$
$$\vec{\nabla}\cdot\vec{B} = 0,$$
$$\vec{\nabla} \times \vec{E} = -\partial\vec{B}/c\partial t,$$
$$\vec{\nabla} \times \vec{B} = 4\pi\vec{J}/c + \partial\vec{E}/c\partial t$$

are not covariant under Galilean transformation.

9 A train is travelling south with velocity $\vec{v}$. When it is located at latitude ϕ, find the force it exerts on the side of the track due to the rotation of the earth.

4
Cartesian tensors at work

4.1. THEORY OF ELASTIC CONTINUA

(a) *Local rotations, compressions, and shears - the strain tensor*

When stresses are applied to an elastic or quasi-elastic medium, distortion occurs. Even in simple cases, for example the twisting of a rod or the bending of a beam, shear takes place. An elastic medium is one in which the atoms return to their original configurations when the stresses are removed. For any given medium, permanent deformations may take place if the stresses exceed the elastic limits.

The elasticity of crystalline solids is of particular interest in physics since sound waves in such solids correspond to elastic waves propagating through the medium, and in modern microscopic theory the various thermodynamic and transport properties such as specific heats and heat conductivity can be analysed to a large extent in terms of quantized wave disturbances called phonons propagating through the crystal.

To begin with, however, we shall not consider the crystalline structure itself, but rather treat the solid as a continuous medium in thermodynamic equilibrium at some fixed temperature T. For us, an infinitesimal volume is one which is macroscopically small but microscopically large (i.e. it still contains an enormous number of atoms). In some given coordinate system, which we take to be Cartesian, a given point in the unstressed medium may be located by the position vector $\vec{r}$. The vector $\vec{r}$ should be regarded as an independent variable with no dependence on the time. Suppose that under moderate stress the atoms in the immediate neighbourhood of $\vec{r}$ displace an amount $\vec{\rho}$

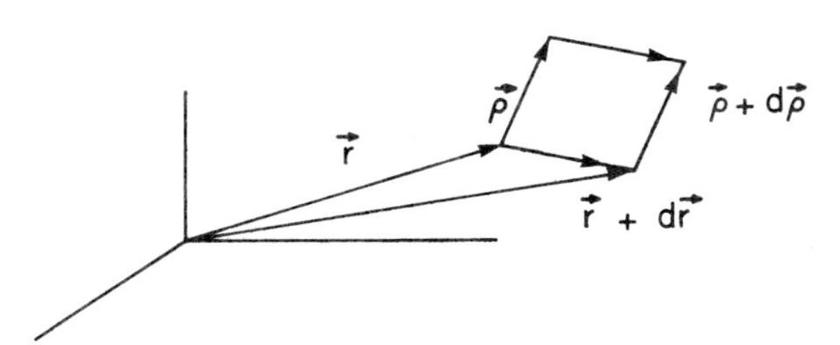

Figure 4.1. Displacements of neighbouring points in the medium due to deformation.

(1) $\vec{r} \to \vec{r} + \vec{\rho}.$

as shown in Figure 4.1.

Similarly, a neighbouring point $\vec{r} + d\vec{r}$ goes over into $\vec{\rho} + \vec{r} + d\vec{r} + d\vec{\rho}$. Let $\vec{r}$ have components x_k and $\vec{\rho}$ components ξ_k. If the deformations are small, as they must be if the stresses are moderate, we can regard $\vec{\rho}$ as a continuous and differentiable function of $\vec{r}$:

(2) $\vec{\rho} = \vec{\rho}(\vec{r})$

or, in tensor notation,

(3) $\xi_k = \xi_k(x_1, x_2, x_3).$

Hence, we can write

$$(4) \quad d\xi_k = \frac{\partial \xi_k}{\partial x_1} dx_1 + \frac{\partial \xi_k}{\partial x_2} dx_2 + \frac{\partial \xi_k}{\partial x_3} dx_3$$

$$= \frac{\partial \xi_k}{\partial x_j} dx_j = \Sigma_{jk} dx_j$$

where Σ_{jk} is defined by these relations; alternatively we may write

(5) $d\vec{\rho} = (d\vec{r} \cdot \nabla)\vec{\rho}.$

To see the significance of these equations for elasticity theory, we rewrite Σ as a sum of symmetric and antisymmetric parts

$$\Sigma_{kj} = \tfrac{1}{2}(\Sigma_{kj} + \Sigma_{jk}) + \tfrac{1}{2}(\Sigma_{kj} - \Sigma_{jk}) = E_{kj} + R_{kj} \tag{6}$$

where $E_{kj} = E_{jk}$ and $R_{kj} = -R_{jk}$. It is important to realize that the tensors Σ_{jk}, E_{jk}, and R_{jk} are really tensor fields since each of the nine components is a continuous, differentiable function of $\vec{r}$.

We can now show quite easily that the differential effect of $R_{kj}(\vec{r})$ is a 'rigid' local rotation of the medium in the neighbourhood of point $\vec{r}$. Since such rotations do not in themselves constitute deformation of the medium at the point $\vec{r}$, we conclude that only the symmetric part E_{kj} of the matrix Σ_{kj} is concerned with true deformation.

Of course, a local rotation at point $\vec{r}$ in the medium is likely to produce a true deformation at some distance away, where shears may occur. For example, if rigid rotations in the same sense were to occur at two separated points $\vec{r}_1$ and $\vec{r}_2$, shear would inevitably occur at some intermediate point $\vec{r}_3$. At this intermediate point, however, the shear would be described by $E_{kj}(\vec{r}_3)$ and not by $R_{kj}(\vec{r}_3)$.

We return now to the problem of showing that R_{kj} by itself implies only local rotation. For convenience, therefore, we set $E_{kj} \equiv 0$. Consider now the point $\vec{r}$ and two neighbouring points $\vec{r} + d\vec{r}$ and $\vec{r} + d\vec{r}\,'$ with angle θ between $d\vec{r}$ and $d\vec{r}\,'$. Hence

$$\cos\theta = \frac{d\vec{r}\cdot d\vec{r}\,'}{|d\vec{r}|\cdot|d\vec{r}\,'|} = \frac{dx_k dx_k'}{|d\vec{r}|\cdot|d\vec{r}\,'|} . \tag{7}$$

After deformation, we have

$$\begin{aligned} \vec{r} &\to \vec{r} + \vec{\rho}, \\ \vec{r} + d\vec{r} &\to \vec{r} + \vec{\rho} + d\vec{r} + d\vec{\rho}, \\ \vec{r} + d\vec{r}\,' &\to \vec{r} + \vec{\rho} + d\vec{r}\,' + d\vec{\rho}\,', \end{aligned} \tag{8}$$

the angle θ' being given by

$$(9) \qquad \cos\theta' = \frac{(dx_k + d\xi_k)(dx'_k + d\xi'_k)}{|d\vec{r} + d\vec{\rho}|\cdot|d\vec{r}' + d\vec{\rho}'|}$$

where k is a running index.

We emphasize now that elastic theory implies small deformation, i.e. in general the $d\xi_k$ are much smaller than the dx_k and we are justified in dropping terms of higher order than the first in the $d\xi_k$. Keeping this in mind, we can write the numerator of the above expression as

$$(10) \qquad dx_k dx'_k + dx'_k d\xi_k + dx_k d\xi'_k = dx_k dx'_k + dx_j dx'_k(R_{jk} + R_{kj})$$
$$= dx_k dx'_k \ ,$$

where we have juggled dummy indices on the right-hand side and used the antisymmetry of R_{jk}.

Consider now the denominator of $\cos\theta'$; in particular,

$$(11) \qquad |d\vec{r} + d\vec{\rho}|^2 \cong |d\vec{r}|^2 + 2d\vec{r}\cdot d\vec{\rho} + \text{higher-order terms}$$
$$\cong |d\vec{r}|^2 + 2dx_k dx_j R_{jk}$$
$$= |d\vec{r}|^2$$

where we have again used the antisymmetry of R_{jk}. Similarly, $|d\vec{r}' + d\vec{\rho}'|^2 \cong |d\vec{r}'|^2$ and hence $\cos\theta' \cong \cos\theta$.

Note that $d\vec{r}\cdot d\vec{\rho}$ and $d\vec{r}'\cdot d\vec{\rho}'$ are zero, i.e. these vector pairs are orthogonal, by virtue of the antisymmetry of R_{jk}. Since the displacement $d\vec{\rho}$ is perpendicular to $d\vec{r}$ for all choices of $d\vec{r}$, it is evident that R_{jk} corresponds to local rotation about point $\vec{r}$.

From the definitions

$$d\xi_k = \frac{\partial \xi_k}{\partial x_j} dx_j = \Sigma_{jk} dx_j$$

we also see that Σ_{jk} and hence also E_{jk} and R_{jk} are second-rank tensor quantities. In the particular case of R_{jk} we can associate with it an axial vector $\vec{K}$ by the usual procedures:

$$(12)\qquad K_i = \frac{1}{2}\,\varepsilon_{imn}\,\Sigma_{mn} = R_{jk}$$

$$= \frac{1}{2}(\Sigma_{jk} - \Sigma_{kj}),\qquad i,j,k \text{ cyclic},$$

$$= \frac{1}{2}\left(\frac{\partial \xi_k}{\partial x_j} - \frac{\partial \xi_j}{\partial x_k}\right) = \frac{1}{2}(\text{curl }\vec{\rho})_i .$$

Therefore

$$(13)\qquad \vec{K} = \frac{1}{2}\text{ curl }\vec{\rho}.$$

It follows that wherever $\vec{\rho}(\vec{r})$ has a non-vanishing curl, local rotation takes place.

Clearly the strain must be described by the symmetric part of Σ_{jk}. For convenience we define the strain tensor operator to be

$$(14)\qquad \delta_{jk} + E_{jk} \equiv \begin{bmatrix} 1 + \dfrac{\partial \xi_1}{\partial x_1} & \dfrac{1}{2}\left[\dfrac{\partial \xi_2}{\partial x_1} + \dfrac{\partial \xi_1}{\partial x_2}\right] & \cdots \\ \dfrac{1}{2}\left[\dfrac{\partial \xi_2}{\partial x_1} + \dfrac{\partial \xi_1}{\partial x_2}\right] & \cdots & \cdots \\ \cdots & \cdots & 1 + \dfrac{\partial \xi_3}{\partial x_3} \end{bmatrix}$$

since then, local rotation being disregarded, it has the interpretation of taking the undeformed vector $d\vec{r}$ into the deformed vector $d\vec{r} + d\vec{\xi}$:

$$(15)\qquad \begin{bmatrix} dx_1 \\ dx_2 \\ dx_3 \end{bmatrix} \Longrightarrow \begin{bmatrix} dx_1 + d\xi_1 \\ dx_2 + d\xi_2 \\ dx_3 + d\xi_3 \end{bmatrix}.$$

(b) *Dilatations*

Let us first note that, in tensor notation, we can write the triple scalar product of three vectors as

(16) $\vec{c}\cdot(\vec{a} \times \vec{b}) = c_k \varepsilon_{k\ell m} a_\ell b_m = \varepsilon_{k\ell m} c_k a_\ell b_m.$

It is also well known that this triple scalar product has a value equal to the volume of a parallelepiped with sides defined by the triad $\vec{a}$, $\vec{b}$, $\vec{c}$.

Now let $\vec{a}$, $\vec{b}$, $\vec{c}$ define a very small parallelepiped in the medium at point $\vec{r}$ (essentially three separate choices of $d\vec{r}$). Suppose that after deformation these vectors become $\vec{a} + \vec{\alpha}$, $\vec{b} + \vec{\beta}$, and $\vec{c} + \vec{\gamma}$ where

(17) $\alpha_k = E_{jk} a_j, \quad \beta_k = E_{jk} b_j, \quad \gamma_k = E_{jk} c_j.$

The new parallelepiped occupies a volume

(18) $(\vec{c} + \vec{\gamma})\cdot[(\vec{a} + \vec{\alpha}) \times (\vec{b} + \vec{\beta})]$

$= \varepsilon_{k\ell m}(c_k + \gamma_k)(a_\ell + \alpha_\ell)(b_m + \beta_m)$

$= \varepsilon_{k\ell m} c_k a_\ell b_m + \varepsilon_{k\ell m}(\gamma_k a_\ell b_m + c_k \alpha_\ell b_m + c_k a_\ell \beta_m)$

+ higher orders in $\gamma\alpha$, $\gamma\beta$, $\alpha\beta$, and $\alpha\beta\gamma$ which we neglect

$= \varepsilon_{k\ell m} c_k a_\ell b_m + \varepsilon_{k\ell m}(E_{jk} c_j a_\ell b_m + E_{j\ell} c_k a_j b_m + E_{jm} c_k a_\ell b_j)$

$= \varepsilon_{k\ell m} c_k a_\ell b_m + (\varepsilon_{j\ell m} E_{kj} + \varepsilon_{kjm} E_{\ell j} + \varepsilon_{k\ell j} E_{mj}) c_k a_\ell b_m$

where we have interchanged (j,k), (j,ℓ), and (j,m) indices in the last three terms.

$= \varepsilon_{k\ell m} c_k a_\ell b_m [1 + (E_{11} + E_{22} + E_{33})]$

$=$ (original volume)(1 + Trace E).

This last result follows from the identity

(19) $$(\varepsilon_{j\ell m}E_{kj} + \varepsilon_{kjm}E_{\ell j} + \varepsilon_{k\ell j}E_{mj}) = \varepsilon_{k\ell m}(E_{11} + E_{22} + E_{33})$$

which can be shown easily (viz., one sees directly that the left-hand side is zero unless $k \neq \ell \neq m$ using the antisymmetry of $\varepsilon_{k\ell m}$; if $k \neq \ell \neq m$, the right-hand side follows at once from the fact that each j-sum has only one non-vanishing term).

Hence

(20) the dilatation = change in volume per unit volume

= Trace E.

It is at least satisfying to see that the dilatation is a scalar quantity independent of the orientation of the coordinate system. Moreover, we can see this result is intuitively correct since $E_{11}dx_1$ is the increase in the length dx_1 in the x_1-direction, etc., so that one is neglecting volumes such as $E_{11}E_{22}dx_1dx_2dx_3$ in comparison with $E_{11}dx_1dx_2dx_3$, etc.

In this way, we see that the diagonal elements have to do with purely volume effects, either expansion or contraction. On the other hand, the off-diagonal elements such as $\partial\xi_2/\partial x_1$ clearly describe the shear strains and local rotations since they predict the incremental 'lateral displacement' $(\partial\xi_2/\partial x_1)dx_1$ which occurs in going from one point to another dx_1 units away in the 1-direction.

(c) *The stress tensor*

When deformation of an elastic medium occurs there is a change in its microscopic structure (i.e. the arrangement of its molecules) away from equilibrium. Short-range molecular forces arise which tend to return the medium to its equilibrium configuration. These forces are called internal stresses, and in the macroscopic theory of elasticity they are taken to be of virtually zero range. Thus the forces exerted on a particular portion of an elastic medium due to the surrounding regions act only over its bounding surface. The internal stresses give rise to a force density σ_j which, by convention, is defined so that the jth component of the force on a small volume element dV is given by $\sigma_j dV$.

Consider now a portion V (bounded by a surface S) of an elastic medium. By Newton's third law (action equals reaction) the total force $\vec{F}$ on V is due to external forces, which act only on the surface S according to the above argument. Since V can be arbitrarily small we anticipate the existence of a stress tensor field σ_{jk} of which the force density is the divergence

$$\text{(21)} \qquad \sigma_k = \partial \sigma_{jk}/\partial x_j .$$

This enables us to express the total force as a surface integral

$$F_k = \int_V \sigma_k dV = \int_V \frac{\partial \sigma_{jk}}{\partial x_j} dV = \int_S \sigma_{jk} da_j ,$$

where $da_k = (da)n_k$, da being an area element and n_k the outward-drawn normal to it.

Clearly σ_{jk} is the kth component of the force/area acting on a small area perpendicular to the jth direction. The components σ_{jj} are considered positive if they tend to expand the volume in question (see Figure 4.2). For $j \neq k$, σ_{jk} over any face is regarded as positive if it contributes a positive torque in the i-direction where j, k, and i are cyclic in that order.

Similarly (if the forces are central) the total torque $\vec{N}$ on V is only due to external forces acting across the surface S. It follows that the kth component of the torque is given by

$$\text{(22a)} \qquad N_k = \int_S (\sigma_{\ell i} x_j - \sigma_{\ell j} x_i) da_\ell , \qquad i,j,k \text{ cyclic.}$$

We can also express the torque as the sum of torques acting on individual volume elements dV, i.e.

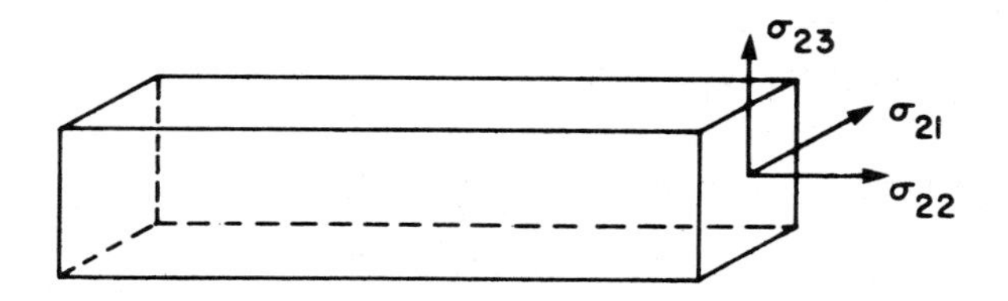

Figure 4.2. Forces per unit area acting positively over the face of a parallelepiped. As drawn, σ_{23} and σ_{21} are positive.

$$N_k = \int_V (x_j\sigma_i - x_i\sigma_j)dV = \int_V\left(x_j\frac{\partial\sigma_{\ell i}}{\partial x_\ell} - x_i\frac{\partial\sigma_{\ell j}}{\partial x_\ell}\right)dV$$

$$= \int_V\frac{\partial}{\partial x_\ell}(\sigma_{\ell i}x_j - \sigma_{\ell j}x_i)dV - \int_V\left(\sigma_{\ell i}\frac{\partial x_j}{\partial x_\ell} - \sigma_{\ell j}\frac{\partial x_i}{\partial x_\ell}\right)dV.$$

Using $\partial x_i/\partial x_k = \delta_{ik}$ we can write

$$\text{(22b)}\quad N_k = \int_S(\sigma_{\ell i}x_j - \sigma_{\ell j}x_i)da_\ell - \int_V(\sigma_{ji} - \sigma_{ij})dV.$$

Subtracting equation (22b) from (22a) then yields

$$\int_V(\sigma_{ji} - \sigma_{ij})dV = 0.$$

Since the volume V is arbitrary the integrand must vanish identically, so that

$$\sigma_{ji} = \sigma_{ij},$$

which proves that the stress tensor is symmetric on its indices.

(d) *Hooke's law - a linear approximation*

Hooke's law supposes linear relationships between the stress and strain coefficients at each point in the elastic medium:

(23) $\sigma_{ij} = S_{ijk\ell}E_{k\ell}$,

where S is the stiffness tensor, or alternatively

(24) $E_{ij} = C_{ijk\ell}\sigma_{k\ell}$,

where C is the compliance tensor. The components of the stiffness tensor have the dimensions of [force]/[area] or [energy]/[volume]; the components of the compliance tensor have the reciprocal of these dimensions.

It is easy to show that the components of C and S at each point in the medium transform like the components of a fourth-rank tensor and hence represent tensor fields. In a homogeneous medium C and S may be regarded as elastic constants, since their $\vec{r}$ dependence then vanishes. In the most general case, it is conceivable that C and S depend not only on space, but also on time and temperature. We shall not consider such cases here. The quantity $S_{ijk\ell}$ has $3^4 = 81$ components. However, since σ_{ij} is symmetric $S_{ijk\ell}$ is also symmetric under interchanges of i and j. The symmetry of $E_{k\ell}$ implies that

(25) $\sigma_{ij} = S_{ijk\ell}E_{k\ell} = \frac{1}{2}(S_{ijk\ell} + S_{ij\ell k})E_{k\ell}$

and so without loss of generality we may assume $S_{ij\ell k}$ is symmetric under interchanges of ℓ and k. Hence there are 6 independent ij and 6 independent $k\ell$ combinations, leaving only 36 independent quantities. Furthermore, if one assumes that the stress energy density U is a function of the state of stress of the material only, then one can write (see problem 1)

(26) $U = \frac{1}{2}S_{ijk\ell}E_{ij}E_{k\ell}$.

Furthermore, we have

(27) $\sigma_{ij} = \frac{1}{2}(\partial U/\partial E_{ij}),\ i \neq j, \quad \sigma_{jj} = \partial U/\partial E_{jj}$

so that

$$(28) \qquad \frac{\partial \sigma_{ij}}{\partial E_{k\ell}} = \frac{\alpha \partial^2 U}{\partial E_{k\ell} \partial E_{ij}} = \frac{\alpha \partial^2 U}{\partial E_{ij} \partial E_{k\ell}} = \frac{\partial \sigma_{k\ell}}{\partial E_{ij}}, \qquad \alpha = \begin{cases} \frac{1}{2}, & i \neq j \\ 1, & i = j \end{cases}.$$

It follows from (23) and (28) that we have the additional symmetry property $S_{ijk\ell} = S_{k\ell ij}$, leaving only 21 fundamental constants to characterize the elastic properties of the medium. Crystal symmetries will further reduce the number of independent constants. For example, a cubic crystal has but three independent constants, and a general homogeneous and isotropic medium is characterized by only two constants. For the latter case the two constants can be chosen to be Young's modulus (the ratio of stress in the direction of a force to tensile strain along it) and Poisson's ratio (the magnitude of the ratio of compressional strain at right angles to a force and the tensional strain along it).

(e) *Principal stresses and strains*

Let $E_{jk}(\vec{r})$ be the strain tensor components as functions of point $\vec{r}$ in the medium. At a particular point $\vec{r}$ the E_{jk} are a set of nine numbers (only six of which are independent) which transform under rotation of coordinate systems according to

$$(29) \qquad E'_{jk} = a_{jn} a_{km} E_{nm}$$

or

$$(30) \qquad E' = AEA^{-1}$$

in matrix notation.

Now it is well known in mathematics that a symmetric matrix with real components can be diagonalized by an orthogonal transformation. It follows that there exists some rotation matrix A, and correspondingly some orientation of coordinate systems, such that E' is a diagonal matrix:

$$(31) \qquad E'_{jk} = 0 \quad \text{if } j \neq k.$$

The diagonal components are then referred to as principal strains. From our previous discussion we see that in this coordinate system it appears as if no shears exist and only dilatation occurs (the dilatation is a scalar and is invariant under such rotations of coordinate systems). Of course, we anticipate that the diagonalizing matrix A will vary as E varies from point to point in the medium. For any given A, certain points in the medium will show no shear, but other points will.

We can also find an orthogonal matrix B at any point in the medium which diagonalizes the stress tensor. Hence we can also talk in terms of principal stresses. In general the same rotation matrix will not simultaneously diagonalize both the stress and strain tensors. Hence the directions of principal stresses and strains are not, in general, coincident.

(f) *Isotropic cubic crystals*

For a coordinate system aligned with the cubic axes, one has

(32a) $S_{1111} = S_{2222} = S_{3333} = \lambda + 2\mu,$

(32b) $S_{2323} = S_{3131} = S_{1212} = \mu,$

(32c) $S_{1122} = S_{2233} = S_{3311} = \lambda,$

where λ and μ are called Lamé's constants. All the components of the stiffness tensor $S_{ijk\ell}$ which cannot be made equal to the above via the symmetry conditions $S_{ijk\ell} = S_{jik\ell}$, $S_{ijk\ell} = S_{ij\ell k}$, and $S_{ijk\ell} = S_{k\ell ij}$ are zero.†

To evaluate the components of the compliance tensor we write equation (23) as a matrix equation and invert†† the S matrix, yielding

---

†See S. Bhagavantam, Crystal Symmetry and Physical Properties (Academic Press, London, 1966), Chapter 11.

††Equations (23) and (24) can be interpreted as matrix equations by regarding $\sigma_{11},\ldots,\sigma_{33}$ and $E_{11},\ldots,E_{33}$ as 9-component column vectors. S and C are then 9 x 9 non-singular matrices and S^{-1} = (

(33a) $$C_{1111} = C_{2222} = C_{3333} = \frac{\lambda + \mu}{\mu(3\lambda + 2\mu)},$$

(33b) $$C_{1122} = C_{2233} = C_{3311} = \frac{-\lambda}{2\mu(3\lambda + 2\mu)},$$

and

(33c) $$C_{2323} = C_{3131} = C_{1212} = 1/\mu.$$

All the components of the compliance tensor $C_{ijk\ell}$ which cannot be made equal to one of the above via the symmetry conditions $C_{ijk\ell} = C_{jik\ell}$, $C_{ijk\ell} = C_{ij\ell k}$, and $C_{ijk\ell} = C_{k\ell ij}$ are zero.

The compressibility K is given by

(34) $$K = \frac{\text{dilatation}}{\text{normal stress}} = \frac{E_{11} + E_{22} + E_{33}}{p} = \frac{3E_{11}}{p}$$

$$= \frac{3(C_{1111}\sigma_{11} + C_{1122}\sigma_{22} + C_{1133}\sigma_{33})}{p}$$

$$= \frac{3p(C_{1111} + 2C_{1122})}{p}$$

$$= \left(\lambda + \frac{2}{3}\mu\right)^{-1},$$

where p is the normal stress or pressure. The bulk modulus B is defined by $B = 1/K$.

Young's modulus E and Poisson's ratio σ are given by†

(35) $$E = \sigma_{11}/E_{11}, \qquad \sigma = -E_{22}/E_{11},$$

and Lamé's constants are related to Young's modulus and

†To explain the minus sign in eq. (35) consider a small cylinder of material oriented along the x_1-axis as in Figure 4.3. A tensile stress $\sigma_{11} > 0$ will produce extension in the cylinder's length and contraction in its cross-section. Thus $E_{11} > 0$, $E_{22} < 0$, $E_{33} < 0$, and σ is positive as required. (Note that $\sigma_{ij} = 0$ unless $i = j = 1$ and $E_{ij} = 0$ for $i \neq j$, whereas $E_{22} = E_{33}$ in eq. (35).)

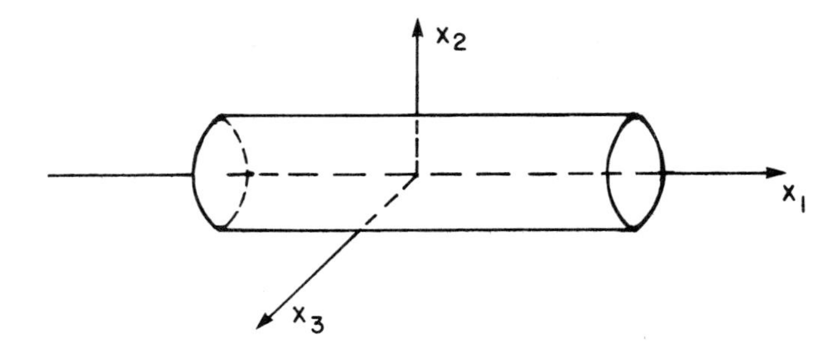

Figure 4.3. Cylinder used to illustrate Young's modulus and Poisson's ratio.

Poisson's ratio via

$$(36)\qquad \mu = \frac{1}{2}\frac{E}{1+\sigma}, \qquad \lambda = \frac{E}{1+\sigma}\frac{\sigma}{1-2\sigma}.$$

4.2. WAVE PROPAGATION IN ELASTIC SOLIDS

Up to this point we have considered static deformations as the result either of permanently stressed media or of imposed external stress forces. We consider now dynamic stresses and strains and show how Newton's equation applied differentially allows solutions describing the propagation of elastic longitudinal and transverse waves in the medium. A more general treatment would include heat transport and heat dissipation through coupling of mechanical and thermal modes of excitation.

Consider a small rectangular element (Figure 4.4) located at point $\vec{r}$ in the medium and bounded by the 6 surfaces corresponding to x_ℓ = constant, $x_\ell + \Delta x_\ell$ = constant, $\ell = 1,2,3$. We neglect body forces (i.e. forces proportional to the volume) and consider the motion of the rectangular element under the action of compressional and shear pressures σ_{ij}. Let the displacement of the point $\vec{r}$ as a function of time be $\vec{\xi}(\vec{r},t)$, or in tensor notation $\xi_\ell(\vec{r},t)$. The acceleration of the element is then $\partial^2\xi_\ell/\partial t^2$ where the partial derivative recognizes that the point $\vec{r}$ is a reference point fixed in space with respect to which the medium displaces $\xi_\ell(\vec{r},t)$. This acceleration is

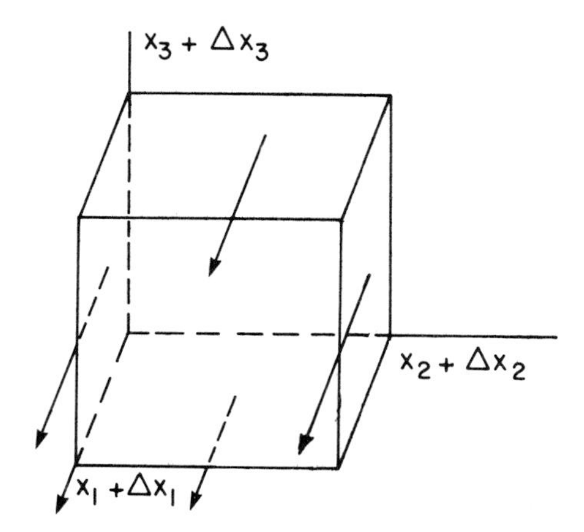

Figure 4.4. Rectangular element at point (x_1, x_2, x_3) and having volume $\Delta x_1 \cdot \Delta x_2 \cdot \Delta x_3$. Shearing stresses in the x_1-direction are indicated.

related to the net force on the element ΔF_ℓ through the mass $m\Delta x_1 \Delta x_2 \Delta x_3$, where m is the mass density at the point $\vec{r}$:

$$m\Delta x_1 \Delta x_2 \Delta x_3 \; \partial^2 \xi_\ell / \partial t^2 = \Delta F_\ell . \tag{37}$$

The force ΔF_ℓ is the algebraic sum of all forces in the ℓth direction transmitted over the six bounding surfaces. Consider the direction $\ell = 1$ or 'x-direction.' The total force ΔF_1 is given by the sum of the compressional forces acting over the ends perpendicular to the x_1-axis, plus the shear forces in the x_1-direction acting over the other four surfaces:

$$\begin{aligned} \Delta F_1 \cong\; & [\sigma_{11}(x_1 + \Delta x_1, x_2, x_3) - \sigma_{11}(x_1, x_2, x_3)]\Delta x_2 \Delta x_3 \\ & + [\sigma_{21}(x_1, x_2 + \Delta x_2, x_3) - \sigma_{21}(x_1, x_2, x_3)]\Delta x_1 \Delta x_3 \\ & + [\sigma_{31}(x_1, x_2, x_3 + \Delta x_3) - \sigma_{31}(x_1, x_2, x_3)]\Delta x_1 \Delta x_2 \\ & \cong \left(\frac{\partial \sigma_{11}}{\partial x_1} + \frac{\partial \sigma_{21}}{\partial x_2} + \frac{\partial \sigma_{31}}{\partial x_3} \right)_{x_1, x_2, x_3} \cdot \Delta x_1 \Delta x_2 \Delta x_3 . \end{aligned}$$

Hence in the ℓth direction we have

$$(38)\qquad \Delta F_\ell \cong \frac{\partial \sigma_{j\ell}}{\partial x_j} \Delta x_1 \Delta x_2 \Delta x_3 \ ^\dagger$$

which, using Hooke's law and assuming that the elastic moduli are constant, can be expressed as

$$\Delta F_\ell \cong S_{j\ell kn} \frac{\partial E_{kn}}{\partial x_j} \Delta x_1 \Delta x_2 \Delta x_3$$

$$= S_{j\ell kn} \frac{\partial^2 \xi_n}{\partial x_k \partial x_j} \Delta x_1 \Delta x_2 \Delta x_3 = m \Delta x_1 \Delta x_2 \Delta x_3 \frac{\partial^2 \xi_\ell}{\partial t^2} .$$

Hence

$$(39)\qquad m \frac{\partial^2 \xi_\ell}{\partial t^2} = S_{j\ell kn} \frac{\partial^2 \xi_n}{\partial x_k \partial x_j} .$$

In this derivation we have kept only the first-order infinitesimals. We now specialize further to cubic crystals and single out the direction $\ell = 1$ for illustration. One then obtains

$$(40)\qquad m \frac{\partial^2 \xi_1}{\partial t^2} = S_{1111} \frac{\partial^2 \xi_1}{\partial x_1^2} + S_{2121} \frac{\partial^2 \xi_1}{\partial x_2^2} + S_{3131} \frac{\partial^2 \xi_1}{\partial x_3^2}$$

$$+ (S_{1122} + S_{2112}) \frac{\partial^2 \xi_2}{\partial x_1 \partial x_2} + (S_{1133} + S_{3113}) \frac{\partial^2 \xi_3}{\partial x_1 \partial x_3}$$

$$= (\lambda + 2\mu) \frac{\partial^2 \xi_1}{\partial x_1^2} + \mu\left(\frac{\partial^2 \xi_1}{\partial x_2^2} + \frac{\partial^2 \xi_1}{\partial x_3^2}\right)$$

$$+ (\lambda + \mu)\left(\frac{\partial^2 \xi_2}{\partial x_1 \partial x_2} + \frac{\partial^2 \xi_3}{\partial x_1 \partial x_3}\right)$$

$$= \mu \nabla^2 \xi_1 + (\lambda + \mu)\left[\frac{\partial}{\partial x_1}\left(\frac{\partial \xi_1}{\partial x_1} + \frac{\partial \xi_2}{\partial x_2} + \frac{\partial \xi_3}{\partial x_3}\right)\right].$$

---

†This result is equivalent to equation (21).

Hence

(41) $m\, \partial^2\xi_\ell/\partial t^2 = \mu\nabla^2\xi_\ell + (\lambda + \mu)\ \partial(\mathrm{div}\ \vec{\rho})/\partial x_\ell$.

To show that these equations support propagating elastic waves, we consider the choice

(42) $\xi_1 = A \sin(kx_1 - \omega t), \quad \xi_2 = \xi_3 = 0.$

Substitution shows that this choice is a solution provided that

(43) $m\omega^2 = (\lambda + 2\mu)k^2.$

Thus the longitudinal waves propagate with a velocity

(44) $v_L = \omega/k = \sqrt{\lambda + 2\mu}/\sqrt{m}.$

Transverse waves in the x_1-direction correspond to the choice

(45) $\xi_2 = B \sin(kx_1 - \omega t).$

Substitution in the wave equation with $\ell = 2$ yields the condition

(46) $m\omega^2 = k^2\mu$

so that the transverse waves propagate with velocity

(47) $v_T = \omega/k = \sqrt{\mu/m}.$

If $\lambda << \mu$, as is approximately true in some instances, $v_L \cong \sqrt{2}v_T$. In seismic work in geophysics, one uses the time interval between the arrival of longitudinal and transverse waves to predict the distance from the detector to the source of disturbance (e.g. earthquake). The failure of transverse waves to propagate through the centre of the earth is used to predict a 'fluid' core and to determine its extent.†

†See G.D. Garland, Introduction to Geophysics - Mantle, Core and Crust (W.B. Saunders, Philadelphia, 1971), Chapters 1-9, and J.B. Macelwane and F.W. Sohon, Introduction to Theoretical Seismology (John Wiley and Sons, New York, 1936), Vol. I.

4.3. HYDRODYNAMICS

In a fixed coordinate system let a particular fluid element be located at a point x_j^o at a time $t = 0$ and at x_j at time t. Clearly x_j depends on the time, and its evolutionary time derivative is the instantaneous velocity of the fluid element

$$v_j = dx_j/dt. \tag{48}$$

However, we are not interested in just a single fluid element, but in all fluid elements throughout some domain D. In order to label fluid elements and distinguish one from another we identify each fluid element by the (constant) value x_j^o of its position at time $t = 0$.

In other words, we define not only a single trajectory $x_j(t)$, but a whole field of trajectories $x_j(\vec{r}^o,t)$ where, for a given fluid element $\vec{r}^o$, the trajectory is $x_j(\vec{r}^o,t)$ and different trajectories are specified by assigning different values of $\vec{r}^o$. The x_j^o are called Lagrangian or convective coordinates, the x_j Eulerian or spatial coordinates.

Consider now the velocity field. By taking two successive snapshots separated by a short time interval, we could in principle find at time t and at each point in space a local velocity of fluid motion. At the point $x_k(\vec{r}^o,t)$ corresponding to the instantaneous location at time t of element $\vec{r}^o$ the fluid would have a velocity

$$v_k(\vec{r}^o,t) = \frac{dx_k}{dt} = \left.\frac{\partial x_k}{\partial t}\right|_{\vec{r}^o}. \tag{49}$$

In order now to exploit the connection with the theory of deformable bodies just concluded, we introduce a displacement field $\vec{\xi}(\vec{r}^o,t)$ such that

$$x_k(\vec{r}^o,t) = x_k^o + \xi_k(\vec{r}^o,t). \tag{50}$$

Clearly, in terms of the displacement field we can also write

(51) $$v_k(\vec{r}^{\,o},t) = \frac{d\xi_k}{dt} = \left.\frac{\partial \xi_k}{\partial t}\right|_{\vec{r}^{\,o}}$$

In hydrodynamics a rather fundamental role is played by the velocity gradient tensor V_{kj} defined as follows. Consider two neighbouring fluid elements at time t separated by a distance δx_k. We can write

(52) $$\delta x_k = x_k(\vec{r}^{\,o} + \delta\vec{r}^{\,o},\ t) - x_k(\vec{r}^{\,o},t)$$

$$= \left(\frac{\partial x_k}{\partial x_j^o}\right)_t \delta x_j^o = J_{kj}\,\delta x_j^o$$

where J_{kj} is the Jacobian tensor at time t expressing the transformation between the Eulerian and Lagrangian coordinates. If we assume invertibility,† we can also write

(53) $$J_{jk}^{-1} = (\partial x_j^o/\partial x_k)$$

and

(54) $$\delta x_j^o = J_{jk}^{-1}\,\delta x_k.$$

Consider now the velocity differential δv_k,

(55) $$\delta v_k = v_k(\vec{r}^{\,o} + \delta\vec{r}^{\,o},\ t) - v_k(\vec{r}^{\,o},t)$$

$$= \left(\frac{\partial v_k}{\partial x_j^o}\right)_t \delta x_j^o = \left(\frac{\partial v_k}{\partial x_j^o}\right)_t J_{j\ell}^{-1}\,\delta x_\ell.$$

Clearly the velocity gradient tensor at time t for fluid element $\vec{r}^{\,o}$ is

†Invertibility requires continuous third derivatives and places 'good behaviour' requirements on the flow, such as: two elements in the same neighbourhood remain in the same neighbourhood; elements cannot bifurcate or occupy each other's positions; and continuous arcs of elements remain continuous.

$$V_{k\ell} = \frac{\delta v_k}{\delta x_\ell} = \left(\frac{\partial v_k}{\partial x_j^o}\right)_t J_{j\ell}^{-1}.$$

Now $$\left(\frac{\partial v_k}{\partial x_j^o}\right)_t = \left[\frac{\partial}{\partial x_j^o}\left(\frac{\partial x_k}{\partial t}\right)_{\vec{r}^o}\right]_t = \left[\frac{\partial}{\partial t}\left(\frac{\partial x_k}{\partial x_j^o}\right)_t\right]_{\vec{r}^o}$$

$$= \frac{d}{dt}\left(\frac{\partial x_k}{\partial x_j^o}\right)_t = \frac{d}{dt}\left[\delta_{kj} + \left(\frac{\partial \xi_k}{\partial x_j^o}\right)_t\right]$$

$$= \frac{d}{dt}\ \Sigma_{kj}^o = \frac{d}{dt} J_{kj},$$

where we have introduced from equation (4) a deformation tensor Σ_{kj}^o, the zero superscript indicating how the displacement field at time t varies as we go from one fluid element to another. Hence

$$(56) \qquad V_{k\ell} = \frac{d}{dt}\Sigma_{kj}^o \ J_{j\ell}^{-1} = \frac{dJ_{kj}}{dt} J_{j\ell}^{-1}.$$

It is now convenient to separate the velocity gradient into two parts: a symmetric part describing sheer (and possibly dissipative flow) and an antisymmetric part describing local rotation (vorticity):

$$(57) \qquad V_{jk} = e_{jk} + f_{jk}$$
$$= e_{jk} - \tfrac{1}{2}\varepsilon_{jk\ell}\omega_\ell,$$

where

$$(58a) \qquad e_{jk} = \tfrac{1}{2}(V_{jk} + V_{kj}),$$

$$(58b) \qquad f_{jk} = \tfrac{1}{2}(V_{jk} - V_{kj}) = -\tfrac{1}{2}\varepsilon_{jk\ell}\omega_\ell.$$

Here $\vec{\omega} = \vec{\nabla} \times \vec{v}$, i.e. $\omega_\ell = \varepsilon_{\ell mn}\partial v_n/\partial x_m$, is called the <u>vorticity</u> vector.

The stress tensor field at a point in a moving fluid is

defined in the same manner as before except that we now separate the tensor σ_{ij} into isotropic and anisotropic parts:

$$\sigma_{ij} = -p(\vec{r},t)\delta_{ij} + d_{ij}(\vec{r},t) \tag{59}$$

where p is the pressure defined as $-\sigma_{ii}/3 = -(\text{Trace } \sigma)/3$. (The negative sign is consistent with the observation that a fluid can have compressional but not tensional strain.) The anistropic part is referred to as the viscosity or deviatoric stress tensor.† It vanishes for a fluid at rest or in uniform motion and has the following properties:

$$d_{ij} = d_{ji}, \quad \text{Trace } d = d_{ii} = 0. \tag{60}$$

The Navier-Stokes equations arise from an attempt to linearize the basic equations of motion. In the Newtonian approximation we assume that a linear relation exists between the deviatoric stress tensor and the velocity gradient tensor

$$d_{ij} = A_{ijk\ell}V_{k\ell} \tag{61}$$

$$= A_{ijk\ell}e_{k\ell} - \tfrac{1}{2}A_{ijk\ell}\varepsilon_{k\ell m}\omega_m \tag{62}$$

at each point $\vec{r}$, t in space-time.

For an isotropic fluid††

$$A_{ijk\ell} = \eta\delta_{ik}\delta_{j\ell} + \eta'\delta_{i\ell}\delta_{jk} + \eta''\delta_{ij}\delta_{k\ell} \tag{63}$$

where η, η', η'' are functions of $\vec{r}$, t. But from equations (60), (61), and (63)

$$A_{ijk\ell} = A_{jik\ell} \text{ implies that } \eta = \eta'$$

†See, for example, L.D. Landau and E.M. Lifshitz, Fluid Mechanics (Pergamon, London, 1959), Section 15.

††See H. Jeffreys, Cartesian Tensors (Cambridge at the University Press, 1963), Chapter VII. However, only monatomic liquids and gases are isotropic.

and

$$A_{ijk\ell} = A_{ji\ell k} \text{ implies that } A_{ijk\ell}\varepsilon_{k\ell m}\omega_m = 0,$$

so that vorticity does not contribute directly to the viscosity. Substitution of these relations into (63) and of (63) into (62) yields

$$(64) \quad d_{ij} = 2\eta e_{ij} + \eta'' e_{kk}\delta_{ij}$$

$$= \eta(2e_{ij} - 2\delta_{ij}e_{kk}/3) + \varepsilon\delta_{ij}e_{kk}$$

where e_{kk} = rate of dilatation = Trace e and $\varepsilon = 2\eta/3 + \eta''$ is the coefficient of bulk viscosity.† The condition Trace D = 0 = d_{ii} then yields

$$(65) \quad \eta'' = -2\eta/3 \text{ or equivalently } \zeta = 0,$$

so that one finally has

$$(66) \quad d_{ij} = 2\eta(e_{ij} - e_{kk}\delta_{ij}/3)$$

where $\eta(\vec{r},t)$ is identified as the shear viscosity field.

From our work on elastic waves we can readily deduce†† the equations of motion for a fluid element as

$$(67) \quad m\frac{Dv_i}{Dt} = mF_i + \frac{\partial\sigma_{ij}}{\partial x_j}$$

where m is the mass density and F_i the body forces per unit mass. The last term on the right gives the pressures and shear gradient which obviously affect the flow.

Substituting the expression

$$(68) \quad \sigma_{ij} = -p\delta_{ij} + 2\eta(e_{ij} - e_{kk}\delta_{ij}/3)$$

†Tisza (Phys. Rev. 61:531 (1942)) points out the important role of bulk viscosity in sound absorption in liquids and gases.

††The velocity field here is to be regarded as a space-time function $v_k = v_k(\vec{r},t)$ where $\vec{r}$ is specified by $x_k = x_k(\vec{r}^{\,o},t)$.

into the equations of motion yields the Navier-Stokes equation†

(69) $$m\frac{Dv_i}{Dt} = mF_i - \frac{\partial p}{\partial x_i} + \frac{\partial}{\partial x_j}[2\eta(e_{ij} - e_{kk}\delta_{ij}/3)].$$

Here the derivative Dv_i/Dt is the so-called hydrodynamic derivative

(70) $$\frac{Dv_i}{Dt} = \frac{\partial v_i}{\partial t} + \frac{\partial v_i}{\partial x_k}\frac{dx_k}{dt} \text{ or } \frac{D\vec{v}}{Dt} = \frac{\partial \vec{v}}{\partial t} + \vec{v}\cdot\vec{\nabla}\vec{v}.$$

The first term on the right expresses the intrinsic rate of change of the velocity field in time, and the second term expresses changes in velocity at a point due to the fact that in time dt new fluid has moved in to the point in question. Using

---

†In this section we have restricted our attention to the simple dynamics of liquid flow. Several practical considerations magnify the complexity of the theory or emphasize its non-linearity. One of these is the development of instabilities in flow patterns leading to turbulence at high Reynolds numbers R. (R is a characteristic number proportional to the flow velocity and the system dimensions, but inversely proportional to the viscosity.) A second consideration has to do with energy transport including heat production, convection, and conduction. A third consideration has to do with mass diffusion, either of one liquid into another or of suspended foreign particles. All of these problems are of great interest, but their solutions are difficult.

A particularly elegant subject is the propagation of sound and its absorption as a function of frequency in viscous fluids where one has to consider the coupling of density variations to hydrodynamic flow and heat conduction. Even more elegant is the theory of sound propagation in superfluids, notably ^{4}He below the λ point. In a picturesque two-fluid model ^{4}He (at saturated vapour pressure) behaves as if it were composed of two interpenetrating fluids below the superfluid transition temperature of 2.18°K. One fluid has zero viscosity while the other has more or less normal properties. The coupling of the two fluids and the macroscopic quantum behaviour of the superfluid fraction combine to give several novel modes of energy propagation (second, third, and fourth sound). The reader interested in pursuing this fascinating subject is referred to the excellent book by Khalatnikov on superfluid hydrodynamics: I.M. Khalatnikov, Introduction to the Theory of Superfluidity (W.A. Benjamin, New York, 1965). See also J. Wilks, The Properties of Liquid and Solid Helium (Clarendon Press, Oxford, 1967).

these expressions, we can write

$$(71)\qquad \frac{\partial}{\partial t}(mv_i) = mF_i - \frac{\partial p}{\partial x_i} - mv_i\,\frac{\partial v_k}{\partial x_k}$$

$$+ \frac{\partial}{\partial x_j}[2\eta(e_{ij} - e_{kk}\delta_{ij}/3) - mv_iv_j].$$

The non-linearity of the hydrodynamic equations is evident.

Finally, if η is a constant, we can write, from (58a) and the definition of the velocity gradient tensor,

$$(72)\qquad m\,\frac{Dv_i}{Dt} = mF_i - \frac{\partial p}{\partial x_i} + \eta\left[\frac{\partial^2 v_i}{\partial x_j \partial x_j} + \frac{1}{3}\,\frac{\partial e_{kk}}{\partial x_i}\right].$$

For an incompressible fluid (see problem 5)

$$(73)\qquad \nabla\cdot\vec{v} = 0$$

and we arrive at the simpler expression

$$(74)\qquad m\,\frac{Dv_i}{Dt} = mF_i - \frac{\partial p}{\partial x_i} + \eta\nabla^2 v_i,$$

or in vector form

$$(75)\qquad m\,\frac{D\vec{v}}{Dt} = m\vec{F} - \vec{\nabla}p + \eta\nabla^2\vec{v}.$$

We note that the hydrodynamic derivative $D\vec{v}/Dt$ refers to a point moving with the fluid, and the equation in this form is called the Lagrange form of the equations of flow. When this derivative is re-expressed using equation (70), the equations are said to be in Euler form. The derivative $\partial\vec{v}/\partial t$ then represents the rate of change of the velocity as 'seen' by an observer fixed in space since $\vec{r}$ is fixed in the expression $v_k = v_k(\vec{r},t)$.

REFERENCES

1 A.E. Green and W. Zerna, Theoretical Elasticity. Oxford at the Clarendon Press, 1968
2 L.D. Landau and E.M. Lifshitz, Theory of Elasticity. Pergamon Press, London, 1970
3 L.D. Landau and E.M. Lifshitz, Fluid Mechanics. Pergamon Press, London, 1959
4 I.S. Sokolnikoff and R.D. Specht, Mathematical Theory of Elasticity. McGraw-Hill, New York, 1946

PROBLEMS

1 Show that if the stress energy density U is a function of the state of stress of the material only, then

$$U = \tfrac{1}{2} S_{ijk\ell}E_{ij}E_{k\ell} \quad \text{and} \quad \sigma_{kj} = \alpha\partial U/\partial E_{kj}, \quad \alpha = \begin{cases} \tfrac{1}{2} & k \neq j, \\ 1 & k = j. \end{cases}$$

Hint: Assume U is expressible in a power series and drop terms of order three and higher.

2 Prove equations (36).

3 Derive the following wave equations for isotropic cubic crystals:

$$\text{(i)} \quad \frac{\partial^2\theta}{\partial t^2} = \frac{\lambda + 2\mu}{m}\nabla^2\theta,$$

$$\text{(ii)} \quad \frac{\partial^2 K_i}{\partial t^2} = \frac{\mu}{m}\nabla^2 K_i,$$

where $K_i = \tfrac{1}{2}(\text{curl}\,\vec{\rho})_i$ and $\theta = \text{div}\,\vec{\rho}$, and justify the conclusion that the divergence and curl of a wave disturbance propagate independently for such crystals.

4 Show that for an isotropic fluid the rate of change of energy density at a fixed point in space is given by

$$\frac{\delta E}{\delta t} = -pe_{kk} + 2\eta\left(e_{ij}e_{ij} - (e_{kk})^2/3\right) + \frac{\partial}{\partial x_i}\left(k\frac{\partial T}{\partial x_i}\right)$$

where T is the temperature and k is the thermal conductivity, provided the velocity of the fluid vanishes at its boundary.

5 Show that for an incompressible fluid $\vec{\nabla}\cdot\vec{v} = 0$.

5
Tensors as a basis of group representations

5.1. INTRODUCTION

We have seen in Chapter 2 how our general notions about the 3-dimensional space continuum, which provides the arena for physical events, leads us to the concept of tensors as the natural mathematical entities or constructs to use for the description of these same events. We may recall that the starting point in arriving at a hierarchical structure of tensors was the simple concept of a 'position vector' together with its transformation properties under coordinate rotation. These transformation properties corresponded to a linear relationship, through an orthogonal matrix, between one set of components and a second 'rotated' set. One is then led in a natural way to the generalization of the tensor concept through the well-known generalization from 3-dimensional position vectors to n-dimensional abstract linear vector spaces defined over the complex number field with unitary (linear) transformations between bases vectors.

In Chapter 2 we also gained some limited insight into symmetry classes of tensors. An arbitrary second-rank proper tensor can be written

$$T_{jk} = \tfrac{1}{2}(T_{jk} + T_{kj}) + \tfrac{1}{2}(T_{jk} - T_{kj}) = S_{jk} + A_{jk}$$

where $S_{jk} = S_{kj}$ is symmetric and $A_{jk} = -A_{kj}$ is antisymmetric on permutation of indices. The three independent components of A_{jk} transform like the components of a first-rank improper tensor (axial vector), the sum of the diagonal components S_{kk} transform as a scalar, and the five remaining independent components of S_{jk} can be chosen so they transform as the components of a second-rank traceless symmetric tensor. (Symmetric, traceless second-rank tensors are said to be irreducible second-rank tensors.) We shall shortly see that this is a special case of a much more general and exciting circumstance, namely that the symmetry classes of tensors provide a unique and interesting categorization of the matrix representations of the groups of importance in mathematical physics. An elaboration of this statement is the primary objective of the present chapter.

Before defining and elaborating on group representations, let us comment briefly on how groups themselves arise so naturally in science from philosophical considerations of the basic problem of describing natural phenomena. Once we have devised orthogonal linear transformations on tensor components which enable an observer O and a (rotated) observer O' to compare physical results, we are entitled to ask about a third observer O". Suppose we know the transformation equations (i.e. the orthogonal matrix) relating tensor components for observers O and O' and again those for O' and O", what can we say about the equations for O and O" directly? It turns out that such successive transformations form a mathematical construct called a group, or as it is sometimes called the group SO(3) of special orthogonal transformations in three real dimensions. Hence, once we are involved in tensors and their transformations, we are also involved in the group theory of the transformations themselves.

Moreover, just as the generalization from 3-dimensional Euclidean space to more general linear vector spaces involves a generalization of the tensor concept, so does the generalization of tensor transformations and their group properties lead us to more general kinds of groups than the one considered above (rotation group in three real dimensions).

Basically, theoretical physics consists of a set of interpreted mathematical structures. One cannot go very far conceptually without the mathematical framework but the framework is meaningless unless it is interpreted. We have partially conceptualized our problem, but before more progress can be made on this front, we need to develop the underlying mathematical structures themselves.

5.2. THE ABSTRACT NOTION OF A GROUP

A set $S = \{a,b,...\}$ is called a group under the operation of 'multipication' denoted $\cdot$ if the following axioms hold:

(i) S is closed under the operation $\cdot$,

(ii) $a\cdot(b\cdot c) = (a\cdot b)\cdot c$ for every $a,b,c \in S$ (associativity),

(iii) there exists an $e \in S$ such that $a\cdot e = e\cdot a = a$ for every $a \in S$ (existence of an identity),

(iv) for every $a \in s$ there exists $a^{-1} \in S$ such that $a\cdot a^{-1} = a^{-1}.a = e$ (existence of an inverse).

Actually the above set of axioms is convenient but considerably more than what is required to characterize a group. The reader might find it interesting to try to discover what the minimum requirements are for S to be a group.

If S' is a subset of S and the axioms (i) to (iv) are satisfied by the elements of S' alone, then S' is called a subgroup of S. Every group S contains two trivial subgroups, itself and $\{e\}$. The number of elements in a group is called the order of the group, and for a group of finite order one can get other subgroups (if they exist) by the following procedure. Pick an element of S, say a, and form its successive powers,† a^2, $a^3,...,a^n$. Eventually, since the order of S is finite, one of the powers must be equal to a lower one, say $a^n = a^k$, $k < n$. Then $a^{n-k} = e$, using the associativity axiom. The set $\{a,a^2,...a^{n-k}\}$ forms a subgroup of S and in fact is the smallest subgroup containing the element a.

To test if a set S' is a subgroup it is only necessary to verify that axioms (i) and (iv) hold in S'. The other axioms hold automatically from the fact that S' is included in S. Further, if the group S is finite, then only axiom (i) needs to be checked for S' to be a subgroup, since for every element

†It is customary to omit the dot symbol in a product and write simply $a\cdot a = aa = a^2$.

a in S' there exists a power, say a^{n-1}, such that $aa^{n-1} = e = a^n$ and hence S' contains the inverse of each of its elements if it is closed.

One convenient method for illustrating a group is by means of a group table. A group table consists of a square array in which the rows and columns are labelled according to the elements of the group. In the box in the nth row and mth column we record the product of the elements labelling the nth row and the mth column, as shown in Figure 5.1. It is easily proved that the same element does not occur twice in any row or column. Thus each row (or column) represents a permutation of the symbols $(a_o,\ldots,a_\ell)$ with no two rows (or columns) representing the same permutation.

An element b of the group S is said to be conjugate to the element a if there exists an element u of S such that

(1) $uau^{-1} = b.$

b is sometimes called the transform of a. Every element is conjugate to itself (i.e. reflexive),

(2) $a = aaa^{-1}.$

If b is conjugate to a, then a is conjugate to b (i.e. conjugation is commutative):

(3) $b = uau^{-1}$ implies that $a = vbv^{-1} = (u^{-1})b(u^{-1})^{-1}.$

Also if a is conjugate to b and b conjugate to c, then a is conjugate to c. Thus conjugation is transitive:

(4) $a = ubu^{-1}$, $b = vcv^{-1}$ implies that $a = uvcv^{-1}u^{-1} = uvc(uv)^{-1}$

Note that conjugation is an equivalence relation (i.e. a relation that is reflexive, commutative, and transitive). An equivalence relation separates a set into classes, the elements of each class being equivalent to one another.†

---

†For a proof of this see J.A. Green, Sets and Groups (Dover Publications, New York, 1965), pp. 17-18.

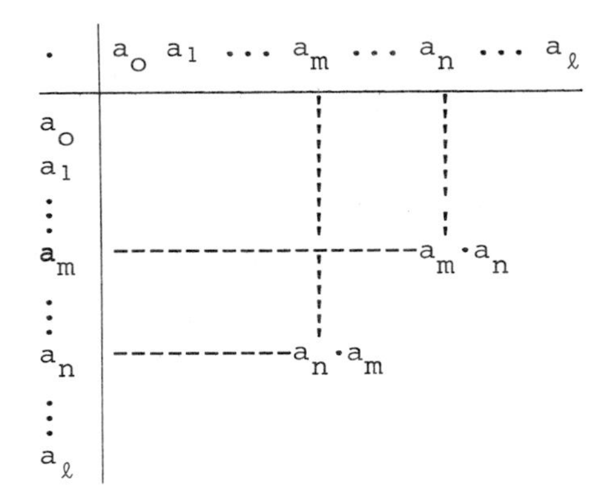

Figure 5.1. Illustration to show the construction of a group table.

Another interesting example of the use of an equivalence relation is given in Appendix C, where it is shown that equivalence relations are useful in defining tensors.

5.3. THE SYMMETRIC GROUP S_n

The symmetric group S_n is the complete group of permutations on n objects. A permutation is a one-to-one mapping of a set onto itself. It is obvious that the set of all permutations on n objects occupying n slots satisfies axioms (i) to (iv) and hence is a group (the operation is, of course, composition; $P_1 \cdot P_2$ means first perform permutation 2 and then permutation 1).

There are several ways of illustrating permutations. Consider for simplicity† a set of three objects which may or may not be distinct but which occupy three distinct slots. The permutation $1 \to 2$, $3 \to 3$, $2 \to 1$, which places the object originally in slot 1 into slot 2 and vice versa but leaves the object in slot 3 alone, may be illustrated by either

$$\text{(i)} \quad \begin{pmatrix} 1 & 2 & 3 \\ 2 & 1 & 3 \end{pmatrix} \quad \text{or} \quad \text{(ii)} \quad (1\ 2)(3).$$

†In the active viewpoint the permutation operators map objects into one another. We adopt the passive viewpoint in which the operation is a permutation of objects from one numbered slot to another.

In (i) each column corresponds to the movement of the object from the indicated slot (upper row) to a designated slot (lower row). The order of the columns is of no consequence whatsoever and hence $\begin{pmatrix} 1 & 3 & 2 \\ 2 & 3 & 1 \end{pmatrix}$ represents the same permutation as (i). We refer to permutations as elements of the group. In (ii) we call each set of brackets a cycle.† The cycle $(1,2,\ldots,n)$ means under permutation $1 \to 2$, $2 \to 3, \ldots, (n-1) \to n$, $n \to 1$. Thus in (ii) $1 \to 2$, $2 \to 1$, $3 \to 3$. To get a feeling for the cycle notation consider the permutation of six objects $\begin{pmatrix} 1 & 2 & 3 & 4 & 5 & 6 \\ 3 & 5 & 1 & 4 & 2 & 6 \end{pmatrix}$. It is easily verified that it has a cycle illustration (13)(25)(4)(6). This permutation is said to have a cycle structure $[22\ 11] \equiv [2^2,1^2]$ since it has two cycles each involving the permutation of two objects and two cycles each involving one object.

Elements of the same conjugate equivalence class have the same cycle structure and vice versa.†† Consider an arbitrary permutation in S_n. It will have a cycle structure $[n^{\nu_n},\ldots,2^{\nu_1},1^{\nu_1}]$ where ν_k is the number of cycles of length k and

(5) $$\nu_1 + 2\nu_2 + \ldots n\nu_n = n$$

since the number of symbols being permuted is n. Hence the number of conjugate equivalence classes in S_n is equal to the number of solutions $\nu_1,\ldots,\nu_n$, for positive integers (or zero), which satisfy equation (5). Putting

(6) $$\sum_{j=i}^{n} \nu_j = \lambda_i, \quad i \in \{1,2,\ldots,n\},$$

then

---

†See B. Baumslag and B. Chandler, Group Theory (Schaum's Outline Series, McGraw-Hill, New York, 1968), p. 167, for a proof that every permutation may be written as the product of disjoint cycles.

††This follows from the result $p^{-1}(a_1,\ldots,a_m)p = (p(a_1),\ldots,p(a_m))$ where $p \in S_n$ and $(a_1,\ldots,a_m)$ is a cycle. For a proof see B. Baumslag and B. Chandler, Group Theory, p. 170.

(7a) $\sum_{i=1}^{n} \lambda_i = n$ with $\lambda_i \geq \lambda_{i+1} \geq 0$ for $i \in \{1,\ldots,n-1\}$.

$(\lambda_1,\lambda_2,\ldots,\lambda_n)$ is called a partition of n since the λ_i satisfy equation (7a) (i.e. a partition of n is a set of non-negative integers which sum to n). Thus to each conjugate equivalence class (i.e. to each cycle structure) there corresponds a partition and to each partition there corresponds a conjugate equivalence class with cycle structure given by

(7b) $\nu_i = \lambda_i - \lambda_{i+1}$, $i \in \{1,\ldots,n\}$, $\lambda_{n+1} = 0$.

Consider the partition (221100) of six. This is usually written $(2^2,1^2)$, the zeros being omitted and repeated numbers expressed as 'powers,' the value of the power being equal to the number of times the number is repeated in the expression for the partition.

The elements of each conjugate equivalence class of S_4 as well as the partition and cycle structure corresponding to each conjugate equivalence class are given in Table 5.1. The symmetric groups play a special role in the study of finite groups since, as the following theorem shows, every finite group of order n has the same structure as some subgroup of the symmetric group S_n.

CAYLEY'S THEOREM: Every group S of order n is isomorphic† with a subgroup of the symmetry group S_n.

PROOF: Let the elements of a group S be $a_1, a_2,\ldots,a_n$. Take any element b of S. With it we associate a row in the group table $ba_1, ba_2,\ldots,ba_n$. The elements of the row are all different (since if $ba_i = ba_j$ then $a_i = a_j$). Thus as was mentioned previously the row represents a permutation of the elements $a_1,\ldots,a_n$. Hence with an arbitrary element b of the

†We say that two groups S and S' are isomorphic if there is a one-to-one mapping of S onto S' such that if $a \to a'$ and $b \to b'$ than $(ab)' = a'b'$.

TABLE 5.1

Correspondence between cycle structure and partition for S_4

Partition	Cycle structure	Elements of the class
(4)	$[1^4]$	(1)(2)(3)(4) = e
(3,1)	$[2,1^2]$	(12)(3)(4), (13)(2)(4), (14)(2)(3), (23)(1)(4), (24)(1)(3), (34)(1)(2)
(2^2)	$[2^2]$	(12)(34), (13)(24), (14)(23)
$(2,1^2)$	[3,1]	(123)(4), (132)(4), (124)(3), (142)(3), (134)(2), (143)(2), (234)(1), (243)(1)
(1^4)	[4]	(1234), (1243), (1324), (1342), (1423), (1432)

group S we associate the permutation P_b given by

$$b \to P_b = \begin{pmatrix} a_1 & \cdots & a_n \\ ba_1 & \cdots & ba_n \end{pmatrix}$$

Similarly,

$$c \to P_c = \begin{pmatrix} a_1 & \cdots & a_n \\ ca_1 & \cdots & ca_n \end{pmatrix} \text{ for } c \in S.$$

Thus

$$P_b \cdot P_c = \begin{pmatrix} a_1 & \cdots & a_n \\ bca_1 & \cdots & bca_n \end{pmatrix} = P_{b \cdot c}$$

and the theorem is proved.

Elements of S_n which may be associated with the elements of a group of order n as was done above are called regular permutations. A regular permutation has the special property that it leaves no element unchanged - i.e. $\begin{pmatrix} a_1 & \ldots & a_n \\ a'_1 & \ldots & a'_n \end{pmatrix}$ is regular if $a'_1 \neq a_1, \ldots, a'_n \neq a_n$ - the identity permutation being excluded, of course.

5.4. REPRESENTATIONS OF GROUPS

In physics, one is often interested not only in the transformation groups themselves, but in their matrix representations, which are, in effect, linear realizations of the group operations. Let S be a group consisting of the set of elements $\{a_1,\ldots,a_n,\ldots\}$. A set of matrices $\{M(a_1),\ldots,M(a_n),\ldots\}$ is called a <u>representation</u> of the group $\{a_1,\ldots,a_n,\ldots\}$ if one can associate with each element of the group a_k a unique matrix $M(a_k)$, $k \in \{1,2,\ldots,n,\ldots\}$, such that matrix multiplication corresponds to group multiplication in the following sense: whenever $a_i \cdot a_j = a_k$

(8) $$M(a_i)M(a_j) = M(a_i a_j) = M(a_k), \quad i,j,k \in \{1,\ldots,n,\ldots\}.$$

The correspondence between group elements and matrices need not be one-to-one. The same matrix may represent several group elements, but each group element must be represented by a specific matrix.

We have already had, incidentally, some examples of matrix representations. In particular, we have seen how rotations in n-dimensional space can be placed in correspondence (actually one-to-one) with the set of orthogonal matrices in n-dimensions with determinant +1. Because the correspondence is one-to-one, we can refer indifferently to the rotation group R_3 (whose elements can be specified by the three Euler angles) or to the associated matrix group SO(3) of the special†

†I.e. orthogonal matrices of determinant +1.

orthogonal matrices. In the latter case one can associate the matrices with linear operators mapping 3-dimensional vector spaces into themselves.

The relevance of matrix representations derives from the importance which is attached in theoretical physics to linear modelling on vector spaces. If a set of physical transformations (e.g. rotations, reflections, Lorentz transformations) induces a mapping of a linear vector space onto itself, such a mapping is always representable in terms of matrices. What is less obvious, but highly significant, is the fact that among all possible matrix representations of a group, certain basic ones, the so-called irreducible representations, play a special role. In quantum theory, for example, a symmetry group (i.e. a group of operators which commute with the Hamiltonian) provides through its irreducible representations quantum numbers for the specification and behaviour of the permissible quantum states of the system. For example, let the non-singular, time-independent operator R act on the Schroedinger equation (time-independent) $R(H\psi) = R(E\psi) = E(R\psi)$. Hence

$$(RHR^{-1})(R\psi) = H(R\psi) = E(R\psi)$$

if H and R commute, so that along with ψ also $R\psi$ is an eigensolution of H with eigenvalue E. Such considerations form the bases of all of molecular, atomic, and nuclear spectroscopy, as well as providing a classification scheme for the so-called elementary particles. Several aspects of this subject will be pursued in the next few chapters of this book.

Some examples of matrix representations are given below:

(a) The identity representation: Every group has an identity representation where every element of the group is associated with the one-dimensional matrix 1.

(b) The antisymmetric representation of S_n: This is also a one-dimensional representation constructed by associating the matrix 1 with every even permutation (i.e. a permutation which can be achieved by an even number (or zero) of transpositions† on the original configuration) and the matrix -1 with every odd permutation (i.e. a permutation which can be achieved by an odd number of transpositions on the original configuration). This is a valid representation since the product of two odd permutations is even, the product of two even permutations is even, and the product of an even permutation and an odd permutation is an odd permutation.

(c) The regular representation: This representation is closely related to the process of left translation of a group in which the group elements are reordered by multiplying a given element into all other group elements from the left. It is constructed as follows. The group $S = \{a_1, \ldots a_n\}$ of order n has a group table

$\cdot$	a_1	a_2	$\ldots$	a_n	→row index
a_1	$a_1 \cdot a_1$	$a_1 \cdot a_2$	$\ldots$	$a_1 \cdot a_n$	
a_2	$a_2 \cdot a_1$	$a_2 \cdot a_2$	$\ldots$	$a_2 \cdot a_n$	
$\vdots$	$\vdots$	$\vdots$		$\vdots$	
a_n	$a_n \cdot a_1$	$a_n \cdot a_2$	$\ldots$	$a_n \cdot a_n$	

Let the elements of the group be considered as ordered n-tuples by putting a 1 in the position where the element occurs in the row index and zeros everywhere else, i.e.,

$$a_1 \Rightarrow \hat{a}_1 = \begin{bmatrix} 1 \\ 0 \\ \vdots \\ 0 \end{bmatrix}, \quad \hat{a}_2 = \begin{bmatrix} 0 \\ 1 \\ 0 \\ \vdots \\ 0 \end{bmatrix}, \ldots, \; a_n \Rightarrow \hat{a}_n = \begin{bmatrix} 0 \\ \vdots \\ 0 \\ 1 \end{bmatrix}.$$

Then the regular matrix representation is formed by associating with the element a_j a matrix $M(a_j)$ such that for all $k \in \{1,2,\ldots,n\}$, if $a_j a_k = a_\ell$, then

†Transpositions are simple interchanges of two numbers or objects in a given arrangement.

(9) $\hat{a}_\ell = \widehat{(a_j \cdot a_k)} = M(a_j)\hat{a}_k$ (left translation).

Thus the matrix $M(a_j)$ takes the column vector $\hat{a}_k$ into the column vector $\hat{a}_\ell$. The $M(a_j)$, $j \in \{1,\ldots,n\}$, as given in (9) form a valid matrix representation of S since if $a_j \cdot a_k = a_\ell$ then under matrix multiplication $M(a_j)M(a_k) = M(a_\ell)$. This can be seen as follows:

$$M(a_j)M(a_k)\hat{a}_\rho = M(a_j)\,\widehat{a_k \cdot a_\rho} = \widehat{(a_j \cdot (a_k \cdot a_\rho))}$$
$$= M(a_j a_k)\hat{a}_\rho, \quad \rho \in \{1,\ldots,n\},$$

where we have used the associative property of group multiplication. Therefore $M(a_j)M(a_k) = M(a_\ell)$, as required. Note that the trace of $M(a_j)$ is zero unless $a_j = e$ and then the trace is n (the order of the group). Other examples of representations appear in the problem set at the end of this chapter.

If $\{A(a_j)\}$ and $\{B(a_j)\}$ are respectively n- and m-dimensional† representations of a group S, then $\begin{bmatrix} A(a_j) & 0 \\ 0 & B(a_j) \end{bmatrix}$ is an (n + m)-dimensional representation of S since

(10) $$\begin{bmatrix} A(a_j) & 0 \\ 0 & B(a_j) \end{bmatrix} \begin{bmatrix} A(a_k) & 0 \\ 0 & B(a_k) \end{bmatrix} = \begin{bmatrix} A(a_j)A(a_k) & 0 \\ 0 & B(a_j)B(a_k) \end{bmatrix}.$$

Two representations (with the same dimension) $[A(a_j)]$ and $[A'(a_j)]$ of a group S are called <u>equivalent</u> representations if there exists some non-singular matrix X such that $A'(a_j) = XA(a_j)X^{-1}$ for all $a_j \in S$.

A representation is called <u>reducible</u> if an equivalent representation exists in which each matrix $M(a_j)$ has the form

†If a matrix $M(a_j)$ is an element of an n-dimensional representation, then $M(a_j)$ is an n x n matrix and can be considered as a linear transformation acting on an n-dimensional vector space.

$$(11)\qquad M(a_j) = \begin{bmatrix} A(a_j) & C_j \\ 0 & B(a_j) \end{bmatrix}$$

where $A(a_j)$ is an n x n matrix $B(a_j)$ is an m x m matrix, C_j is an n x m matrix, and 0 is the m x n zero matrix. The matrices $A(a_j)$ and $B(a_j)$ are also representations of the group since

$$(12)\qquad M(a_j)M(a_k) = M(a_\ell) \text{ implies that } A(a_j)A(a_k) = A(a_\ell) \text{ and } B(a_j)B(a_k) = B(a_\ell).$$

The representations $[A(a_j)]$ and $[B(a_j)]$ may also be reducible but if not they are called irreducible representations. If $[A(a_j)]$ and $[B(a_j)]$ are reducible, then by repeated similarity transformations one may get an equivalent representation in which every matrix has the same block structure:

$$(13)\qquad M(a_j) = \left[\begin{array}{c|c|c|c} A^1(a_j) & & & \\ \hline & A^2(a_j) & & C_j \\ \hline & 0 & \ddots & \\ \hline & & & A^k(a_j) \end{array}\right],$$

where the matrix sets $\{A^1(a_j)\}$, $\{A^2(a_j)\}$, ..., $\{A^k(a_j)\}$ are separately irreducible representations. In such a reduction process, it may happen that some irreducible representations occur more than once: for example, the matrix sets $\{A^n(a_j)\}$ and $\{A^m(a_j)\}$ may differ only by a similarity transformation.

A representation is said to be <u>fully reducible</u> if an equivalent representation exists in which each matrix $M(a_j)$ has the form

$$(14)\qquad M(a_j) = \begin{bmatrix} A^1(a_j) & & 0 \\ & \ddots & \\ 0 & & A^k(a_j) \end{bmatrix}$$

(i.e. $C_j = 0$) where $[A^1(a_j)],\ldots,[A^k(a_j)]$ are all irreducible representations.

We write each matrix in this case as a direct sum:

(15) $$M = \sum_{i=1}^{k} A^i = A^1 \oplus A^2 \oplus \ldots \oplus A^k.$$

If some of the representations $[A^i(a_j)]$ are equivalent to each other the direct sum can be expressed as

(16) $$M = \sum_i \alpha_i A^i$$

where α_i is a positive integer equal to the number of times the representation A^i occurs† in the reducible representation $[M(a_j)]$. For example, for a fully reduced representation the matrices can all be expressed in the block diagonal form

$$M(a_j) = \left[\begin{array}{c|c|c} A^1(a_j) & & \\ \hline & A^1(a_j) & \\ \hline & & A^2(a_j) \end{array}\right]$$

which we write symbolically as

$$M = 2A^1 \oplus A^2.$$

If we consider the matrices of a group representation $[M(a_j)]$ as acting on a linear vector space V (whose dimension is the same as that of the representation) then V is called a basis of the representation $[M(a_j)]$. If the representation $[M(a_j)]$ is fully reducible then the matrices will transform certain subspaces of V among themselves. For example, suppose all matrices of some fully reduced representation have the form

$$M = \begin{bmatrix} a & b & 0 \\ c & d & 0 \\ 0 & 0 & e \end{bmatrix}$$

and V is the space of ordered 3-tuples (x_1, x_2, x_3); the matrices M map the 2-dimensional subspace of vectors $(x_1, x_2, 0)$ into

†One does not distinguish between a representation or its equivalent in this counting process.

itself and the one-dimensional subspace of vectors $(0,0,x_3)$ into itself as follows:

$$\begin{bmatrix} a & b & 0 \\ c & d & 0 \\ 0 & 0 & e \end{bmatrix}\begin{bmatrix} x_1 \\ x_2 \\ 0 \end{bmatrix} = \begin{bmatrix} ax_1 + bx_2 \\ cx_1 + dx_2 \\ 0 \end{bmatrix} \quad \text{and} \quad \begin{bmatrix} a & b & 0 \\ c & d & 0 \\ 0 & 0 & e \end{bmatrix}\begin{bmatrix} 0 \\ 0 \\ x_3 \end{bmatrix} = \begin{bmatrix} 0 \\ 0 \\ ex_3 \end{bmatrix},$$

i.e. the subspaces spanned by $(x_1,x_2,0)$ and $(0,0,x_3)$ are invariant subspaces. Similarly, the basis vectors of an irreducible representation of a symmetry group are eigenfunctions of the Hamiltonian belonging to the same energy eigenvalue.

It can be shown that any unitary representation (i.e. a representation consisting of unitary matrices) is fully reducible. For groups of finite order all representations are equivalent to a unitary representation† and hence fully reducible.

A simple criterion for distinguishing one irreducible representation from another is found in the traces of the matrices of the representation. We first note that the trace is invariant under similarity transformations:

$$\text{Trace}(UMU^{-1}) = \sum_k \sum_\ell \sum_i U_{ik}M_{k\ell}U^{-1}_{\ell i} = \sum_k \sum_\ell M_{k\ell}\left(\sum_i U_{ik}U^{-1}_{\ell i}\right)$$

$$= \sum_k \sum_\ell M_{k\ell}\delta_{\ell k} = \text{Trace } M$$

so that equivalent representations have equal traces, matrix by corresponding matrix. Thus if $[M(a_j)]$ and $[M'(a_j)]$ are equivalent representations of a group S then Trace $M(a_j)$ = Trace $M'(a_j)$ for every a_j in S. Secondly, the trace is also a class function, matrices corresponding to elements of the same conjugate equivalence class having the same trace since they differ from each other only by a similarity transformation (see equation (1)). The character $\chi^{(\beta)}_j$ is defined to be the

---

†This result is proved in many books on group theory. See for example M. Hamermesh, Group Theory (Addison-Wesley, Reading, Mass., 1962), p. 92, and L.M. Falicov, Group Theory and Its Physical Applications (University of Chicago Press, 1966), p. 22.

trace of any matrix in the βth irreducible representation which corresponds to any element in the jth class of the group.

For representations of a group of finite order the number of non-equivalent irreducible representations is equal to the number of conjugate equivalence classes of the group (see problem 6(i)). Thus it is possible to construct a table of characters as is shown below.

		conjugacy classes				
		K_1	...	K_i	...	K_k
irreducible representations	A^1	$\chi_1^{(1)}$		$\chi_i^{(1)}$		$\chi_k^{(1)}$
	⋮					
	A^μ	$\chi_1^{(\mu)}$		$\chi_i^{(\mu)}$		$\chi_k^{(\mu)}$
	⋮					
	A^k	$\chi_1^{(k)}$		$\chi_i^{(k)}$		$\chi_k^{(k)}$

It can be shown that each irreducible representation μ is uniquely characterized by the k numbers (characters) $\chi_j^{(\mu)}$, $j = 1,2,\ldots,k$.

5.5. IRREDUCIBLE REPRESENTATIONS OF THE SYMMETRIC GROUP S_n

The regular representation of the symmetric group S_n is fully reducible. We shall describe here a very simple method for deducing the dimensions of the irreducible representations of S_n. With each partition $(n^{\nu_n},\ldots 2^{\nu_2},1^{\nu_1})$ we associate a graph which is an arrangement of boxes:

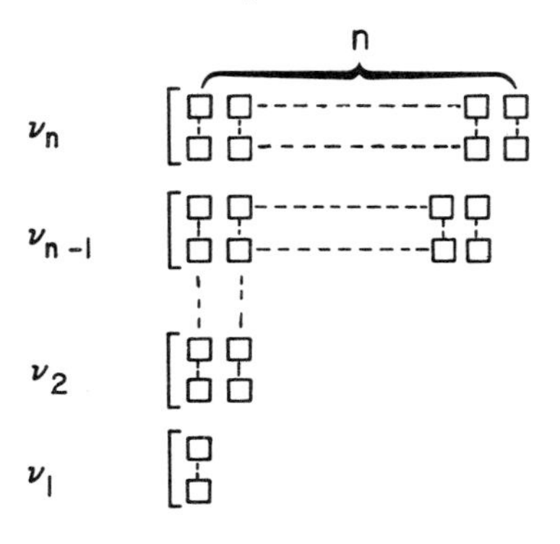

where ν_k is the number of rows with k boxes. Thus a graph consists of n boxes arranged in a configuration descriptive of the corresponding partition. A graph with the numbers 1,2,...,n assigned to the n boxes is called a Young tableau.† One says that the graph has been filled by a regular application of the numbers 1,...,n if the magnitude of the numbers increases down any column and across any row (moving from left to right). A graph filled by a regular application of 1,...,n is called a standard tableau.

Since the number of irreducible representations (non-equivalent) equals the number of classes (or the number of distinct partitions) we may also associate with each partition an irreducible representation. It can be shown that the dimension of an irreducible representation is equal to the number of standard tableaux which can be constructed by filling the graph associated with that representation.

Consider, for example, the group S_4. The standard tableaux associated with each partition (i.e. class) are given in the Table 5.2 (boxes omitted).

It can also be shown that for any finite group each representation occurs in the regular representation a number of times equal to the dimension of the irreducible representation.†† Hence, by constructing standard tableaux, we are able to assign the numbers α_i which occur in (16).

The representations of the symmetric groups††† play a special role in physics and, indeed, in mathematics itself. This special role stems partly from the fundamental connection between the statistics of identical elementary particles in

†The Collected Papers of Alfred Young (University of Toronto Press, 1977).

†See, for example, M. Hamermesh, Group Theory, p. 107.

†For an excellent exposition on the structure of the irreducible representations of the symmetric groups see A.J. Coleman, The Symmetric Group Made Easy, Advances in Quantum Chemistry 4, 83 (1968).

TABLE 5.2
Standard Tableaux for S_4

Partition	Standard tableaux			Dimension of representation
(4)	1 2 3 4			1
(3,1)	1 2 3 4	1 3 4 2	1 2 4 3	3
(2^2)	1 2 3 4	1 3 2 4		2
$(2,1^2)$	1 2 3 4	1 3 2 4	1 4 2 3	3
(1^4)	1 2 3 4			1

quantum theory and their spin dynamics. As we shall discuss in Chapters 6 and 9, particles whose spin is characterized by integral quantum numbers obey Bose-Einstein statistics, whereby the wave function of the many-particle system is required to be a completely symmetric function of the particle coordinates; such a function for an n-particle system forms the basis of the 1-dimensional identity representation with a Young tableau consisting of a single row of n boxes. For Fermi-Dirac particles corresponding to spin quantum numbers having half-odd integral values, the statistics require that the wave function be completely antisymmetric in the particle coordinates; such a function for an n-particle system forms the basis of the 1-dimensional alternating representation of

S_n corresponding to a Young tableau consisting of a single column.

Partly, the symmetric group plays a fundamental role because symmetry tensors (tensors with permutation symmetry on their indices) characterize the irreducible representations of all the unitary groups, including the rotation group R_3 (or SO(3)) as we shall see shortly.

Finally, it can be shown that the symmetry functions forming a basis of the irreducible representations of the symmetry groups S_n, n = 0,1,2,..., form a complete set of functions in the sense that an unsymmetrized function can be expanded uniquely in terms of them. A well-known example of this theorem is the statement that any function of two arguments can be written as a linear combination of a symmetric function (belonging to tableau $\Box\Box$) and an antisymmetric function (belonging to tableau $\begin{array}{c}\Box\\\Box\end{array}$). Explicitly

$$f(x,y) = \tfrac{1}{2}[f(x,y) + f(y,x)] + \tfrac{1}{2}[f(x,y) - f(y,x)].$$

Similarly, an unsymmetrized function of three variables f(x,y,z) can be written as a linear combination of the six symmetry functions, 1 belonging to $\Box\Box\Box$, one to $\begin{array}{c}\Box\\\Box\\\Box\end{array}$, two to the 2-dimensional representation $\begin{array}{ll}\Box&\Box\\\Box&\end{array}$, and two to the second (and independent) 2-dimensional representation $\begin{array}{ll}\Box&\Box\\\Box&\end{array}$ which also occurs in the decomposition of the regular representation of S_3.

5.6. PRODUCT REPRESENTATIONS

If $[M^{\mu}(a)]$ and $[M^{\nu}(a)]$ are two representations of the group $S = \{a_1, a_2, \ldots, a_k, \ldots\}$ with dimensions n_{μ} and n_{ν} respectively, we can construct a product representation of dimension $n_{\mu}n_{\nu}$

denoted by $M^{(\mu x \nu)}(a)$ where the matrix elements are defined by

$$(17) \qquad M^{(\mu x \nu)}_{ik,j\ell}(a) = M^{\mu}_{ij}(a) M^{\nu}_{k\ell}(a).$$

Thus the matrix $M^{(\mu x \nu)}(a)$ is obtained by taking all possible products of elements of $M^{\mu}(a)$ with elements of $M^{\nu}(a)$. The rows and columns are doubly indexed but could be renamed with a single index running from 1 to $n_{\mu} n_{\nu}$.

That the set $M^{(\mu x \nu)}(a)$ forms a valid representation follows from successive applications of equations (17) and (8):

$$\begin{aligned} M^{(\mu x \nu)}_{ik,j\ell}(a_1 a_2) &= M^{\mu}_{ij}(a_1 \cdot a_2) M^{\nu}_{k\ell}(a_1 \cdot a_2) \\ &= \sum_{m=1}^{n_\nu} \sum_{n=1}^{n_\mu} M^{\mu}_{in}(a_1) M^{\mu}_{nj}(a_2) M^{\nu}_{km}(a_1) M^{\nu}_{m\ell}(a_2) \\ &= \sum_{n=1}^{n_\nu} \sum_{m=1}^{n_\mu} M^{(\mu x \nu)}_{ik,nm}(a_1) M^{(\mu x \nu)}_{nm,j\ell}(a_2). \end{aligned}$$

The characters of the product representation are simply related to those of the original representations. We recall that the characters are class functions. Let element a belong to class s and let the characters of the original representations for this class be χ^{μ}_{s} and χ^{ν}_{s}. For this same class in the product representation $(\mu \times \nu)$ the character is again given by the trace:

$$(18) \qquad \chi_s^{(\mu x \nu)} = \sum_{j=1}^{n_\nu} \sum_{k=1}^{n_\mu} M^{(\mu x \nu)}_{jk,jk}(a) = \sum_{j=1}^{n_\nu} \sum_{k=1}^{n_\mu} M^{\mu}_{jj}(a) M^{\nu}_{kk}(a)$$

$$= \chi^{\mu}_{s} \chi^{\nu}_{s}.$$

Let the ordered tuples $(x_1, \ldots, x_{n_\mu})$ for a basis for the representation $[M^{\mu}(a)]$ and the ordered tuples $(y_1, \ldots, y_{n_\nu})$ form a basis for the representation $[M^{\nu}(a)]$ such that

$$x_i' = \sum_{j=1}^{n_\mu} M^\mu_{ij}(a)x_j \quad \text{and} \quad y_k' = \sum_{\ell=1}^{n_\nu} M^\nu_{k\ell}(a)y_\ell.$$

Then the quantities $x_i y_k$ form the components of a vector in the $n_\mu n_\nu$-dimensional space on which the representation $[M^{(\mu x\nu)}(a_k)]$ acts, since

$$x_k' y_k' = \sum_{j=1}^{n_\mu} \sum_{\ell=1}^{n_\nu} M^{(\mu x\nu)}_{ik,j\ell} x_j y_\ell.$$

The matrices defined by equation (17) are rather hard to visualize. For example, let $n_\nu = 2$ and $n_\mu = 2$. Writing

$$M^\mu(a) = \begin{bmatrix} M^\mu_{11}(a) & M^\mu_{12}(a) \\ M^\mu_{21}(a) & M^\mu_{22}(a) \end{bmatrix} \quad \text{and} \quad M^\nu(a) = \begin{bmatrix} M^\nu_{11}(a) & M^\nu_{12}(a) \\ M^\nu_{21}(a) & M^\nu_{22}(a) \end{bmatrix},$$

then

$$M^{(\mu x\nu)}(a) = \begin{bmatrix} M^{(\mu x\nu)}_{11,11}(a) & M^{(\mu x\nu)}_{11,12}(a) & M^{(\mu x\nu)}_{11,21}(a) & M^{(\mu x\nu)}_{11,22}(a) \\ M^{(\mu x\nu)}_{12,11}(a) & M^{(\mu x\nu)}_{12,12}(a) & M^{(\mu x\nu)}_{12,21}(a) & M^{(\mu x\nu)}_{12,22}(a) \\ M^{(\mu x\nu)}_{21,11}(a) & M^{(\mu x\nu)}_{21,12}(a) & M^{(\mu x\nu)}_{21,21}(a) & M^{(\mu x\nu)}_{21,22}(a) \\ M^{(\mu x\nu)}_{22,11}(a) & M^{(\mu x\nu)}_{22,12}(a) & M^{(\mu x\nu)}_{22,21}(a) & M^{(\mu x\nu)}_{22,22}(a) \end{bmatrix}$$

$$= \begin{bmatrix} M^\mu_{11}(a)M^\nu_{11}(a) & M^\mu_{11}(a)M^\nu_{12}(a) & M^\mu_{12}(a)M^\nu_{11}(a) & M^\mu_{12}(a)M^\nu_{12}(a) \\ M^\mu_{11}(a)M^\nu_{21}(a) & M^\mu_{11}(a)M^\nu_{22}(a) & M^\mu_{12}(a)M^\nu_{21}(a) & M^\mu_{12}(a)M^\nu_{22}(a) \\ M^\mu_{21}(a)M^\nu_{11}(a) & M^\mu_{21}(a)M^\nu_{12}(a) & M^\mu_{22}(a)M^\nu_{11}(a) & M^\mu_{22}(a)M^\nu_{12}(a) \\ M^\mu_{21}(a)M^\nu_{21}(a) & M^\mu_{21}(a)M^\nu_{22}(a) & M^\mu_{22}(a)M^\nu_{21}(a) & M^\mu_{22}(a)M^\nu_{22}(a) \end{bmatrix}$$

The representation $[M^{(\mu x\nu)}(a)]$ is called the Kronecker product of the representations $\{M^\mu(a)\}$ and $\{M^\nu(a)\}$.

The notion of product representations may be generalized to products of an arbitrary number of representations. For example, if $[M^{\nu_1}(a)]$, $[M^{\nu_2}(a)]$, ..., $[M^{\nu_n}(a)]$ are representations of the group S with dimensions n_{ν_1}, n_{ν_2}, ..., n_{ν_n} respectively then $[M^{(\nu_1 x \nu_2 x \ldots x \nu_n)}(a)]$ is a representation of S of dimension $n_{\nu_1} \cdot n_{\nu_2} \cdots n_{\nu_n}$ defined by

$$(19) \qquad M^{(\nu_1 x \nu_2 x \ldots x \nu_n)}_{i_1 i_2 \ldots i_n, j_1 j_2 \ldots j_n}(a_k) = M^{\nu_1}_{i_1 j_1}(a_k) M^{\nu_2}_{i_2 j_2}(a_k) \ldots M^{\nu_n}_{i_n j_n}(a_k)$$

and the character of the product representation is the product of the characters of each individual representation, all referred to the same group class:

$$(20) \qquad \chi^{(\nu_1 x \nu_2 x \ldots x \nu_n)}(a) = \chi^{\nu_1}(a) \chi^{\nu_2}(a) \ldots \chi^{\nu_n}(a).$$

Of course any number of the representations $[M^{\nu_j}(a_k)]$, $j \in \{1,\ldots,n\}$, may be identical. If all of them are identical, we call the representation $[M^{(\nu x \nu x \ldots x \nu)}(a)]$ defined by equation (19) the nth Kronecker product of the representation $[M^{\nu}(a)]$.

5.7. REPRESENTATIONS OF THE GENERAL LINEAR GROUP

In the previous chapter we defined a tensor as a set of components which transform in a certain way under coordinate rotations. This definition may be generalized by considering a tensor as a set of components which transform in a certain fashion under a given group of linear transformations (not necessarily rotations). A quantity which is a tensor under one group of transformations need not be a tensor under another group of transformations. In this section we associate tensors with the group of all invertible linear transformations in n-dimensional space over the field R, denoted GL(n):

$$(21) \qquad GL(n) = \left\{ A = \begin{bmatrix} a_{11} & \ldots & a_{1n} \\ \vdots & & \vdots \\ a_{n1} & \ldots & a_{nn} \end{bmatrix} \middle| \det A \neq 0,\ a_{ij} \in R \right\}.$$

If we regard the matrices A as a set of transformations on a linear vector space, we can again look for quantities which transform multilinearly in the coefficients a_{ij}. In general, if we can find a set of n^r quantities

$$F_{i_1 \ldots i_r} \quad \text{with all } i_k \in \{1,2,\ldots,n\},$$

which transform under A as

$$(22) \qquad F'_{i_1 \ldots i_r} = a_{i_1 j_1} a_{i_2 j_2} \cdots a_{i_r j_r} F_{j_1 j_2 \ldots j_r}$$

(summation convention used), then the set is called an rth-rank tensor in n-dimensions under the group GL(n). It is clear that the rth-rank tensors form a basis for the rth Kronecker product of the matrices of the group GL(n).

It should be noted that for any abstract group one can find a matrix representation, and in fact many matrix representations and with various dimensions. The group GL(n) is, however, defined as a set of matrices, so that its own matrices form a representation which is isomorphic to the group itself. We call this representation by matrices A the self-representation. Other representations of GL(n) exist, of course, including the trivial one-dimensional representation in which each group element (matrix) is represented by unity.

Consider a permutation p of r objects: $p = \begin{pmatrix} 1 & \ldots r \\ p(1) & \ldots p(r) \end{pmatrix}$. Associated with p and with an rth-rank tensor we define a permutation operator P which acts on the tensor by permuting its indices, successive indices occupying slots 1,...r:

$$(23) \qquad PF_{i_1 \ldots i_r} = F_{p^{-1}(i_1 \ldots i_r)} = F_{i_{p(1)} \ldots i_{p(r)}}$$

where in order to have a representation† for P we must interpret the action of P on F as the operation of p^{-1} on the

†Note that to compose successive operations one must use inverses. If P acts on f(x) through its argument we must have $Pf(x) = f'(x) = f(p^{-1}x)$ since then $QPf(x) = Qf'(x) = f'(q^{-1}x) = f(p^{-1}q^{-1}x) = f((qp)^{-1}x)$.

indices of F: $P^{-1}(i_1 \ldots i_r) = i_{p(1)} \ldots i_{p(r)}$. The operator P is conveniently denoted by the permutation it represents written in cycle form. Thus, for example,

(24a) $$(1\ 2)(3)F_{i_1 i_2 i_3} = F_{i_2 i_1 i_3}.$$

It is common practice to delete the elements which are left unpermuted when writing P. Thus we write the permutation operator corresponding to the permutation (1 2)(3) as just (1 2) so that equation (24a) is written

(24b) $$(1\ 2)F_{i_1 i_2 i_3} = F_{i_2 i_1 i_3}.$$

If $(jk)F_{i_1 i_2 \ldots i_j \ldots i_k \ldots i_r} = F_{i_1 i_2 \ldots i_j \ldots i_k \ldots i_r}$ we say that $F_{i_1 \ldots i_r}$ is symmetric on interchange of i_j with i_k and if $(jk)F_{i_1 \ldots i_j \ldots i_k \ldots i_r} = -F_{i_1 \ldots i_j \ldots i_k \ldots i_r}$ we say that $F_{i_1 \ldots i_r}$ is antisymmetric on interchange of i_j with i_k. Tensors of a given symmetry type retain that symmetry under transformation since

$$\begin{aligned} PF'_{i_1 \ldots i_r} &= F'_{i_{p(1)} \ldots i_{p(r)}} \\ &= a_{i_{p(1)} j_1} a_{i_{p(2)} j_2} \ldots a_{i_{p(r)} j_r} F_{j_1 j_2 \ldots j_r} \\ &= a_{i_{p(1)} j_{p(1)}} \ldots a_{i_{p(r)} j_{p(r)}} F_{j_{p(1)} \ldots j_{p(r)}} \end{aligned}$$

where we have renamed the j-indices, since they are all dummy (i.e. summed) indices. Rearranging the product of a coefficients we have

$$\begin{aligned} PF'_{i_1 \ldots i_r} &= a_{i_1 j_1} \ldots a_{i_r j_r} F_{j_{p(1)} \ldots j_{p(r)}} \\ &= a_{i_1 j_1} \ldots a_{i_r j_r} PF_{j_1 \ldots j_r} = (PF)'_{i_1 \ldots i_r}. \end{aligned}$$

Tensors of a given symmetry type are transformed among themselves,[†] and hence the rth Kronecker product representation can be reduced to representations each having tensors of some symmetry type for a basis.

General linear transformations have matrix elements on which no constraints have been placed; thus the only technique for reducing the rth Kronecker product representation is symmetrization, and the bases for the irreducible representations are tensors of the 'highest possible symmetry.'

With each standard tableau formed by filling the graph corresponding to a partition[††] $(\lambda_1,\dots,\lambda_m)$ we may associate a symmetrized rth-rank tensor in the following fashion. First form the operator P:

$$(25)\qquad P = \sum_{\text{all } \tilde{p}} \tilde{P}$$

where $\tilde{p}$ is any permutation[†††] which does not exchange elements belonging to different rows of the tableau. Then the tensor

$$(26)\qquad PF_{i_1\dots i_r} = \left(\sum_{\text{all } \tilde{p}} \tilde{P}\right) F_{i_1\dots i_r} = \sum_{\text{all } \tilde{p}} (\tilde{P}F_{i_1\dots i_r})$$

is simultaneously symmetric on all transpositions among the indices i_{a_1} where a_1 are elements of the first row of the tableau, on all transpositions among the indices i_{a_2} where the a_2 are elements of the second row of the tableau,..., and finally on all transpositions among the indices i_{a_m} where the a_m are elements of the last row of the tableau. (Note that $PF_{i_1\dots i_r}$ is not symmetric on interchange of i_{a_j} with i_{a_k} if $k \neq j$.) Thus, for example, with the standard tableau

---

†If $(PF) = -F$, for example, then $PF' = (PF)' = -F'$, showing that F' has the same symmetry as F.

†Here we adopt the convention that zeros are omitted but repeated numbers are not written as powers. Thus the number of rows in the graph associated with the partition $(\lambda_1,\dots,\lambda_m)$ is m.

†Note that $\tilde{P}$ is the operator associated with permutation $\tilde{p}$.

1	2
3	4

corresponding to the partition (2,2) (i.e. (2^2)) of 4 we have the operator

$$P = e + (1\ 2) + (3\ 4) + (1\ 2)(3\ 4)$$

(where e is the identity permutation)

and hence the tensor

$$PF_{i_1i_2i_3i_4} = F_{i_1i_2i_3i_4} + F_{i_2i_1i_3i_4} + F_{i_1i_2i_4i_3} + F_{i_2i_1i_4i_3}$$

is symmetric on interchange of i_2 with i_1 and on interchange of i_3 with i_4.

This process of symmetrization does not yield tensors of the 'highest possible symmetry,' however, for one can also partially antisymmetrize the tensor. Antisymmetrization on indices which have been symmetrized yields the zero tensor (i.e. if $F_{i_1i_2} = F_{i_2i_1}$ then $F_{i_1i_2} - F_{i_2i_1} = 0$). Thus the anti-symmetrizing operator is

$$(27)\qquad Q = \sum_{\text{all } \tilde{q}} (\text{sgn } \tilde{q})\tilde{Q} = \sum_{\text{all } \tilde{q}} \delta_q \tilde{Q}$$

where $\tilde{q}$ is any permutation which does not exchange elements belonging to different columns of the tableau. sgn $\tilde{q}$ or δ_q is called the sign of the permutation $\tilde{q}$ and is defined by

$$(28)\qquad \text{sgn } \tilde{q} = \begin{cases} -1 & \text{if } \tilde{q} \text{ is an odd permutation,} \\ +1 & \text{if } \tilde{q} \text{ is an even permutation.} \end{cases}$$

Then tensors† of the 'highest possible symmetry' are given by

†See, for example, H. Weyl, Group Theory and Quantum Mechanics (Dover Publications, New York, 1931).

$$(29) \quad YF_{i_1 \ldots i_r} = \mathcal{Q} \cdot \mathcal{P} \, F_{i_1 \ldots i_r}$$

$$= \sum_{\tilde{q}} \delta_q \tilde{Q} \sum_{\tilde{p}} \tilde{P} \, F_{i_1 \ldots i_r}$$

$$= \sum_{\tilde{q}, \tilde{p}} \delta_q \tilde{Q} \cdot \tilde{P} \, F_{i_1 \ldots i_r} .$$

Y is called a Young operator and the tensors $YF_{i_1 \ldots i_r}$ are antisymmetric on transposition of any two indices i_j and i_k where j and k are in the same column of the standard tableau associated with Y.

Some standard tableaux yield Young operators which give $YF_{i_1 \ldots i_r} \equiv 0$. For example, consider the general linear group of dimension n, GL(n). Then the indices can take on the values 1,2,...,n. If we have a partition $(\lambda_1, \ldots, \lambda_m)$ of r such that the number of rows of its graph exceeds n (i.e. $m > n$) then certainly some of the indices corresponding to elements in the first column of a standard tableau associated with the partition $(\lambda_1, \ldots, \lambda_m)$, $m > n$, must take on the same value (for example one might have $i_j = i_k = 2$ where j and k appear in the first column of the standard tableau). But the tensor $YF_{i_1 \ldots i_r}$ is antisymmetric on interchange of all indices corresponding to elements in the first column; hence $YF_{i_1 \ldots i_r} \equiv 0$. (Y is the Young operator in this case for any standard tableau formed by filling the graph associated with the partition $(\lambda_1, \ldots, \lambda_m)$, $m > n$, of r in a regular fashion.)

It can be shown† rigorously that the space of tensors $YF_{i_1 \ldots i_r}$ forms a basis for an irreducible representation of GL(n), where Y is a Young operator associated with any standard tableau corresponding to a partition $(\lambda_1, \ldots, \lambda_m)$ of r such that $m \leq n$. Young operators associated with standard tableaux corresponding to the same partition give rise to equivalent irreducible representations, and Young operators associated with standard

†See H. Weyl, Group Theory and Quantum Mechanics.

tableaux corresponding to different partitions give rise to non-equivalent irreducible representations.

It can also be shown† that the number of times an irreducible representation associated with the partition $(\lambda_1,\ldots,\lambda_m)$ (such that $m \leq n$) of r occurs in the rth Kronecker product of the self-representation A of GL(n) (see equation (21)) is equal to the number of possible standard tableaux formed by filling the graph of the partition $(\lambda_1,\ldots,\lambda_m)$ in a regular fashion. The number of independent components in the tensor $YF_{i_1\ldots i_r}$ is the dimension of the irreducible representation whose basis is the space of tensors $YF_{i_1\ldots i_r}$.

Consider, for example, the third Kronecker product of the self-representation A of GL(n) ($n \geq 3$). Associated with the partition $(1,1,1) = (1^3)$ is the standard tableau

1
2
3

The associated Young operator is given by

$$Y = [e - (1\ 2) - (2\ 3) - (1\ 3) + (1\ 2\ 3) + (1\ 3\ 2)]\cdot[e]$$

$$= e - (1\ 2) - (2\ 3) - (1\ 3) + (1\ 2\ 3) + (1\ 3\ 2),$$

so that the space of tensors,

$$YF_{i_1i_2i_3} = F_{i_1i_2i_3} - F_{i_2i_1i_3} - F_{i_1i_3i_2} - F_{i_3i_2i_1} + F_{i_2i_3i_1} + F_{i_3i_1i_2}, \tag{30}$$

$i_k \in \{1,2,\ldots,n\}$, forms a basis for the irreducible representation (1^3) of GL(n).

---

†This follows from the results on the regular representations of the symmetric group quoted in Section 5 of this chapter.

The tensor $YF_{i_1i_2i_3}$ as given by equation (30) is antisymmetric on any interchange among the three indices i_k.

With the partition (2,1) we have two possible standard tableaux:

1	2
3	

and

1	3
2	

The two Young operators corresponding to the tableaux are

$$Y^1 = [e - (1\ 3)]\cdot[e + (1\ 2)] = e - (1\ 3) + (1\ 2) - (1\ 2\ 3)$$

and

$$Y^2 = [e - (1\ 2)]\cdot[e + (1\ 3)] = e - (1\ 2) + (1\ 3) - (1\ 3\ 2).$$

The space of tensors (see equation (23))

$$(31)\quad Y^1F_{i_1i_2i_3} = F_{i_1i_2i_3} - F_{i_3i_2i_1} + F_{i_2i_1i_3} - F_{i_2i_3i_1}$$

forms the basis of the irreducible representation (2,1) of GL(n). Similarly, the space of tensors

$$(32)\quad Y^2F_{i_1i_2i_3} = F_{i_1i_2i_3} - F_{i_2i_1i_3} + F_{i_3i_2i_1} - F_{i_3i_1i_2}$$

is the basis of an equivalent representation (2,1) of GL(n). The tensor $Y^1F_{i_1i_2i_3}$ is antisymmetric on interchange of i_1 and i_3. The tensor $Y^2F_{i_1i_2i_3}$ is antisymmetric on interchange of i_2 and i_1.

With the partition (3) we have the standard tableau

1	2	3

and Y = [e]·[e + (1 2) + (2 3) + (1 3) + (1 2 3) + (1 3 2)

= e + (1 2) + (2 3) + (1 3) + (1 2 3) + (1 3 2).

Therefore, the tensors

$$(33)\quad YF_{i_1 i_2 i_3} = F_{i_1 i_2 i_3} + F_{i_2 i_1 i_3} + F_{i_1 i_3 i_2} + F_{i_3 i_2 i_1} + F_{i_2 i_3 i_1} + F_{i_3 i_1 i_2}$$

which are symmetric on all interchanges among the indices i_1, i_2, i_3 form a basis for still another irreducible representation of GL(n), viz. (3).

5.8. REPRESENTATIONS OF THE FULL ORTHOGONAL GROUP

Let O(n) represent the group of orthogonal transformations in n-dimensions. This is a subgroup of the group of all linear transformations GL(n). The matrix elements of representations of O(n), unlike those for GL(n), have a constraint placed on them, namely

$$(34)\quad a_{ij}a_{ik} = a_{ji}a_{ki} = \delta_{jk}.$$

The orthogonal group O(n) is not irreducible when broken up into representations by symmetry as was done for GL(n) since there exists another operation, contraction, which is preserved under transformation. The contraction of the tensor $F_{i_1 \ldots i_r}$ on the first two indices is denoted $F^{(1,2)}_{i_3 \ldots i_r}$ and defined by

$$(35)\quad F^{(1,2)}_{i_3 \ldots i_r} = \delta_{i_1 i_2} F_{i_1 i_2 i_3 \ldots i_r} \qquad \text{(summation convention).}$$

Then, using equation (21) but considering $\{a_{ij}\}$ as members of the matrix representation of O(n) (i.e. satisfying equation (34)), we get

$$F'^{(1,2)}_{i_3\ldots i_r} = \delta_{i_1 i_2} F'_{i_1 i_2 \ldots i_r}$$

$$= \delta_{i_1 i_2} a_{i_1 j_1} a_{i_2 j_2} \ldots a_{i_r j_r} F_{j_1 j_2 \ldots j_r}$$

$$= a_{i_1 j_1} a_{i_2 j_2} a_{i_3 j_3} \ldots a_{i_r j_r} F_{j_1 \ldots j_r}$$

$$= \delta_{j_1 j_2} a_{i_3 j_3} \ldots a_{i_r j_r} F_{j_1 \ldots j_r}$$

$$= a_{i_3 j_3} \ldots a_{i_r j_r} F^{(12)}_{j_3 \ldots j_r}$$

$$= (F^{(12)}_{i_3 \ldots i_r})'.$$

Thus tensors which yield zero on contraction of a given set of indices retain that property under transformation† (by elements of O(n)).

It can be shown†† by such considerations that bases for irreducible representations of the full orthogonal group are obtained by applying Young operators to tensors which yield zero upon contraction of any pair of indices (i.e. traceless tensors).

REFERENCES

1 H. Boerner, Representations of Groups. North-Holland, Amsterdam, 1963
2 L.M. Falicov, Group Theory and Its Physical Applications. University of Chicago Press, Chicago, 1966
3 J.W. Leech and D.J. Newman, How to Use Groups. Methuen, London, 1969
4 M. Hamermesh, Group Theory. Addison-Wesley, Reading, Mass., 1962
5 A.E. Littlewood, The Theory of Group Characters. Oxford at the Clarendon Press, 1958
6 F. Murnaghan, The Theory of Group Representations. Dover Publications, New York, 1963
7 G. de B. Robinson, Representation Theory of the Symmetric Group. University of Toronto Press, Toronto, 1961

†I.e. if $F^{(j,k)} = 0$, then $F'^{(j,k)} = 0$.
††See, for example, M. Hamermesh, Group Theory.

8 E.P. Wigner, Group Theory. Academic Press, New York, 1959
9 H. Weyl, The Classical Groups. Princeton University Press, Princeton, 1953
10 H. Weyl, Group Theory and Quantum Mechanics. Dover Publications, New York, 1931

PROBLEMS

1 (i) Prove that the complex numbers excluding zero form a group under multiplication.
(ii) Show that the set of all matrices of the form

$$\begin{bmatrix} \alpha & -\beta \\ \beta & \alpha \end{bmatrix} \equiv \alpha I + \beta J, \qquad \alpha, \beta \text{ real},$$

where $I = \begin{bmatrix} 1 & 0 \\ 0 & 1 \end{bmatrix}$ and $J = \begin{bmatrix} 0 & -1 \\ 1 & 0 \end{bmatrix}$,

is a representation of the group of complex numbers under multiplication.

2 A quaternion Z is a four-dimensional vector written as

$$Z = (Z_0, Z_1, Z_2, Z_3) = Z_0\hat{e} + Z_1\hat{\imath} + Z_2\hat{\jmath} + Z_3\hat{k}$$

where Z_0, Z_1, Z_2, Z_3 are real. The sum of two quaternions is defined by

$$Z + Y = (Z_0 + Y_0)\hat{e} + (Z_1 + Y_1)\hat{\imath} + (Z_2 + Y_2)\hat{\jmath} + (Z_3 + Y_3)\hat{k}.$$

(i) Prove that quaternions form a group under addition.
One can define a product for quaternions by

$$Z \odot Y = (Z_0\hat{e} + Z_1\hat{\imath} + Z_2\hat{\jmath} + Z_3\hat{k}) \odot (Y_0\hat{e} + Y_1\hat{\imath} + Y_2\hat{\jmath} + Y_3\hat{k})$$

$$= (Z_0Y_0)\hat{e} \odot \hat{e} + (Z_0Y_1)\hat{e} \odot \hat{\imath} + (Z_0Y_2)\hat{e} \odot \hat{\jmath}$$

$$+ (Z_0Y_3)\hat{e} \odot \hat{k} + (Z_1Y_0)\hat{\imath} \odot \hat{e} + \ldots + (Z_3Y_3)\hat{k} \odot \hat{k}$$

where the multiplicative properties of $\hat{e}, \hat{\imath}, \hat{\jmath}$, and $\hat{k}$ are given in the following table:

	$\hat{e}$	$\hat{i}$	$\hat{j}$	$\hat{k}$
$\hat{e}$	$\hat{e}$	$\hat{i}$	$\hat{j}$	$\hat{k}$
$\hat{i}$	$\hat{i}$	$-\hat{e}$	$\hat{k}$	$-\hat{j}$
$\hat{j}$	$\hat{j}$	$-\hat{k}$	$-\hat{e}$	$\hat{i}$
$\hat{k}$	$\hat{k}$	$\hat{j}$	$-\hat{i}$	$-\hat{e}$

Thus
$$\begin{aligned} Z \odot Y = {} & (Z_0Y_0 - Z_1Y_1 - Z_2Y_2 - Z_3Y_3)\hat{e} \\ & + (Z_0Y_1 + Z_1Y_0 + Z_2Y_3 - Z_3Y_2)\hat{i} \\ & + (Z_0Y_2 + Z_2Y_0 + Z_3Y_1 - Z_1Y_3)\hat{j} \\ & + (Z_0Y_3 + Z_3Y_0 + Z_1Y_2 - Z_2Y_1)\hat{k}. \end{aligned}$$

(ii) Show that $(\hat{i} \odot \hat{j}) \odot \hat{k} = -\hat{e} = \hat{i} \odot (\hat{j} \odot \hat{k})$.

(iii) One may write a quaternion as $Z = Z_0\hat{e} + \vec{Z}$ where $\vec{Z}$ is a vector in 3-dimensional space $\vec{Z} = (Z_1, Z_2, Z_3) = Z_1\hat{i} + Z_2\hat{j} + Z_3\hat{k}$. Show that the product of $Z = Z_0\hat{e} + \vec{Z}$ and $Y = Y_0\hat{e} + \vec{Y}$ may be written as

$$Z \odot Y = (Z_0Y_0 - \vec{Z}\cdot\vec{Y})\hat{e} + Z_0\vec{Y} + Y_0\vec{Z} + \vec{Z} \times \vec{Y}.$$

(iv) Show that quaternion multiplication is associative but not commutative.

Conjugation is defined for quaternions (in a manner similar to that for complex numbers) by

$$Z^* = (Z_0, Z_1, Z_2, Z_3)^* = (Z_0, -Z_1, -Z_2, -Z_3).$$

(v) Show that $(X + Y)^* = X^* + Y^*$ for any two quaternions X and Y.

We define the inner product of two quaternions Z and X by

$$\langle Z, X\rangle = Z_0X_0 + Z_1X_1 + Z_2X_2 + Z_3X_3;$$

then $Z \odot Z^* = \langle Z, Z\rangle\hat{e}$. The quantity $\sqrt{\langle Z, Z\rangle}$ is called the magnitude of Z and denoted by $|Z|$.

(vi) Prove that $\langle Z \odot Y, Z \odot X\rangle = |Z|^2\langle Y, X\rangle = \langle Y \odot Z, X \odot Z\rangle$.

(vii) Prove that the set of all quaternions excluding zero forms a group under multiplication.

(viii) Show that the set of all two-by-two matrices of the form

$$\begin{bmatrix} Z_0 - Z_3 i & -Z_1 i - Z_2 \\ -Z_1 i + Z_2 & Z_0 + Z_3 i \end{bmatrix} \equiv Z_0 I_1 + Z_1(-i)\sigma_1 + Z_2(-i)\sigma_2 + Z_3(-i)\sigma$$

where Z_0, Z_1, Z_2, Z_3 are real and $\sigma_1, \sigma_2, \sigma_3$ are the Pauli spin matrices defined by

$$\sigma_1 = \begin{bmatrix} 0 & 1 \\ 1 & 0 \end{bmatrix}, \quad \sigma_2 = \begin{bmatrix} 0 & -i \\ i & 0 \end{bmatrix}, \quad \sigma_3 = \begin{bmatrix} 1 & 0 \\ 0 & -1 \end{bmatrix},$$

is a representation of the group of quaternions under multiplication.

(ix) Show that the Pauli spin matrices obey the following rules:

$$\sigma_i \sigma_j + \sigma_j \sigma_i = 0 \quad \text{(anticommutation; } i \neq j)$$

$$\sigma_i \sigma_j - \sigma_j \sigma_i = 2i\sigma_k \quad \text{(commutation; } i,j,k \text{ cyclic)}.$$

(x) Let $\hat{n}$ be a unit vector, $Q_1 = \cos\theta_1 \hat{e} + \sin\theta_1 \hat{n}$, and $Q_2 = \cos\theta_2 \hat{e} + \sin\theta_2 \hat{n}$. Show that

$$Q_1 \odot Q_2 = \cos(\theta_1 + \theta_2) + \sin(\theta_1 + \theta_2)\hat{n}.$$

Thus one writes the quaternion $\cos\theta\hat{e} + \sin\theta\hat{n}$ as $e^{\hat{n}\theta}$.

(xi) Prove that any quaternion $Z = Z_0\hat{e} + \vec{Z}$ may be written as $|Z|e^{\hat{n}\theta}$.

(xii) Prove that if $o\hat{e} + \vec{v}_1 = e^{\hat{n}\theta/2} \odot (o\hat{e} + \vec{v}_0) \odot e^{-\hat{n}\theta/2}$, then $\vec{v}_1$ is the vector $\vec{v}_0$ rotated counterclockwise through an angle θ (see diagram below)

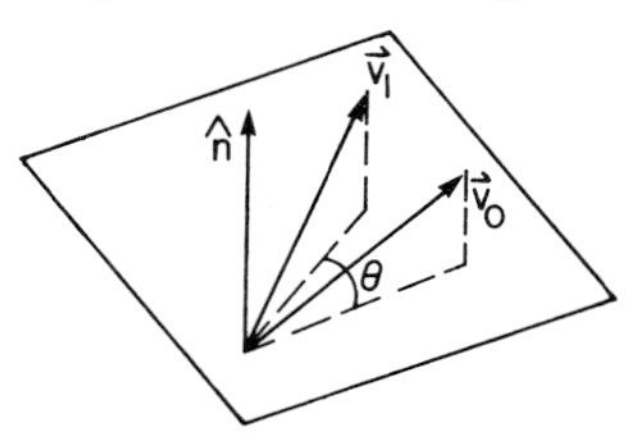

3 Prove that each class in a commutative (i.e. Abelian) group contains only one element.

4 Construct the group table for the rigid-body transformations of a square into itself.

5 Prove that the number of distinct elements in the class of S_n associated with the cycle structure $[1^{\nu_1},\ldots,n^{\nu_n}]$ is

$$\frac{n!}{(1^{\nu_1}\ \nu_1!)(2^{\nu_2}\ \nu_2!)\ldots(n^{\nu_n}\ \nu_n!)}$$

6 The characters for a unitary representation of a finite group satisfy the following relations:

$$\sum_i \chi_i^{(\mu)}\chi_i^{(\nu)*}g_i = g\delta_{\mu\nu} \qquad \text{(orthogonality relations)}$$

and

$$\sum_\nu \chi_i^{(\nu)}\chi_j^{(\nu)*} = (g/g_i)\delta_{ij}$$

where g is the order of the group, g_j is the number of elements (i.e. order) in the jth class, and $\chi_j^{(\beta)}$ is the character for the jth class βth irreducible representation.

(i) From the above relations deduce that the number of non-equivalent irreducible representations of a finite group is equal to the number of classes in the group.

(ii) For a finite group $\sum_\mu n_\mu^2 = g$ where n_μ is the dimension of each non-equivalent irreducible representation. From this prove that a commutative group of finite dimension has only one-dimensional irreducible representations. (Hint: Use the results of problem 3.)

7 Let G be a finite group and H be a subgroup of G. If $x \in G$ then the set of all elements xh where $h \in H$ is denoted xH and called a left coset of G.

(i) Show that for every $x,y \in G$ either $xH = yH$ or $xH \cap yH = \emptyset$, the null set.

(ii) Prove that if g is the order of G and h the order of H then $g = mh$ where m is a positive integer. This is called

Lagrange's Theorem. (Hint: Break the group up into the sum of left cosets.)

8 Construct the standard tableaux for all the partitions of five and from these calculate the dimensions of the irreducible representations of S_5 and the number of times they occur in the regular representation.

9 Construct the Young operators for all the standard tableaux associated with partitions of four.

10 Show that a 3-dimensional 3rd-rank irreducible tensor† which is symmetric on every pair of indices and traceless on every pair of indices has seven independent components.

11 For the irreducible rank-two tensor

$$\begin{bmatrix} (1/3)r^2 - x & xy & xz \\ xy & (1/3)r^2 - y & yz \\ xz & yz & (1/3)r^2 - z \end{bmatrix}$$

formed from the tensor $x_j x_k$ show that the components can be reorganized through linear combinations as the five spherical harmonics of order two times r^2. The spherical harmonics are given by

$$Y_{2,0} = \left(\frac{5}{16\pi}\right)^{\frac{1}{2}} (3\cos^2\theta - 1),$$

$$Y_{2,\pm 1} = \left(\frac{15}{8\pi}\right)^{\frac{1}{2}} \sin\theta\cos\theta e^{\pm i\phi},$$

$$Y_{2,\pm 2} = \left(\frac{15}{32\pi}\right)^{\frac{1}{2}} \sin^2\theta e^{\pm 2i\phi},$$

where θ, ϕ, r are spherical polar coordinates (see diagram below).

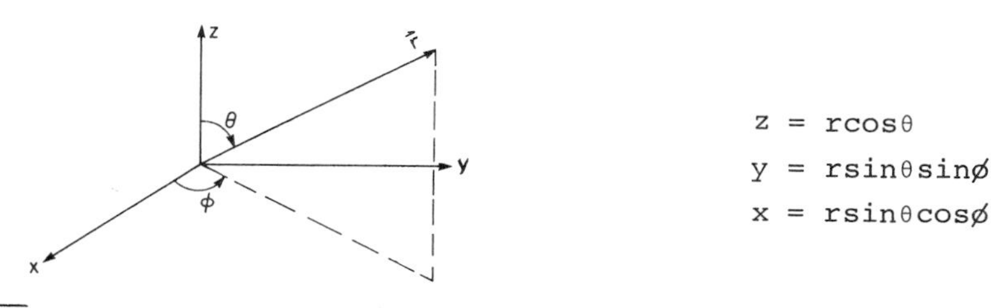

†A tensor which forms a base for an irreducible representation of a group is said to be irreducible.

12 Let $G = \{T, T^2, \ldots, T^n\}$, where $T^n = e$, the identity, be a a cyclic group of transformations on 3-dimensional space (i.e., $T: R^3 \to R^3$). Show that the general solution to the equation

$$g(\vec{x}) = g(T(\vec{x}))$$

is

$$g(\vec{x}) = \sum_{k=1}^{n} f(T^k(\vec{x}))$$

where f is an arbitrary function.

6
Quantum mechanics as a linear theory

6.1. INTRODUCTION

Quantum mechanics presents a striking example of a highly developed and successful linear theory. In the sense of Thomas Kuhn† it forms an integral part of the scientists' current paradigm. It is a pragmatic theory in that it has given birth to many wonders of technology of the present era; but it is also an artistic creation, incorporating as it does a beautiful and sophisticated conceptual framework articulated in elegant mathematics, replete with a rather elaborate and appealing theory of measurement.

However, the student of quantum physics should not expect quantum mechanics to tell him why the world is the way it is, only to describe it the way it is! The 'why' question is basically metaphysical and largely inappropriate to the subject of physics; in physics proper, the objective is to find an intellectually and aesthetically satisfying description of natural physical phenomena. This does not mean one should not attempt to axiomatize the theory, and indeed we shall attempt to do so; but the axioms should be regarded as convenient foundations to describe phenomenology rather than as obvious or acutely logical entities. In the last analysis, a theory is validated by a set of experimentally verifiable propositions, rather than by a set of a priori axioms. It is only after the consequences of a set of propositions or conjectures have been tested by experience that one has confidence in an axiomatic

---

†T. Kuhn, The Structure of Scientific Revolutions (University of Chicago Press, Chicago, 1970).

base which encompasses them.

A further caution regarding axiomatic bases arises from the celebrated theorem of Gödel† which may be paraphrased in the statement that no mathematical structures exist which have a complete and satisfactory axiomatic basis: for a given set of axioms, questions can always be posed which cannot be answered without additional axioms. Even the brilliant approaches of Von Neumann and others to axiomatize quantum mechanics are incomplete, and perhaps not free from contradiction. Even within conventional logic, there is no agreement; Reichenbach,†† for example, has argued that quantum mechanics requires three-valued logic for its proper explication. Aside from these questions, there are also deep and perplexing issues at the interface between the philosophical foundations of quantum mechanics and relativity, as pointed out by Wigner§ and others.

Although researchers disagree, almost violently at times, as to what constitutes a satisfactory axiomatic basis for quantum mechanics, almost everyone agrees in standard practice on how actual calculations are to be carried out and results verified by experiment for specific systems of interacting quantum particles. From this perspective, it is useful to write down some rules for calculation which at the same time amount to a set of propositions because they involve a new conceptual framework as well as providing an operational structure.

The point of view is basically this: all possible states of an isolated microscopic system are describable in terms of the vectors called state vectors of a linear vector space (LVS). If the LVS is realized in terms of Lebesgue square-integrable functions in some manifold, one speaks of wave mechanics because the functions can usually be interpreted as amplitudes of probability waves. All information which can possibly be

---

†E. Nagel and J. Newman, 'Godel's proof,' Scientific American, June 1956.

††Hans Reichenbach, Elements of Symbolic Logic (Macmillan, New York, 1952).

§E.P. Wigner, Rev. Mod. Phys. 29: 255 (1957).

obtained about a system is supposed to be contained in the state vector or wave function, as the case may be. The information is extracted by means of linear operators on the state vectors, with appropriate operators representing in effect a measurement on the system of some physical quantity (e.g. momentum, energy, position, mass, spin). The state vector itself is not observable; the 'observables' of the system are associated with appropriate extraction operators whose eigenvalues constitute the possible results of measurement.

For simplicity, and because appeal can be made to intuitive notions concerning wave motion, we shall first use the wave mechanical description, bearing in mind that everything can be translated into terms involving the state vector formalism. In modern terms, this amounts to the transition from Schroedinger's theory to Heisenberg's matrix mechanics. Also for simplicity we limit ourselves at the outset to the wave mechanics of a single, one-dimensional non-relativistic point particle without internal structure. The 'position variable' x may be regarded as generic in the sense that it stands now for an arbitrary point in one-dimensional space, but later it may stand for a point in 3-dimensional space, or even for several points in space $\vec{r}_1$, $\vec{r}_2$, $\vec{r}_3$, etc. corresponding to the possible locations of the centres of mass of several particles, or even to such variables plus some internal variables like spin or isospin or both. Later we shall deal briefly with some of the modifications which have to be made to accommodate internal structure and also relativistic covariance.

Finally, it should be emphasized that we are concerned primarily with the structure of quantum mechanics as an example of a linear theory. Exposition of detailed methodology including approximation techniques is left to textbooks on quantum mechanics. Nonetheless, the definitions and rules outlined below are relatively complete, and may be quickly adapted to formal calculations for particular physical problems.

6.2. EIGENFUNCTION EXPANSIONS FOR SELF-ADJOINT OPERATORS

Fundamental to both the conceptual and functional aspects of quantum mechanics is the notion of complete sets of orthonormal state vectors or wave functions associated with self-adjoint operators in a Hilbert space. We leave to more detailed treatments the difficult task of proving the necessary theorems, and of excluding or including, as the case may be, the various exotic circumstances which can arise in the solution of eigenvalue problems associated with linear operators.

Quantum kinematics is concerned with the possible states of a system, and we have already remarked that the state of a system is described in wave mechanics by a complex-valued wave function representing a probability amplitude. Since probability is real and normalized to unity, the modulus squared of the wave function must be Lebesgue integrable.† The class $L_2(-\infty,\infty)$ of Lebesgue square-integrable functions $\psi(x)$ on the real line $(-\infty,\infty)$ is known to form a Hilbert space (see Appendix F), i.e. a complete, normed LVS with an inner product defined. For $L_2(-\infty,\infty)$ we define the inner product of two vectors $\psi(x)$ and $\phi(x)$ by the Lebesgue integral

$$\langle\psi|\phi\rangle = \int_{-\infty}^{\infty} \psi^*(x)\phi(x)dx.$$

This notation is consistent with the Dirac bracket notation to be introduced later.

A linear operator A mapping its domain of definition D(A) (which is some subspace of the Hilbert space $L_2(-\infty,\infty)$) into $L_2(-\infty,\infty)$ is called self-adjoint (actually only symmetric) if it satisfies

$$\text{(1a)} \quad \langle\psi|A\phi\rangle = \int_{-\infty}^{\infty} \psi^*(x)(A\phi)(x)dx = \int_{-\infty}^{\infty} (A\psi)^*(x)\phi(x)dx = \langle A\psi|\phi\rangle$$

†The mathematical framework required for a systematic explication of quantum mechanics is treated briefly in Appendices D, E, and F.

for every $\psi(x)$ and $\phi(x)$ in D(A) with the closure of its domain of definition $\bar{D}$ given by

(1b) $\bar{D}(A) = L_2(-\infty,\infty)$.†

Strictly speaking physicists use the term self-adjoint when they really mean symmetric. For example, the operator $-i\hbar\partial/\partial x$ with once-differentiable functions†† as the domain of definition is not self-adjoint. This poses no real difficulty, however, since the definitions of symmetric operators occuring in the quantum mechanics may be extended (in a straightforward fashion) to make them self-adjoint.

For every self-adjoint operator A with a pure point spectrum (see Appendix F) there exists an orthonormal set of eigenvectors $\{\phi_k(x)\}$ with real eigenvalues a_k,

$$(A\phi_k)(x) = a_k\phi_k(x), \quad \phi_k(x) \in D(A),$$

such that any vector $\psi(x)$ in the Hilbert space $L_2(-\infty,\infty)$ can be expanded in terms of them. That is,

$$\psi(x) = \sum_k C_k\phi_k(x)$$

where the expansion coefficients C_k are given by

$$C_k = \int_{-\infty}^{\infty} \phi_k^*(x)\psi(x)dx.$$

The orthonormality is expressed as

$$\langle\phi_k|\phi_j\rangle = \int_{-\infty}^{\infty} \phi_k^*(x)\phi_j(x)dx = \delta_{kj}.$$

Unfortunately most self-adjoint operators occuring in quantum mechanics do not have a pure point spectrum. A self-adjoint operator A is called an 'observable' if there exists a

†I.e. the closure of the domain of definition of A is the entire Hilbert space $L_2(-\infty,\infty)$. In other words D(A) is dense in $L_2(-\infty,\infty)$.
††Whose derivatives also lie in the Hilbert space.

set of (proper) eigenvectors $\{\phi_k(x)\}$ with real eigenvalues a_k

(2a) $$(A\phi_k)(x) = a_k\phi_k(x), \quad \phi_k(x) \in D(A),$$

as well as a set of 'improper' eigenvectors $\{\phi(\lambda;x)\}$ with real eigenvalues† $a(\lambda)$

(2b) $$A\phi(\lambda;x) = a(\lambda)\phi(\lambda;x), \quad \phi(\lambda;x) \notin L_2(-\infty,\infty),$$

such that any vector $\psi(x)$ in the Hilbert space $L_2(-\infty,\infty)$ may be expanded in terms of them. That is,

(3a) $$\psi(x) = \sum_k C_k\phi_k(x) + \int C(\lambda)\phi(\lambda;x)d\lambda$$

where now

(3b) $$C_k = \int_{-\infty}^{\infty} \phi_k^*(x)\psi(x)dx$$

and

(3c) $$C(\lambda) = \int_{-\infty}^{\infty} \phi^*(\lambda;x)\psi(x)dx.$$

In physics, one refers to the a_k as the discrete part of the spectrum of A and $a(\lambda)$ as the continuous part. The hydrogen atom Hamiltonian, for example, has discrete spectra for bound states and continuous spectra for ionized states.

The improper eigenvectors are not square-integrable and thus the definition of A must be extended to give meaning to the operation of A on an improper eigenvector. Taking the inner product of ψ with the proper eigenvector $\phi_k(x)$ and using equations (3) gives

(4a) $$0 = \sum_\ell C_\ell\left[\int_{-\infty}^{\infty} \phi_k^*(x)\phi_\ell(x)dx - \delta_{\ell k}\right] + \int C(\lambda)\left[\int_{-\infty}^{\infty} \phi_k^*(x)\phi(\lambda;x)dx\right]d\lambda$$

†Mathematicians refer to the eigenvalues corresponding to improper eigenvectors as eigen-numbers.

whereas taking the 'inner product' of $\psi(x)$ with the improper eigenvector $\phi(\lambda;x)$ and again using equations (3) gives

$$\text{(4b)}\qquad 0 = \sum_{\ell} C_{\ell}\left[\int_{-\infty}^{\infty} \phi^*(\lambda;x)\phi_{\ell}(x)\,dx\right] + \int C(\lambda')\left[\int_{-\infty}^{\infty} \phi^*(\lambda;x)\phi(\lambda';x)\,dx - \delta(\lambda - \lambda')\right]d\lambda',$$

where $\delta(\lambda - \lambda')$ is the Dirac delta function defined by

$$\text{(5)}\qquad \delta(\lambda - \lambda') = 0 \text{ for } \lambda \neq \lambda', \qquad \int_{-\infty}^{\infty} C(\lambda')\delta(\lambda - \lambda')\,d\lambda' = C(\lambda)$$

for any $C(\lambda')$ continuous in some neighbourhood of λ. It is clear from equation (5) that $\delta(\lambda - \lambda')$ is not a proper function since it must be infinite at the point $\lambda = \lambda'$. Although a rigorous mathematical treatment of the Dirac delta function involves the theory of distributions, the physical content of equation (5) is easily justified.

Since equations (4) are true for arbitrary coefficients $C(\lambda)$ and C_k, we have

$$\text{(6a)}\qquad \int_{-\infty}^{\infty} \phi_k^*(x)\phi_{\ell}(x)\,dx = \delta_{\ell k},$$

$$\text{(6b)}\qquad \int_{-\infty}^{\infty} \phi^*(\lambda;x)\phi(\lambda';x)\,dx = \delta(\lambda - \lambda'),$$

and

$$\text{(6c)}\qquad \int_{-\infty}^{\infty} \phi_k^*(x)\phi(\lambda;x)\,dx = 0.$$

These are the orthonormality conditions for the set of all proper and improper eigenvectors $\{\phi_k(x),\ \phi(\lambda;x)\}$.

General theorems on eigenfunction expansion are difficult to prove and involve such exotic devices as equipped and rigged Hilbert spaces. For our purposes, however, it is sufficient to assume that the self-adjoint operators occurring in quantum mechanics are observables.

Some observables have only improper eigenvectors. The eigenfunction expansion for the momentum operator $-i\hbar d/dx$ is provided by theorems on Fourier analysis, which guarantee that any $\psi(x) \in L_2(-\infty,\infty)$ may be expanded in terms of the improper eigenvectors $\phi(k;x) = (2\pi)^{-\frac{1}{2}} e^{ikx}$ (with eigenvalues $a(k) = \hbar k$) in the following manner:

$$\psi(x) = \text{l.i.m.} \frac{1}{(2\pi)^{\frac{1}{2}}} \int_{-\infty}^{\infty} C(k)e^{ikx}dk$$

(where l.i.m. implies that the improper integral is to be interpreted as a limit in the mean). One can explicitly avoid the use of improper eigenvectors and eigenvalues by enclosing the system under consideration in a large, but finite box, in which case a pure point spectrum for the momentum operator can be assured. The box must be large enough so that the physical phenomenon under study is not affected by the presence of the box. In practice it always turns out that the box parameters drop out of any expressions referring to physical measurements.

In the Schroedinger equation the potential energy has a smoothing effect so that the total Hamiltonian usually has both proper and improper eigenvectors. In the case of the harmonic oscillator the smoothing effect of the potential is so great that the Hamiltonian has a pure point spectrum (i.e. only proper eigenvectors occur).

Our remarks essentially establish the diagonalizability of self-adjoint operators (occurring in quantum mechanics) in an orthonormal basis. Arising from the expansion properties of the eigenfunctions (proper and improper) they are said to constitute a complete set.

6.3. POSTULATES OF QUANTUM MECHANICS

We shall take the following set of postulates as a basis for our wave mechanics, motivated more by clarity of exposition than by mathematical rigour:

I. For a system comprising a point particle its state is completely specified by a complex-valued (wave) function $\psi(x,t)$ which is to be interpreted as a probability-density amplitude in the sense that

$$|\psi(x,t)|^2 dx$$

is the probability that a position measurement at time t will reveal the particle to be in the closed interval between x and x + dx. Since the total probability must be unity, we have the normalization condition:

$$\int_{-\infty}^{\infty} |\psi(x,t)|^2 \, dx = 1. \tag{7}$$

Remark: Some physicists prefer a strictly statistical interpretation in which one speaks of the relative number of particles to be found between x and x + dx when measurement is carried out on a large number of identical systems.

II. The possible states of the system referred to in postulate I form the set $L_2(-\infty,\infty)$ of all square-integrable functions defined on $(-\infty,\infty)$ and normalized to unity.

Remark: The theorem of linear superposition, which is of paramount importance, follows immediately from postulate II, viz. any normalized linear combination of two possible states of the system at a time t is itself a possible state of the system at time t. Let $\psi_1(x,t)$ and $\psi_2(x,t)$ be kinematically possible states; then if c_1 and c_2 are in the complex number field† and satisfy

$$|c_1|^2 + |c_2|^2 = 1, \tag{8}$$

then

---

†Here c_1 and c_2 are constants independent of position x and time t.

(9) $\psi_3(x,t) = c_1\psi_1(x,t) + c_2\psi_2(x,t)$

is also a kinematically possible state.

III. All possible information about the system is contained in the wave function and can be extracted by a suitable set of linear self-adjoint operators. Measurement on a system in state $\psi(x,t)$ of a physical quantity represented by operator A yields the value a_k with probability $|c_k|^2$, where a_k is the eigenvalue of A corresponding to eigenfunction $\phi_k(x)$ in the discrete spectrum,

(2a) $A\phi_k(x) = a_k\phi_k(x)$,

and c_k is the expansion coefficient for $\psi(x)$ in terms of the eigenfunctions of A at the time t:

(3a) $\psi(x,t) = \sum_\ell c_\ell\phi_\ell(x) + \int c(\lambda)\phi(\lambda;x)d\lambda$;

similarly a measurement of A yields a value in the continuous spectrum between $a(\lambda)$ and $a(\lambda + d\lambda)$ with probability $|c(\lambda)|^2 d\lambda$, where $a(\lambda)$ is the eigenvalue of A corresponding to the improper eigenfunction $\phi(\lambda;x)$:

(2b) $A\phi(\lambda;x) = a(\lambda)\phi(\lambda;x)$,

and $c(\lambda)$ is the expansion coefficient of $\phi(\lambda;x)$ for $\psi(x)$ in equation (3a) at time t.

Remark: This postulate implies that the probability of finding the system exactly in a given improper eigenstate of a self-adjoint operator A is zero. Thus the improper eigenvectors are not possible states of the system in accord with postulate II. Note, however, that wave packets which are continuous, normalized linear superpositions of improper eigenvectors are possible states of the system.

Remark: Simultaneous measurements on a large number of identical systems do not all lead to the same result.

Rather, the probability $|C_k|^2$ of the measurement yielding a_k varies with a_k, such as in the manner shown in Figure 6.1. Only if ψ is an eigenstate of A is the result of measurement strictly determined.

Remark: For a large number of simultaneous measurements on identical systems, the average or expectation value is denoted by $\langle A\rangle$† and given by

$$(10) \quad \langle A\rangle = \sum|C_k|^2 a_k + \int |C(\lambda)|^2 a(\lambda)d\lambda$$

$$= \int_{-\infty}^{\infty} \psi^*(x,t)A\psi(x,t)dx$$

as can be easily verified by expanding $\psi(x,t)$ and $\psi^*(x,t)$ on the right-hand side and using the orthonormality of eigenfunctions (equation (6)).

IV. In analogy with Hamilton's formulation of classical mechanics, operators representing dynamical quantities in quantum mechanics are self-adjoint operator functions of the basic position and momentum operators

$$(11a) \quad x_{op} = x \text{ (algebraic multiplication by x)}$$

$$(11b) \quad p_{op} = -i\hbar\partial/\partial x$$

with a complete set of eigenfunctions (i.e. observables).

Remark: The momentum operator p (we drop the 'op' designation for simplicity) derives its form from the supposition that de Broglie waves, i.e. plane waves of the form

$$(12) \quad \psi(x,t) = a e^{ikx-i\omega t},$$

where $\hbar\omega = \hbar^2k^2/2m$ and a is a constant amplitude, correspond

---

†Note that $\langle A\rangle$ may be a function of time, since the expansion coefficients C_ℓ and $C(\lambda)$ in equation (3a) vary with the time t, in general.

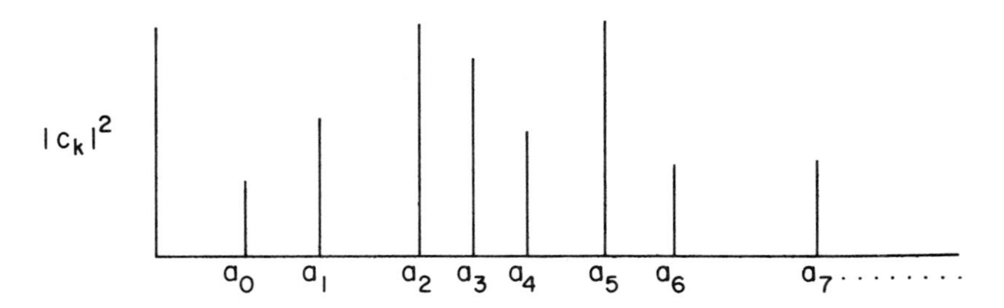

Figure 6.1. A possible spread of results for measurement of A (a self-adjoint operator with a pure point spectrum) on a large number of identical systems. The possible eigenvalues of A are plotted along the abscissa and the relative frequency of occurrence along the ordinate.

to free particles moving with fixed momentum. Hence de Broglie waves must be improper eigenfunctions, according to postulate III, of the momentum operator:

$$(13) \qquad -i\hbar \frac{\partial}{\partial x} a e^{ikx-i\omega t} = \hbar k a e^{ikx-i\omega t}.$$

De Broglie waves are also eigenfunctions of the kinetic energy operator $T = p^2/2m$

$$(14) \qquad T a e^{ikx-i\omega t} = \frac{-\hbar^2}{2m} \frac{\partial^2}{\partial x^2} a e^{ikx-i\omega t}$$

$$= \frac{\hbar^2 k^2}{2m} a e^{ikx-i\omega t}.$$

If the potential energy V depends only on the spatial location of the particles, then the Hamiltonian operator is given by

$$(15) \qquad H_{op} = H(p_{op}, x_{op}) = T_{op} + V_{op}$$

$$= \frac{-\hbar^2}{2m} \frac{\partial^2}{\partial x^2} + V(x).$$

Again, for simplicity in notation, we drop the 'op' designation on the Hamiltonian and write simply H.

Remark: The plane wave states $a e^{ikx-i\omega t}$ are improper eigenvectors of the momentum operator since they are not Lebesgue square-integrable. The normalization condition for improper eigenvectors gives

$$|a|^2 = (2\pi)^{-1}.$$

Postulate II implies that in a strict mathematical sense it is not possible to interpret such functions as representing possible states of the system. Physically, however, such complications are of no real consequence since it is always possible to find a proper vector (i.e. Lebesgue square-integrable function) which uniformly approximates a plane wave to any desired accuracy over any finite interval. For example, consider the following sequence $\{\psi_n(x,t)\}$ of square-integrable functions:

$$\psi_n(x,t) = e^{ikx-i\omega t}/\sqrt{2\pi}, \qquad |x| < n,$$

$$= 0, \qquad |x| > n.$$

The sequence converges pointwise to

$$\psi(x,t) = e^{ikx-i\omega t}/\sqrt{2\pi}$$

over the whole real line $(-\infty,\infty)$. The fact that the sequence $\{\psi_n(x,t)\}$ converges uniformly to $\psi(x,t)$ only on a finite interval places no essential restriction in practice since we can always envisage the system under study to be contained in a box and take the size of the box very much larger than the system, yet still finite.

Remark: From mathematics we know that the necessary and sufficient condition for two self-adjoint operators to have a complete set of simultaneous eigenfunctions is that they commute. We write the commutator of two operators A and B as $[A,B] = AB - BA$. If the commutator of two self-adjoint operators is non-zero then they may have simultaneous eigenfunctions, but not a complete set of them.

The operators x and p do not commute since

$$(16) \quad [x,p] = i\hbar \neq 0$$

so that they cannot have a complete set of simultaneous eigenfunctions. In fact they do not have any simultaneous eigenfunctions, for if they did have a simultaneous eigenfunction $\psi(x)$:

$$x\psi(x) = a\psi(x) \quad \text{and } p\psi(x) = b\psi(x),$$

this would imply that

$$\psi(x) = \frac{1}{i\hbar}[x,p]\psi(x) = \frac{ab}{i\hbar}\,\psi(x) - \frac{ba}{i\hbar}\,\psi(x) = 0.$$

A system cannot be in a state of definite position and momentum at the same time. This is the origin of Heisenberg's celebrated uncertainty principle $\Delta x \Delta p \geq \hbar/2$. One can show that fundamentally it is 'action' (e.g. erg-sec), the dimension of the product xp, that is quantized in the quantum theory.† Any quantum process can only alter its amount of action by an integral number of units of $\hbar$.

Remark: A set of n commuting observables $\{A_j\}$ (i.e. $[A_i,A_j] = 0$, $i,j = 1,\ldots,n$) is called maximal if any other observable B that commutes with every A_j is a function of the A_j (i.e. $B = F(A_1,A_2,\ldots,A_n)$). A maximal set of observables has an orthonormal basis of simultaneous eigenvectors. The vectors in this basis may be labelled unambiguously by their eigenvalues; for example a typical eigenvector in the basis would have the form

$$\phi(x)_{a^{(1)}_{\lambda_1},\, a^{(2)}_{\lambda_2},\ldots,a^{(n)}_{\lambda_n}} \equiv \phi_a(x)$$

†R.P. Feynman and A.R. Hibbs, Quantum Mechanics and Path Integrals (McGraw Hill, New York, 1965).

with

$$A_k \phi_a = a^{(k)}_{\lambda_k} \phi_a, \quad k = 1,2,\ldots,n.$$

V. <u>The dynamics of a quantum system are specified by Schroedinger's equation</u>

(17) $$H\psi(x,t) = i\hbar \partial\psi(x,t)/\partial t.$$

<u>Remark</u>: We see that in quantum mechanics just as in classical theory the dynamics relates time rates of change to spatial gradients. Of fundamental importance is the fact that the Schroedinger equation is first order in time, since this implies that initial conditions require only specifications of $\psi(x,t_0)$ at some initial time t_0 and not its derivatives. Hence if we know the state of the system at $t = t_0$, in principle at least, we can integrate Schroedinger's equations to find the state at any later time. For example, let us expand $\psi(x,t)$ in a Taylor series about t_0:

(18) $$\psi(x,t) = \psi(x,t_0) + (t - t_0)\left(\frac{\partial\psi}{\partial t}\right)_{t_0} + \ldots$$

$$= 1 - \frac{i(t - t_0)}{\hbar} H \psi + \ldots$$

Now let $\Delta t = t - t_0$ and $\tau = n(\Delta t)$ for some finite interval τ where n is a large integer. Repeating the first-order Taylor expansion n times, we get

(19) $$\psi(x, t_0 + \tau) = \lim_{n\to\infty} (1 - \frac{i\tau}{n\hbar})^n \psi(x,t_0)$$

$$= e^{-iH\tau/\hbar}\psi(x,t_0).$$

The operator

(20) $$U(\tau) = e^{-iH\tau/\hbar},$$

which is clearly a unitary operator,† is called the <u>time development</u> or <u>time translation</u> operator of the system.

Note that the operator H may not be defined on the entire Hilbert space of square-integrable functions. Hence equation (17) does not give the time development of all state vectors. This handicap is not serious, however, since the operator $e^{-iH\tau/\hbar}$ may be defined on the entire Hilbert space (see Appendix F) and so equation (19) does give the time development of every state vector.

<u>Remark</u>: In equation (19), let us choose the time scale $t_o = 0$ for convenience, so that the time development for an arbitrary state $\psi(x,t)$ of the system is given by

$$\text{(19a)} \quad \psi(x,t) = e^{-iHt/\hbar}\psi(x,0).$$

We can expand $\psi(x,0) \equiv \psi(x)$ in a complete set of energy eigenfunctions, as in equations (2a), (2b), and (3a), where $A = H$ is the Hamiltonian and $a_k = E_k$ and $a(\lambda) = E(\lambda)$ are the discrete and continuous energy eigenvalues. It follows that equation (19) takes the form

$$\text{(19b)} \quad \psi(x,t) = \sum C_k\phi_k(x)e^{-iE_k t/\hbar} + \int C(\lambda)\phi(\lambda;x)e^{-iE(\lambda)t/\hbar}d\lambda$$

Thus a knowledge of the energy eigenvalue spectrum and of the expansion coefficients at time t = 0 enables one to determine completely the evolved state $\psi(x,t)$ at any later time t.

6.4. MATRIX MECHANICS

Historically, quantum mechanics developed along two parallel but separate channels, which were eventually shown to lead to equivalent formulations. The two formulations are those of Heisenberg and Schroedinger. Following de Broglie's hypothesis concerning matter waves, Schroedinger endeavoured to find a

---

†Since H is self-adjoint, $H = H^+$ and hence $U(\tau)U^+(\tau) = e^{-iH\tau/\hbar}e^{iH\tau/\hbar} = 1$.

wave equation to describe their propagation. His attempts at a relativistically invariant formulation were not successful (success was achieved by Dirac several years later), but his non-relativistic equation became the centrepiece of all subsequent developments in quantum theory. Heisenberg's matrix mechanics, however, derived directly from his experience in trying to understand the relationships among transition amplitudes in two-stage quantum processes. Today, the equivalence between the wave and matrix formulations seems entirely natural, and one goes from the function space formulation of wave mechanics to the matrix formulation by expanding Schroedinger functions in a complete orthonormal basis. For convenience we treat the basis as if it consisted only of proper vectors. This involves no real loss of generality since the extension to improper vectors is achieved merely by replacing sums by integrals.

Let $\phi_k(x)$ be such a set of complete orthonormal functions belonging to some appropriate self-adjoint operator W with eigenvalues ω_k:

$$W\phi_k(x) = \omega_k\phi_k(x). \tag{21}$$

Further, let $\psi(x)$ be the Schroedinger 'wave function' describing some particular state of the system, and let B be a linear (algebraic or differential) operator corresponding to some dynamical quantity of interest. Further, let the action of B on ψ produce the quantum state χ:

$$B\psi(x) = \chi(x). \tag{22}$$

Using the theorem of Section 5.2, we expand both $\psi(x)$ and $\chi(x)$ in terms of the complete set $\phi_k(x)$:

$$\begin{aligned} \psi(x) &= \sum_k c_k\phi_k(x), \qquad c_k = \int_{-\infty}^{\infty} \phi_k^*(x)\psi(x)\,dx, \\ \chi(x) &= \sum_j a_j\phi_j(x), \qquad a_j = \int_{-\infty}^{\infty} \phi_j^*(x)\chi(x)\,dx. \end{aligned} \tag{23}$$

Substitution in the linear equation $\chi = B\psi$ then yields

$$(24) \qquad \sum_j a_j\phi_j(x) = \sum_k c_k B\phi_k(x)$$

which, upon multiplication by $\phi_n^*(x)$ and integration, yields the following linear (matrix) equations relating the a_n and the c_k:

$$(25) \qquad a_n = \sum_k b_{nk}c_k$$

where

$$(26) \qquad b_{nk} = \int_{-\infty}^{\infty} \phi_n^*(x) B\phi_k(x)\,dx$$

is called the <u>matrix element</u> of the operator B in the representation $\phi_k(x)$.

Clearly, these linear equations can be arranged in matrix form:

$$(27) \qquad \begin{bmatrix} a_1 \\ a_2 \\ a_3 \\ \vdots \end{bmatrix} = \begin{bmatrix} b_{11} & b_{12} & b_{13}\cdots \\ b_{21} & b_{22} & b_{23}\cdots \\ b_{31} & b_{32} & b_{33}\cdots \\ \vdots & \vdots & \vdots \end{bmatrix} \begin{bmatrix} c_1 \\ c_2 \\ c_3 \\ \vdots \end{bmatrix}.$$

Moreover it is easy to show, using the same expansion technique, that successive linear differential operators, in their matrix form, compose by matrix multiplication. For example, if in Schroedinger's theory we have the operator relation H = BG, then in Heisenberg's formulation we have

$$(28) \qquad h_{ij} = \sum_k b_{ik}g_{kj},$$

where the matrix elements of H and G are defined in a manner entirely analogous to those of B.

We now introduce the Dirac bracket notation by rephrasing the relation $\chi = B\psi$ as

(29) $|\chi\rangle = B|\psi\rangle$

which translates these functional relations into state vector notation. Vectors like $|\chi\rangle$ and $|\psi\rangle$ are called <u>ket vectors</u>, and their complex conjugates, arranged as row matrices, are called <u>bra vectors</u>. Scalar products in the Hilbert space correspond to closing the <u>bracket</u>; for example, if

$$|\psi\rangle = \begin{bmatrix} c_1 \\ c_2 \\ c_3 \\ \vdots \end{bmatrix}, \quad \langle\chi| = \begin{bmatrix} a_1^* & a_2^* & a_3^* & \cdots \end{bmatrix},$$

then

(30) $$\langle\chi|\psi\rangle = \sum_k a_k^* c_k = \begin{bmatrix} a_1^* & a_2^* & a_3^* & \cdots \end{bmatrix} \begin{bmatrix} c_1 \\ c_2 \\ c_3 \\ \vdots \end{bmatrix}$$

$$= \int_{-\infty}^{\infty} \chi^*(x)\psi(x)\,dx.$$

If one now considers the most general 'matrix element' of some dynamical operator G between two states $\chi(x)$ and $\psi(x)$ of the system, we can write

(31) $$\langle\chi|G|\psi\rangle = \sum_{kj}\sum a_k^* c_j \langle\phi_k^*|G|\phi_j\rangle = \sum_{kj}\sum a_k^* c_j g_{kj}.$$

Matrix elements of this type are called <u>transition matrix elements</u>, because the operator G 'causes' the state $|\psi\rangle$ to transform into $|\chi\rangle$. In the usual applications where G is the perturbing term in the Hamiltonian, the probability of transition is proportional to the modulus squared: $|\langle\chi|G|\psi\rangle|^2$.

6.5. AN EXAMPLE - THE SIMPLE HARMONIC OSCILLATOR

We consider as an example the simple harmonic oscillator (SHO), in order to illustrate these formal ideas. Although our example is modest, it is also basic to theoretical developments throughout the whole range of mathematical physics, including such diverse problems as the thermal properties of crystalline solids, nuclear collective motions, and the theory of the laser. We shall not, however, solve equations but only quote results; the formal solution of SHO problems is given in every standard book on quantum mechanics.

In one-dimension the Schroedinger equation for the SHO becomes

$$\left(-\frac{\hbar^2}{2m}\frac{\partial^2}{\partial x^2}+\frac{1}{2}\kappa x^2\right)\psi(x,t) = i\hbar\frac{\partial\psi}{\partial t} \tag{32}$$

where the first term on the left represents the kinetic energy and the second term the potential energy. For solutions of definite energy E, we can substitute

$$\Psi = \psi(x)e^{-iEt/\hbar} \tag{33}$$

to obtain the time-independent equation

$$\left(-\frac{\hbar^2}{2m}\frac{\partial^2}{\partial x^2}+\frac{1}{2}\kappa x^2\right)\psi(x) = E\psi(x). \tag{34}$$

The possible eigenvalues of the SHO correspond to those solutions which are square-integrable (i.e. which vanish at $\pm\infty$). For simplicity, we write the classical frequency as $\omega_0 = \sqrt{\kappa/m}$ and introduce a dimensionless length parameter ξ and a dimensionless energy parameter λ as follows:

$$E = \lambda\hbar\omega_0/2, \quad x = \sqrt{\hbar/m\omega_0}\,\xi.$$

With these substitutions, the SHO equation becomes an equation for $\psi(\xi)$:

$$\psi'' + (\lambda - \xi^2)\psi = 0 \tag{35}$$

where primes indicate differentiation with respect to ξ. Using a series method of solution, one finds that the only square-integrable solutions are of the form

$$\psi_n(x) = (2^n n!\pi)^{-\frac{1}{2}} e^{-\xi^2/2} H_n(\xi), \quad n = 0,1,2,\ldots, \tag{36}$$

where the $H_n(\xi)$ are standard Hermite polynomials of which the first four are

$$H_0 = 1, \quad H_2 = 4\xi^2 - 2,$$
$$H_1 = 2\xi, \quad H_3 = 8\xi^3 - 12\xi.$$

Corresponding to ψ_n the eigenvalue for the (n + 1)st level is $E_n = (n + \frac{1}{2})\hbar\omega_0$, so that the energy levels are evenly spaced beginning with the ground state energy $\hbar\omega_0/2$. The eigenfunctions $\psi_n(x)$ form a complete orthonormal set, so that an arbitrary function f(x) obeying the same boundary conditions can be expanded in them:

$$f(x) = \sum_n c_n\psi_n(x) \tag{37}$$
$$= \sum_n c_n(2^n n!\pi)^{-\frac{1}{2}} e^{-\xi^2/2} H_n(\xi)$$

with

$$c_n = (2^n n!\pi)^{-\frac{1}{2}} \int_{-\infty}^{\infty} f(x)e^{-\xi^2/2} H_n(\xi)dx; \tag{38}$$

moreover

$$\int_{-\infty}^{\infty} \psi_n(x)\psi_m(x)dx = \delta_{nm}. \tag{39}$$

One can use the SHO eigenfunctions as a basis to obtain a matrix representation of any dynamical operator. The simplest examples are given by the position operator x and the momentum operator $-i\hbar\partial/\partial x$. Using the simple recursion relations for the Hermite polynomials

$$\text{(40)}\qquad \begin{aligned} H_n' &= 2nH_{n-1}, \\ H_{n+1} &= 2\xi H_n - H_n', \end{aligned}$$

and

$$H_{n+1} - 2\xi H_n + 2nH_{n-1} = 0,$$

one can easily show that x has matrix elements

$$\text{(41)}\qquad \langle\psi_n|x|\psi_m\rangle = \langle n|x|m\rangle = \sqrt{\frac{\hbar}{m\omega_0}}\left(\sqrt{\frac{n}{2}}\,\delta_{m,n-1} + \sqrt{\frac{n+1}{2}}\,\delta_{m,n+1}\right)$$

and $p = -i\hbar\partial/\partial x$ has matrix elements

$$\text{(42)}\qquad \langle\psi_n|p|\psi_m\rangle = \langle n|p|m\rangle = i\sqrt{\hbar m\omega_0}\left(-\sqrt{\frac{n}{2}}\,\delta_{m,n-1} + \sqrt{\frac{n+1}{2}}\,\delta_{m,n+1}\right).$$

Thus

$$\text{(43)}\qquad x = \sqrt{\frac{\hbar}{2m\omega_0}}\begin{bmatrix} 0 & \sqrt{1} & 0 & 0 & \cdots \\ \sqrt{1} & 0 & \sqrt{2} & 0 & \cdots \\ 0 & \sqrt{2} & 0 & \sqrt{3} & \cdots \\ \vdots & \vdots & \vdots & \vdots & \end{bmatrix},$$

$$\text{(44)}\qquad p = i\sqrt{\frac{\hbar m\omega_0}{2}}\begin{bmatrix} 0 & -\sqrt{1} & 0 & 0 & \cdots \\ \sqrt{1} & 0 & -\sqrt{2} & 0 & \cdots \\ 0 & \sqrt{2} & 0 & -\sqrt{3} & \cdots \\ 0 & 0 & \sqrt{3} & 0 & \\ \vdots & \vdots & \vdots & \vdots & \end{bmatrix}.$$

The reader can easily check that if one forms x^2 and p^2 by matrix multiplication of the x and p matrices respectively, one obtains the matrix for the Hamiltonian

$$(45) \quad H = (\frac{p^2}{2m} + \frac{\kappa x^2}{2}) = \frac{\hbar\omega_0}{2} \begin{bmatrix} 1 & & & & & & \\ & 3 & & & & \text{zeros} & \\ & & 5 & & & & \\ & & & 7 & & & \\ & \text{zeros} & & & \cdot & & \\ & & & & & \cdot & \\ & & & & & & \cdot \end{bmatrix}$$

which is, of course, diagonal because we used eigenfunctions of H as the basis of the matrix representation.

As an example of a transition operator, we consider the dipole moment operator ex, where e is the charge on the oscillating particle and x is its position variable. The matrix for x above immediately tells us that transitions occur only between states which differ by one oscillator quantum $\hbar\omega_0$, which comprises the selection rule $n \rightarrow n \pm 1$ for dipole transitions.

6.6. PARTICLES WITH INNER STRUCTURE

In the previous sections of the chapter we have assumed that the quantum systems under consideration were point particles, i.e. particles without internal structure. We now raise the question of how internal structure is to be accommodated. One aspect of internal structure is parity, the behaviour of the wave function at a point in space under coordinate inversion, i.e. under change from a right-handed to a left-handed system or vice versa. Since, in quantum mechanics, only the square modulus of the wave function, not the wave function itself, is observable, the wave function (regarded for the moment as a scalar field) could be either a true scalar field

$$(46) \quad I\psi(\vec{r},t) = \psi(-\vec{r},t)$$

or a pseudoscalar field

(47) $I\psi(\vec{r},t) = -\psi(-\vec{r},t)$,

where I is the inversion operator.† An example of the scalar is the ε particle (resonance in π-π scattering); the π-meson itself belongs to the pseudoscalar class.††

In the sense that the scalar-pseudoscalar diversity reflects the structure of 3-dimensional space, one can say that the parity (inversion) quantum number ± 1 is a purely geometric property within the framework of quantum mechanics.

Other internal properties of elementary particles, such as spin, charge, magnetic moment, isospin, strangeness, may or may not have purely geometric interpretations. At the present time this must be regarded as an open question. What is clear is the convenience of using multi-component wave functions for describing these properties in the simplest possible fashion. For example, in Dirac's theory of the electron, which we shall return to in Chapter 9, a four-component wave function is used, i.e. at each point in space one must specify four functions to describe the state of the particle, two of these having to do with charge and two with spin angular moment. It is as though one had to specify four separate fields, one describing electrons 'spin-up,' one describing electrons 'spin-down,' and two corresponding fields for the positron (anti-particle).

For the present, however, we shall restrict ourselves to non-relativistic models, and to the consideration of intrinsic angular momentum (spin) as the only quantum property having to do with internal structure. If we assume a quantum analogue of the well-known classical theorem in mechanics that the energy can be separated into centre-of-mass motion and internal dynamics, we can write the Hamiltonian as

(48) $$H = H_{ext} + H_{int} = -\left(\frac{\hbar^2}{2m}\nabla^2 + V(\vec{r})\right) + H_{int}$$

---

†The inversion operator I commutes with the Hamiltonian operator and so the wave function can be chosen to be simultaneous eigenfunctions of H and I.

††Electrons, neutrons, and protons have a more complicated internal structure and will be discussed later.

where $V(\vec{r})$ is the externally imposed potential energy and H_{int} depends only on internal variables. For simplicity we designate the total collection of internal variables by the symbol τ. From the linear properties of the separation into internal and external parts, it is clear that the time-independent Schroedinger equation splits into two separate equations, if we assume a product solution of the form

(49) $$\psi(\vec{r},\tau) = \phi(\vec{r})\chi(\tau).$$

Substituting this product function into the Schroedinger equation

$$H\psi = E\psi$$

and dividing out by ψ yields

(50) $$\frac{1}{\phi} H_{ext}\phi + \frac{1}{\chi} H_{int}\chi = E.$$

Since the first term depends only on the independent variable $\vec{r}$ and the second only on the independent variables τ, and E is a constant, we conclude that each term is separately equal to a constant. This observation yields the separated equations:

(51) $$H_{int}\chi = \varepsilon\chi,$$

(52) $$\left(-\frac{\hbar^2}{2m}\nabla^2 + V\right)\phi = (E - \varepsilon)\phi,$$

where the 'separation constant' ε has the interpretation of internal energy.

In the non-relativistic theory of the electron,† the two spin states corresponding to quantum number $S = \frac{1}{2}$ (total spin momentum $\sqrt{S(S+1)}\hbar$) but to z-components $S_z = \pm\hbar/2$ are energy degenerate choices of the internal wave function $\phi(\tau)$, viz. $\phi_{\frac{1}{2},\frac{1}{2}} = \alpha(\tau)$ (spin-up state) and $\phi_{\frac{1}{2},-\frac{1}{2}} = \beta(\tau)$ (spin-down state).

†I.e. of the electron moving in a potential $V(\vec{r})$ which is independent of the spin variable τ.

In a detailed treatment of the quantum theory of angular momentum it is shown that the dynamical spin operator $\vec{S}$ has components obeying the commutation rules

$$[S_x, S_y] = i\hbar S_z \quad \text{(and cyclic permutations)}$$

and

$$(53) \qquad [S^2, S_x] = [S^2, S_y] = [S^2, S_z] = 0,$$

where $S^2 = S_x{}^2 + S_y{}^2 + S_z{}^2$. Hence a maximal set of commuting operators in spin space is S^2 and S_z, say, which can have simultaneous eigenfunctions, viz. the ϕ functions introduced above:

$$(54) \qquad \begin{aligned} S^2\phi_{\frac{1}{2},\pm\frac{1}{2}} &= \tfrac{1}{2}(\tfrac{1}{2} + 1)\hbar^2\phi_{\frac{1}{2},\pm\frac{1}{2}}, \\ S_z\phi_{\frac{1}{2},\pm\frac{1}{2}} &= \pm\tfrac{1}{2}\hbar\phi_{\frac{1}{2},\pm\frac{1}{2}}, \end{aligned}$$

or in Dirac notation

$$(55) \qquad \begin{aligned} S^2|\tfrac{1}{2},\pm\tfrac{1}{2}\rangle &= \tfrac{1}{2}(\tfrac{1}{2} + 1)\hbar^2|\tfrac{1}{2},\pm\tfrac{1}{2}\rangle, \\ S_z|\tfrac{1}{2},\pm\tfrac{1}{2}\rangle &= \pm\tfrac{1}{2}\hbar|\tfrac{1}{2},\pm\tfrac{1}{2}\rangle. \end{aligned}$$

A realization of these operator equations (see problem 3) is given by the Pauli spin matrices $\vec{\sigma}$, viz.

$$(56) \qquad \begin{aligned} S_x &= \frac{\hbar}{2}\sigma_x = \frac{\hbar}{2}\begin{bmatrix} 0 & 1 \\ 1 & 0 \end{bmatrix}, \quad & S_y &= \frac{\hbar}{2}\sigma_y = \frac{\hbar}{2}\begin{bmatrix} 0 & -i \\ i & 0 \end{bmatrix}, \\ S_z &= \frac{\hbar}{2}\sigma_z = \frac{\hbar}{2}\begin{bmatrix} 1 & 0 \\ 0 & -1 \end{bmatrix}, \quad & S^2 &= \frac{3\hbar^2}{4}\begin{bmatrix} 1 & 0 \\ 0 & 1 \end{bmatrix}, \end{aligned}$$

with the spin-up and spin-down states represented by

$$(57) \qquad |\tfrac{1}{2},\tfrac{1}{2}\rangle = \begin{bmatrix} 1 \\ 0 \end{bmatrix} \quad \text{and} \quad |\tfrac{1}{2},-\tfrac{1}{2}\rangle = \begin{bmatrix} 0 \\ 1 \end{bmatrix}.$$

An arbitrary spin state of the electron is then represented by the matrix

$$(58)\quad a\begin{bmatrix}1\\0\end{bmatrix} + b\begin{bmatrix}0\\1\end{bmatrix} = \begin{bmatrix}a\\b\end{bmatrix}$$

with $|a|^2 + |b|^2 = 1$ for normalization.

More generally, a particle of spin S is described by a (2S + 1)-component wave function consisting of a linear combination of the product functions

$$\phi(\vec{r})\chi_{S,M_S}(\tau), \quad M_S = S,\ S - 1,\ \ldots,\ -S + 1,\ -S.$$

Again, the χ could be represented by the base vectors of a (2S + 1)-dimensional vector space

$$(59)\quad \chi_{S,S} = \begin{bmatrix}1\\0\\\vdots\\0\\0\end{bmatrix}, \quad \chi_{S,S-1} = \begin{bmatrix}0\\1\\0\\\vdots\\0\end{bmatrix}, \ldots, \chi_{S,-S} = \begin{bmatrix}0\\0\\\vdots\\0\\1\end{bmatrix},$$

while the spin operators would obey the same quantum commutation relations as the electron spin operators, but the matrices corresponding to these operators would be (2S + 1)-dimensional square matrices.

It is a legitimate and important question to ask how one knows that a particular particle has a (2S + 1)-dimensional internal structure, and how these states are isolated in laboratory measurements. It is a matter of experience that particles have magnetic moments $\vec{\mu} = g\vec{S}$ associated with their spin dynamics, where the gyromagnetic ratio g varies from particle to particle depending on its electric charge substructure, but is a constant for a given particle. In the absence of a magnetic field, the (2S + 1)-dimensional substructure is degenerate in energy but shows up in the statistical behaviour of the particle. In the presence of an inhomogeneous magnetic field (Stern-Gerlach experiment), not only is the energy degeneracy removed through the interaction $\vec{\mu}\cdot\vec{B}$ ($\vec{B}$ = magnetic field intensity), but the bean corresponding to the quantized values of μ, viz. gS,...,-gS, along the direction of $\vec{B}$ are split apart so that they can be separately detected.

6.7. QUANTIZATION OF ORBITAL MOMENTUM

In this section we take a brief look at the quantum theory of orbital angular momentum. In contrast to spin angular momentum, where the detailed inner dynamics is not known and one is contented with the enumeration of states and their associated angular momentum eigenvalues, here we can deal with the detailed dynamics as well, if we so wish, since this involves only the translational motion of the centre of mass. Hence, we face two choices: one is to develop a detailed description of the orbital motion through appropriate differential equations; the second is to derive formal results that give us information about state enumeration in terms of angular momentum quantum numbers only. We shall pursue the first briefly for the perspective gained, but switch to the formal method to achieve our results.

If we are concerned only with centre-of-mass motion, the inner structure can be temporarily set aside. We consider then the second Schroedinger equation arising from separation† into external and internal energies of the orbiting electron,

$$\left(-\frac{\hbar^2}{2m}\nabla^2 + V(r)\right)\phi(\vec{r}) = (E - \varepsilon)\phi = \mathcal{E}\phi \tag{60}$$

where $\mathcal{E}$ is the energy of centre-of-mass motion of the electron, i.e., the difference between the total and the internal energies, and $V(r)$ is a central field such as that experienced by an electron in the hydrogen atom. Substitution of the product wave function

$$\phi(\vec{r}) = R(r)Y(\theta,\phi) \tag{61}$$

into equation (60) and division by RY again leads to two separate equations, the radial equation

$$\frac{1}{r^2}\frac{d}{dr}\left(r^2\frac{dR}{dr}\right) + \frac{2m}{\hbar^2}\left[(\mathcal{E} - V(r)) - \frac{\lambda}{r^2}\right]R = 0 \tag{62}$$

---

†Note that external and internal here mean with respect to the electron per se.

and the angular equation

$$\text{(63)}\qquad L^2Y(\theta,\phi) = \lambda\hbar^2Y(\theta,\phi),$$

where L^2 is the angular momentum operator squared:

$$\text{(64)}\qquad L^2 = -\hbar^2\left[\frac{1}{\sin\theta}\frac{\partial}{\partial\theta}\left(\sin\theta\frac{\partial}{\partial\theta}\right) + \frac{1}{\sin^2\theta}\frac{\partial^2}{\partial\phi^2}\right].$$

The identification of L^2 in equation (64) follows from the construction of the self-adjoint vector operator

$$\text{(65)}\qquad \vec{L} = (\vec{r} \times \vec{p}) = -i\hbar\,\vec{r} \times \vec{\nabla}$$

expressed in spherical polar coordinates and the formation of the self-adjoint scalar product $L^2 = \vec{L}\cdot\vec{L}$.

We only quote here the results of solving equation (63) with the definition (64). Applying the condition of square integrability and requiring that the probability density be single-valued and continuous leads to the eigensolutions

$$\text{(66)}\qquad Y_{\ell m} = (-1)^m\sqrt{\frac{2\ell+1}{4\pi}\frac{(\ell-|m|)!}{(\ell+|m|)!}}\,P_\ell^{|m|}(\cos\theta)e^{im\phi},$$

satisfying the eigenvalue equations

$$\text{(67)}\qquad \begin{aligned} L^2Y_{\ell m} &= \ell(\ell+1)\hbar^2Y_{\ell m}, & \ell &= 0,1,2,\ldots, \\ L_zY_{\ell m} &= -i\hbar\frac{\partial}{\partial\phi}Y_{\ell m} = m\hbar Y_{\ell m}, & m &= \ell,\ell-1,\ldots,-\ell. \end{aligned}$$

Here z has been chosen as the so-called axis of quantization and the $P_\ell^{|m|}$ are the standard associated Legendre polynomials. The $Y_{\ell m}$ form a complete orthonormal set on the unit sphere in consonance with the expectations of the mathematical theorem outlined in Section 5.2: viz. for an arbitrary angular function $f(\theta,\phi)$ satisfying the same boundary and regularity conditions, we can expand

$$\text{(68)}\qquad f(\theta,\phi) = \sum_{\ell=0}^{\infty}\sum_{m_\ell=-\ell}^{\ell} g_{\ell m}Y_{\ell m}(\theta,\phi)$$

where the $g_{\ell m}$ are coefficients given by

$$(69) \qquad g_{\ell m} = \int_0^{2\pi} d\phi \int_0^{\pi} \sin\theta d\theta \, \{Y^*_{\ell m}(\theta,\phi) f(\theta,\phi)\}.$$

This last result follows from the orthonormality of the $Y_{\ell m}$:

$$(70) \qquad \int_0^{2\pi}\int_0^{\pi} Y^*_{\ell m}(\theta,\phi) Y_{\ell' m'}(\theta,\phi) \sin\theta d\theta d\phi = \delta_{\ell\ell'}\delta_{mm'}.$$

If we use the $Y_{\ell m}$ eigenfunctions as a basis of a matrix representation, it is evident that L^2 and L_z are already diagonal; for example,

$$(71) \qquad L^2 = \hbar^2 \begin{bmatrix} 0 & & & & & & & & & & & \\ & 2 & & & & & & & & & & \\ & & 2 & & & & & & & & & \\ & & & 2 & & & & & & & & \\ & & & & 6 & & & & & & & \\ & & & & & 6 & & & & & & \\ & & & & & & 6 & & & & & \\ & & & & & & & 6 & & & & \\ & & & & & & & & 6 & & & \\ & & & & & & & & & \cdot & & \\ & & & & & & & & & & \cdot & \\ & & & & & & & & & & & \cdot \end{bmatrix}, \qquad L_z = \hbar \begin{bmatrix} 0 & & & & & & & & & & & \\ & 1 & & & & & & & & & & \\ & & 0 & & & & & & & & & \\ & & & -1 & & & & & & & & \\ & & & & 2 & & & & & & & \\ & & & & & 1 & & & & & & \\ & & & & & & 0 & & & & & \\ & & & & & & & -1 & & & & \\ & & & & & & & & -2 & & & \\ & & & & & & & & & \cdot & & \\ & & & & & & & & & & \cdot & \\ & & & & & & & & & & & \cdot \end{bmatrix}$$

Since L_x and L_y do not commute with L_z they cannot be diagonalized in this representation. It is not difficult to show that the operator combinations $L_\pm = L_x \pm iL_y$, the so-called raising and lowering operators, satisfy the equations

$$(72) \qquad (L_x \pm iL_y)Y_{\ell,m} = \sqrt{\ell(\ell+1) - m(m\pm 1)}\,\hbar Y_{\ell,m\pm 1}$$

so that the corresponding matrices vanish except for a set of values immediately above (or below) the main diagonal.

6.8. ROTATION OPERATOR AND REPRESENTATION OF THE SPECIAL ORTHOGONAL (ROTATION) GROUP

Given some general distribution in space, say $f(r,\theta,\phi)$, which in our context is interpreted as a wave amplitude, we can ask what this distribution is like in a rotated reference frame. Let the reference frame undergo rotation $\delta\phi$ about the z-axis. The new angles are now $\theta' = \theta$ and $\phi' = \phi - \delta\phi$. To first order in a Taylor's expansion†

$$(73) \quad f'(\theta,\phi) = f(\theta, \phi + \delta\phi) = \left(1 + \delta\phi\frac{\partial}{\partial\phi}\right) f(\theta,\phi) + \ldots$$

$$= \left(1 + \frac{i\delta\phi}{\hbar} L_z\right) f(\theta,\phi).$$

If we now rotate in n successive such steps in order to achieve a finite rotation through angle $\phi_z = n\delta\phi$, we have††

$$(74) \quad (R_z(\phi_z)f)(\theta,\phi) = \lim_{n\to\infty}\left(1 + \frac{i\phi_z L_z}{n\hbar}\right)^n f(\theta,\phi)$$

$$= \exp\left(\frac{i\phi_z}{\hbar} L_z\right) f(\theta,\phi) = f'(\theta,\phi).$$

The exponential is a unitary operator which expresses the function f' in terms of the original variables θ,ϕ. If the rotation has a value $\vec{\alpha}$, where $|\vec{\alpha}|$ is the rotational angle and $\hat{\alpha}$ the rotation axis, equation (74) is generalized to

$$(75) \quad (R(\vec{\alpha})f)(\theta,\phi) = e^{i\vec{\alpha}\cdot\vec{L}/\hbar} f(\theta,\phi).$$

Since $\vec{\alpha}$ encompasses all possible rotations of axes, the operator $R(\vec{\alpha})$ in fact gives a realization of the rotation group. It also follows that we can form a matrix representation of the rotation group by choosing a set of base functions and casting the exponential operator $R(\vec{\alpha})$ in matrix form.

The natural choice to make of base functions is the set $Y_{\ell m}(\theta,\phi)$. We observe that L^2 commutes with L_x, L_y, and L_z and hence with $\vec{L}$ and $R(\vec{\alpha})$. Thus L^2 commutes with all group operators. Such an operator is called a Casimir operator and, by a

† $f'(\theta',\phi') = f(\theta,\phi) \Rightarrow f'(\theta,\phi - \delta\phi) = f(\theta,\phi) \Rightarrow f'(\theta,\phi) = f(\theta,\phi + \delta\phi$

†† The notation on the left-hand side of (74) signifies that after the action of $R_z(\phi_z)$ on f the result is a function of θ and ϕ.

lemma due to Schur, its eigenvalues label irreducible representations of the rotation group. But the eigenvalues of L^2 are $\ell(\ell + 1)\hbar^2$ where ℓ is the angular momentum quantum number with possible values $\ell = 0,1,2,\ldots$ Hence the irreducible representation of R_3 (or SO(3)) consists of just the submatrices corresponding to definite ℓ-values in equations (71) and (72) exponentiated according to the definition of $R(\vec{\alpha})$ in equation (75).

We give two examples. For $\ell = 0$, L_z is the 1-dimensional zero matrix and $L_\pm$ are zero as well. For $\ell = 1$, we get a 3-dimensional irreducible representation whose rows and columns may be labelled by the quantum numbers $m_\ell = 1,0,-1$. The $L_x \pm iL_y$ matrices, and hence the L_x and L_y matrices, may be obtained using equation (72). With the choice

$$Y_{11} = \begin{bmatrix} 1 \\ 0 \\ 0 \end{bmatrix}, \quad Y_{10} = \begin{bmatrix} 0 \\ 1 \\ 0 \end{bmatrix}, \quad \text{and } Y_{1,-1} = \begin{bmatrix} 0 \\ 0 \\ 1 \end{bmatrix}.$$

the results for $\ell = 1$ are

$$L^2 = 2\hbar^2 \begin{bmatrix} 1 & 0 & 0 \\ 0 & 1 & 0 \\ 0 & 0 & 1 \end{bmatrix}, \quad L_z = \hbar \begin{bmatrix} 1 & 0 & 0 \\ 0 & 0 & 0 \\ 0 & 0 & -1 \end{bmatrix},$$

(76)

$$L_+ = \hbar \begin{bmatrix} 0 & \sqrt{2} & 0 \\ 0 & 0 & \sqrt{2} \\ 0 & 0 & 0 \end{bmatrix}, \quad L_- = \hbar \begin{bmatrix} 0 & 0 & 0 \\ \sqrt{2} & 0 & 0 \\ 0 & \sqrt{2} & 0 \end{bmatrix},$$

since

$$L_+Y_{11} = 0, \quad L_+Y_{10} = \sqrt{2}\begin{bmatrix} 1 \\ 0 \\ 0 \end{bmatrix} = \sqrt{2}\, Y_{11}, \text{ etc.}$$

Hence, the designation raising and lowering operators.

Although the above results were obtained from solving the angular part of the Schroedinger equation for centre-of-mass motion, the results are more general and can be derived strictly from the commutation relations for self-adjoint angular momentum operators:

(77) $[J_i, J_j] = i\hbar J_k$, i,j,k cyclic and equal to 1,2,3,

and the definition

(78) $$J^2 = J_x{}^2 + J_y{}^2 + J_z{}^2.$$

We use here $\vec{J}$ rather than $\vec{L}$ to emphasize that the results apply to any angular momentum, whether orbital, spin, or total. The steps are briefly as follows:

1. We observe that the commutation relations may be rewritten

(79) $$[J^2, J_\pm] = 0,$$
$$[J_z, J_\pm] = \pm\hbar J_\pm,$$
$$[J_+, J_-] = 2\hbar J_z.$$

Let the simultaneous eigenfunctions, as yet unknown, of J^2 and J_z be designated $|j,m\rangle$ where the designations j,m are also unknown. We have, however,

(80) $$J_z|j,m\rangle = \hbar m|j,m\rangle,$$
$$J^2|j,m\rangle = \hbar^2\lambda_j|j,m\rangle,$$

where m and λ_j are the appropriate eigenvalues.

2. Define states

$$|jm\rangle_\pm = J_\pm|jm\rangle.$$

Then, since J^2 commutes with $J_\pm$, we have

(81) $$J^2|jm\rangle_\pm = J^2J_\pm|jm\rangle = J_\pm J^2|jm\rangle = \hbar^2\lambda_j|jm\rangle_\pm.$$

Hence $|jm\rangle_\pm$ have the same eigenvalues as $|jm\rangle$ with respect to J^2.

3. However,

(82) $$J_z|jm\rangle_\pm = J_zJ_\pm|jm\rangle = (J_\pm J_z + [J_z,J_\pm])|jm\rangle = \hbar(m \pm 1)|jm\rangle_\pm.$$

Hence $|jm\rangle_\pm$ are eigenstates of J_z, but with eigenvalues raised and lowered by $\hbar$, or are null states.

4. Consider the following relation:

$$J_\mp J_\pm = (J_x \mp iJ_y)(J_x \pm iJ_y) = J_x{}^2 + J_y{}^2 \pm i[J_x, J_y] \tag{83}$$

$$= J_x{}^2 + J_y{}^2 \mp \hbar J_z = J^2 - J_z{}^2 \mp \hbar J_z.$$

5. Consider now the length (squared) of the vector $|jm\rangle_\pm$

$${}_\pm\langle jm|jm\rangle_\pm = \langle jm|J_\mp J_\pm|jm\rangle = \langle jm|J^2 - J_z{}^2 \mp \hbar J_z|jm\rangle \tag{84}$$

$$= \hbar^2(\lambda_j - m^2 \mp m)\langle jm|jm\rangle = \hbar^2(\lambda_j - m^2 \mp m),$$

since $|jm\rangle$ is assumed normalized. Hence $\lambda_j \geq m^2 \pm m$. Hence there must be a largest value of m, say m_ℓ, and a smallest value, m_s. By successively applying J_+ to the state $|jm_s\rangle$, we get a sequence of states with m values,

$$m_s,\ m_s + 1,\ \ldots,\ m_\ell\ .$$

Hence $m_\ell = m_s + n$ where n is an integer.

6. Now $J_+|jm_\ell\rangle = 0$ and $J_-|jm_s\rangle = 0$ by assumption. By calculating the lengths of these vectors, we get

$$\begin{aligned} &\lambda_j - m_\ell(m_\ell + 1) = 0, \\ &\lambda_j - m_s(m_s - 1) = 0. \end{aligned} \tag{85}$$

Combining this result with that in step 5 we get $\lambda_j = j(j + 1)$, where j is any integer or half-integer, so that the possible eigenvalues of J^2 are $j(j + 1)\hbar^2$, with $j = 0, 1/2, 1, 3/2, \ldots$ and, for a given j, m taking on the values $j,\ j - 1,\ j - 2, \ldots, -j$.

Finally, we remark that orbital momenta can have only integral values for j. The half-integral values occur only for particles with inner structure (e.g. electron spin) and perhaps can be regarded as corresponding to the quantum analogue of rigid body rotation in classical mechanics.

6.9. MANY-PARTICLE SYSTEMS

In this section we shall be concerned with systems of identical particles and the symmetry properties of the many-particle wave functions. Again our considerations will be restricted to particles without internal structure, for simplicity, and we simplify the notation by setting $\vec{r} = x$. At first we shall set aside statistical considerations and consider only the structure of the wave functions themselves.

In the quantum mechanics of many-body systems, we regard the wave function as a many-particle probability amplitude;

$$|\Psi(x_1,x_2,\ldots,x_n,t)|^2 dx_1 dx_2 \ldots dx_n,$$

represents the probability that at time t particle 1 will be found in interval dx_1 about x_1, and simultaneously particle 2 will be found in dx_2 about x_2, etc. The wave function Ψ obeys a many-particle Schroedinger equation

$$H\Psi = i\hbar\partial\Psi/\partial t$$

where

$$H = \sum_{k=1}^{n} - \frac{\hbar^2}{2m} \nabla_k^2 + \sum_{j\neq k} V(x_j,x_k), \text{ for example.}$$

Let us, for simplicity, consider the free-particle case where all the interparticle interactions are shut off. The energy operators are then additive and one can separate the Schroedinger equation into n single-particle equations by assuming a solution of the form

$$\Psi(x_1,x_2,\ldots,x_n,t) = \prod_{j=1}^{n} \phi_{k_j}(x_j) e^{-iEt/\hbar} \tag{86}$$

where the ϕ_k satisfy

(87) $$-\frac{\hbar^2}{2m}\nabla_k^2\,\phi_k = \varepsilon_k\phi_k \quad\text{and}\quad E = \sum_{j=1}^{n}\varepsilon_{k_j},$$

with individual particle energies given by $\varepsilon_k = \hbar^2k^2/2m$.

Since the ϕ_k individual functions form a complete orthonormal set in one-particle space, the product $\prod_{j=1}^{n}\phi_{k_j}(x_j)$ forms a complete set in the product space of n particles. Effectively such a space is a tensor space since it corresponds to the outer product of n single-particle vector spaces.

Imagine a unitary transformation on the base vectors $\phi(x)$ corresponding to one particle. The base vectors $\phi(x)$ transform linearly 'like vectors.' Hence, under the action of such unitary groups, the products $\prod_{j=1}^{n}\phi_{k_j}(x_j)$ transform linearly as the components of an nth-rank tensor. Since the basis is complete, we can regard it as the basis for a tensor space consisting of all the solutions of the n-particle problem including the potential energy terms in the Hamiltonian.

Consider now some solution $\psi(x_1,x_2,\ldots,x_n)$ of the n-particle Schroedinger equation with interaction terms included. Given one such solution, we can find n! solutions, since the particles are identical, by taking the n! permutations of the n variables x_j:

$$\psi(x_1,x_2,\ldots,x_n),$$

$$P_{12}\psi(x_1,x_2,\ldots,x_n) = (12)\psi(x_1,x_2,\ldots x_n),\ \text{etc.}$$

The question then arises as to how Nature knows which of these n! solutions to choose, or whether they are all valid. The answer is interesting and almost unique: Nature divides the particle zoo up into two groups. For the particle group having integral values of the spin quantum numbers S = 0,1,2, etc., Nature prescribes that the correct wave function is the completely symmetrized wave function

(88) $$\psi_S(x_1,x_2,\ldots,x_n) = \sum_P P\psi(x_1,x_2,\ldots,x_n)$$

where p goes over all n! permutations of the x_j coordinates. This unique function corresponds to a one-dimensional irreducible representation of the symmetric permutation group corresponding to the Young pattern

1	2	...	n

On the other hand, for particles with spin quantum number j equal to half an odd integer (j = 1/2,3/2,5/2, etc.) Nature requires the wave function to be completely antisymmetric:

$$\psi_A(x_1,x_2,\ldots,x_n) = \sum_p \delta_p \, P\psi(x_1,x_2,\ldots,x_n) \tag{89}$$

where δ_p, called the signature, is ±1 according as p is an even or odd permutation. Again the function is unique and corresponds to the one-dimensional representation of the symmetric group with Young pattern

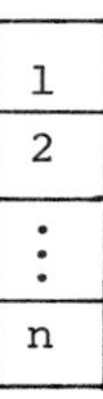

Finally, we remark in passing that these symmetry requirements on the wave functions of systems of identical particles have far-reaching consequences both in terms of the statistical behaviour of the particles and in terms of their contribution to the structure of matter. The symmetrized particles are said to obey Bose (or Bose-Einstein) statistics, and the antisymmetrized particles are said to obey Fermi (or Fermi-Dirac) statistics. The fermions tend to constitute what one normally regards as substantial matter, since electrons, protons, and neutrons are all fermions. The bosons, however, mediate interactions between fermions and thus provide the glue that holds the material world together at the nuclear and atomic levels. The deep connection between spin and statistics first investigated by Pauli can be shown to arise from the basic requirements of relativity, microscopic causality, and unitarity (conservation of probability).

REFERENCES

1 E.E. Anderson, Modern Physics and Quantum Mechanics. Saunders, Toronto, 1971
2 D. Bohm, Quantum Theory. Prentice-Hall, Englewood Cliffs, New Jersey, 1951
3 P.A.M. Dirac, The Principles of Quantum Mechanics. Clarendon Press, Oxford, 1959
4 M. Jammer, The Conceptual Development of Quantum Mechanics. McGraw-Hill, New York, 1966
5 E. Merzbacher, Quantum Mechanics. Wiley, New York, 1970
6 A. Messiah, Quantum Mechanics. North Holland, Amsterdam, 1961
7 A.B. Migdal, Approximation Methods in Quantum Mechanics. Benjamin, New York, 1961
8 W. Pauli, Quantentheorie. Handbuch der Physik, Vol. 23, Springer, Berlin, 1926, pp. 1-278
9 E. Prugovecki, Quantum Mechanics in Hilbert Space. Academic Press, New York, 1971
10 L.I. Schiff, Quantum Mechanics. McGraw-Hill, New York, 1968

PROBLEMS

1 (i) Prove that $\lim_{n\to\infty} f_n(x-y) = \delta(x-y)$ where

(a) $f_n(x-y) = \begin{cases} 0 \text{ for } |x-y| \geq 1/n, \\ n/2 \text{ for } |x-y| < 1/n, \end{cases}$

(b) $f_n(x-y) = n/2\, e^{-n|x-y|}$,

(c) $f_n(x-y) = \dfrac{n}{\pi(1+n^2|x-y|^2)}$,

(d) $f_n(x-y) = n\pi^{-\frac{1}{2}}e^{-n^2(x-y)^2}$.

(ii) Show that

$$\int_{-\infty}^{\infty} \left[\frac{d^n}{dx^n}\,\delta(x-y)\right] f(x)\,dx = (-1)^n \frac{d^n}{dy^n} f(y).$$

(Hint: Integrate by parts n times.)

2 Using the Schroedinger equation for a particle of mass m in a potential V(r) show that a conservation theorem holds:

$$\frac{\partial}{\partial t} Q + \vec{\nabla}\cdot\vec{S} = 0$$

where

$$Q = \psi^*(\vec{r},t)\,\psi(\vec{r},t) = \text{probability density},$$

$$\vec{S} = \frac{\hbar}{2im}\left[\psi^*(\vec{r},t)\,\vec{\nabla}\psi(\vec{r},t) - \left(\vec{\nabla}\psi(\vec{r},t)\right)^*\psi(\vec{r},t)\right].$$

Interpret $\vec{S}\cdot d\vec{a}$ as a probability flux through area $d\vec{a}$.

3 Show that the matrices given by equation (56) are a realization of the operator equations (53), (54), and (55) with spin states given by equation (57).

4 Show that the general solution to the time-dependent Schroedinger equation for a free particle of mass m can be written in terms of its initial value at a time t_o in the following way:

$$\psi(\vec{r},t) = \int d\vec{r}_o G(\vec{r}-\vec{r}_o;\ t-t_o)\,\psi(\vec{r}_o,t_o), \quad t > t_o ,$$

where the Green function

$$G(\vec{r}-\vec{r}_o;\ t-t_o) = \left(\frac{m}{2\pi i\hbar|t-t_o|}\right)^{3/2} \exp\left(\frac{im|\vec{r}-\vec{r}_o|^2}{2\hbar(t-t_o)}\right).$$

5 A particle of mass m is confined by rigid walls to the interior of a rectangular box, $0 \le x \le a_1$, $0 \le y \le a_2$, $0 \le z \le a_3$. Show that the allowed energy eigenvalues are

$$E_{(n_1,n_2,n_3)} = \frac{\hbar^2\pi^2}{2m}\left(\frac{n_1^2}{a_1^2} + \frac{n_2^2}{a_2^2} + \frac{n_3^2}{a_3^2}\right)$$

where $n_j = 1,2,3,4,\ldots$ and $j = 1,2,3$.

6 Show that

$$\psi(\xi,t) = Ne^{-\frac{1}{2}[\xi-\xi_o\cos\omega_o t]^2}e^{-i[\xi\xi_o\sin\omega_o t-\frac{1}{4}\xi_o^2\sin2\omega_o t]},$$

where ξ_o is a constant, is a solution of the time-dependent Schroedinger equation for a 1-dimensional harmonic oscillator. Comment on the probability density of such a state.

7 (a) Using the commutation relations $[p_i,x_j] = \frac{\hbar}{i}\,\delta_{ij}$ show that

$$\frac{i}{\hbar}[F(\vec{p},\vec{x}),\vec{x}] = \frac{\partial F}{\partial \vec{p}},$$

$$\frac{i}{\hbar}[F(\vec{p},\vec{x}),\vec{p}] = -\frac{\partial F}{\partial \vec{x}}.$$

Be careful about the ordering of operators in F.

(b) Use the above relations to show that for any Hamiltonian $H(\vec{p},\vec{x})$

$$\frac{d\langle\vec{x}\rangle}{dt} = \left\langle\frac{\partial H}{\partial \vec{p}}\right\rangle,$$

$$\frac{d\langle\vec{p}\rangle}{dt} = \left\langle-\frac{\partial H}{\partial \vec{x}}\right\rangle.$$

Under what conditions are these the classical Hamiltononian equations for particle motion?

8 Consider a system consisting of two spin 1/2 particles labelled 1,2. A maximal set of commuting operators in spin space is $\{S^2, S_z, S_1^2, S_2^2\}$ where

$$\vec{S} = \vec{S}_1 + \vec{S}_2$$

is the total spin. Show that the states

$$|1,1\rangle = |\tfrac{1}{2},\tfrac{1}{2}\rangle_1 \otimes |\tfrac{1}{2},\tfrac{1}{2}\rangle_2,$$

$$|1,0\rangle = \frac{1}{\sqrt{2}}(|\tfrac{1}{2},\tfrac{1}{2}\rangle_1 \otimes |\tfrac{1}{2},-\tfrac{1}{2}\rangle_2 + |\tfrac{1}{2},-\tfrac{1}{2}\rangle_1 \otimes |\tfrac{1}{2},\tfrac{1}{2}\rangle_2),$$

$$|1,-1\rangle = |\tfrac{1}{2},-\tfrac{1}{2}\rangle_1 \otimes |\tfrac{1}{2},-\tfrac{1}{2}\rangle_2,$$

$$|0,0\rangle = \frac{1}{\sqrt{2}}(|\tfrac{1}{2},\tfrac{1}{2}\rangle_1 \otimes |\tfrac{1}{2},-\tfrac{1}{2}\rangle_2 - |\tfrac{1}{2},-\tfrac{1}{2}\rangle_1 \otimes |\tfrac{1}{2},\tfrac{1}{2}\rangle_2)$$

are an orthonormal basis of eigenstates of the operators S^2, S_z, such that

$$S^2|s,m_s\rangle = \hbar^2 s(s+1)|s,m_s\rangle, \qquad S_z|s,m_s\rangle = \hbar m_s|s,m_s\rangle.$$

This is an illustration of the addition of angular momentum rule mentioned in Chapter 3.

7
Generalized tensors in Riemannian geometry

7.1. INTRODUCTION

In the first four chapters we dealt either implicitly or explicitly with vector spaces and tensor maps in a Euclidean world. In what follows, a more precise meaning will be given to the term 'Euclidean world,' but basically we means a space (manifold) which can be uniquely parametrized by means of n Cartesian coordinates representing distances from an origin in n mutually perpendicular directions. Our conscious experience that we live in a 3-dimensional Euclidean world provides an example. In using Euclidean worlds, our conceptual framework is basically linear. The tensors we have introduced are in fact linear maps, as are the underlying transformation groups.

We have now reached a milestone in our development, and further progress is awkward without extending our conceptual framework. In the present chapter we introduce spaces (geometries) which are basically non-linear, but which are sufficiently non-pathological that linear theory can be locally applied. We are not yet ready to abandon linear theory, however. In fact, two basic aspects of modern physics, viz. the special theory of relativity and quantum theory, are beautifully creative examples of how far linearity goes in satisfactorily modelling phenomena at the microscopic level. Nonetheless, the theory of general relativity is couched in decisively non-linear terms as we shall see in Chapter 10. A more immediate argument for going beyond the geometry of Euclid at this time is related to the need for a Lorentz metric in the space-time manifold of special relativity.

We emphasize, however, that special relativity is a

linear theory; although the geometry of space-time is no longer Euclidean, it is still 'flat.' Since we are in any case at a departure point with respect to Euclidean geometry, our technique is to introduce Riemannian geometry at this point, but defer its full development until the explicit need arises. In Chapter 8 we shall only require the special circumstances of 'flat,' but non-Euclidean geometry.

7.2. GAUSS'S INTRODUCTION TO NON-EUCLIDEAN GEOMETRY

One of the fundamental concepts in geometry is 'distance or length' and its invariance under coordinate transformations. Once the concept of distance is introduced, the problem of how to determine the path of shortest length between two points becomes meaningful. Such a path is usually called a geodesic.† In Euclidean space geodesics are, of course, 'straight' lines. There are, however, spaces where the shortest distance between two points is not the length of a 'straight' line joining the points. An example is the 2-dimensional space comprising the surface of a sphere where geodesics are segments of great circles.

Gauss took the first step toward a non-Euclidean geometry by considering a 'curved space' embedded in a Euclidean space of higher dimension. Consider, for example, any 2-dimensional surface defined in 3-dimensional Euclidean space by the relation

$$f(x_1,x_2,x_3) = 0. \tag{1}$$

Under certain conditions we may solve this equation to obtain

$$x_3 = \phi_3(x_1,x_2). \tag{2}$$

The above equation makes it clear that the surface requires only two parameters to specify points on it.

We may introduce an independent parametrization†† u^1,u^2

---

†The definition of a geodesic will be modified to a more workable one.

††The reason for using superscripts instead of subscripts will become apparent later.

such that every point on the 2-dimensional surface is specified by the intersection of curves u^1 = constant, u^2 = constant. Hence one can write for points on the surface

(3) $$x_\nu = \phi_\nu(u^1,u^2), \qquad \nu = 1,2,3.$$

Consider now a curve traced out on the surface. Let the curve be parametrized by a single variable τ which runs from 0 to 1. The 'length' L of the curve is given by

(4) $$L = \int_0^L ds = \int_0^1 \sqrt{\left(\frac{dx_1}{d\tau}\right)^2 + \left(\frac{dx_2}{d\tau}\right)^2 + \left(\frac{dx_3}{d\tau}\right)^2}\, d\tau$$

$$= \int_0^1 \sqrt{\frac{dx_\nu}{d\tau}\frac{dx_\nu}{d\tau}}\, d\tau \qquad \text{(summation on } \nu = 1,2,3).$$

Since

$$\frac{dx_\nu}{d\tau} = \sum_{j=1}^{2} \left[\frac{\partial\phi_\nu}{\partial u^j}\right] \frac{du^j}{d\tau} = \left[\frac{\partial\phi_\nu}{\partial u^j}\right]\frac{du^j}{d\tau},$$

it follows that the length can be expressed by the integral

(5) $$L = \oint \sqrt{\frac{\partial\phi_\nu}{\partial u^j}\frac{\partial\phi_\nu}{\partial u^k}\, du^j du^k} = \oint \sqrt{g_{jk}du^j du^k},$$

where

(6) $$g_{jk}(u^1,u^2) = \sum_{\nu=1}^{3} \frac{\partial\phi_\nu}{\partial u^j}\frac{\partial\phi_\nu}{\partial u^k}.$$

From equations (4) and (5) it is clear that the square of the differential element of arc length ds is given by

(7) $$(ds)^2 = g_{jk}du^j du^k,$$

which implies that the differential quadratic form $g_{jk}du^j du^k$ is positive definite. The four quantities g_{jk}, $j = 1,2$, $k = 1,2$, are called the components of the metric tensor.

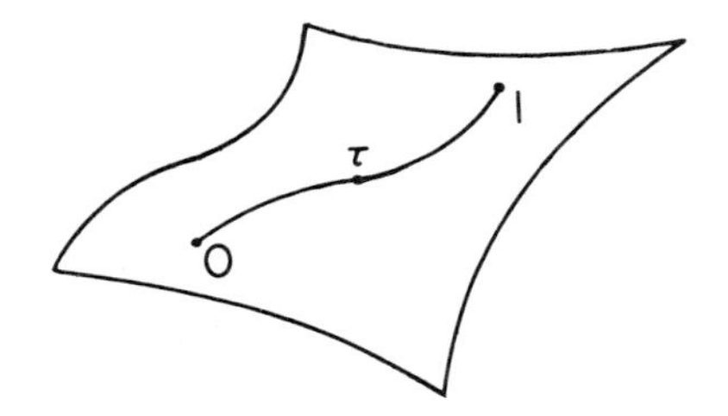

Figure 7.1. Curve parametrized by the variable τ.

The parameters u^1, u^2 do not determine the nature of the 2-dimensional space. They are simply a parametrization of it. The curved nature of the space is determined by the functions g_{jk}.

7.3. CURVILINEAR COORDINATES

(a) *General considerations*

As an example of the utility of Gauss' idea we consider briefly the nature of curvilinear coordinates. In usual practice, one is still dealing with 3-dimensional Euclidean space, but using a non-Cartesian parametrization of it, such as spherical polar coordinates. However, the ideas are still closely related to Gaussian geometry because the 2-dimensional surfaces generated by fixing one parameter and letting the remaining two take on their full range of values usually constitute a genuine non-Euclidean 2-dimensional manifold. For example, if the parameter r in spherical coordinates is fixed, the 2-dimensional world generated by θ and ϕ is a sphere which is inherently a curved manifold in 2-dimensions.

The value of curvilinear coordinates arises in practice when, because of some symmetry in the problem, the curved surfaces are equipotential sheets or define the shape of a mass or charge distribution, or otherwise coincide with some structural or boundary feature of the problem at hand (see problem 4).

Let $u^1(x_1,x_2,x_3)$, $u^2(x_1,x_2,x_3)$, and $u^3(x_1,x_2,x_3)$ (where

(x_1,x_2,x_3) are Cartesian coordinates of a point with position vector $\vec{r} = x_1\hat{i} + x_2\hat{j} + x_3\hat{k}$) be three sets of well-behaved, single-valued functions, which can be solved uniquely for

(8) $$x_\nu = x_\nu(u^1,u^2,u^3)$$

so that any space point can be described by the intersection of the three curved surfaces

(9) $$u^1 = \text{constant}, \quad u^2 = \text{constant}, \quad u^3 = \text{constant}.$$

We call u^1,u^2,u^3 a set of curvilinear coordinates. It follows that

(10) $$d\vec{r} = \sum_{k=1}^{3} \frac{\partial\vec{r}}{\partial u^k} du^k = \sum_{k=1}^{3} du^k\, \hat{e}_k$$

and

(11) $$(ds)^2 = d\vec{r}\cdot d\vec{r} = \hat{e}_k\cdot\hat{e}_j du^k du^j.$$

Our conditions of uniqueness ensure that the $\hat{e}_k$ at any point in space do not all lie in a plane. (Note that although any vector position $\vec{r}$ can be expressed as $\vec{r} = \hat{e}_k r^k$, in general $r^k \neq \vec{r}\cdot\hat{e}_k$ because the $\hat{e}_k$ are not orthogonal.)

Any vector $\vec{v}$ can be written as

(12) $$\vec{v} = v^k\hat{e}_k.$$

The quantities v^k are called the contravariant components of a vector $\vec{v}$. This definition is consistent with the more general definition of contravariance to be introduced in the next section. In a new (barred) set of curvilinear coordinates $\bar{u}^k = \bar{u}^k(u^1,u^2,u^3)$, $k = 1,2,3$, the basic vectors $\bar{\hat{e}}_k$, $k = 1,2,3$, are given by

(13) $$\bar{\hat{e}}_k = \frac{\partial\vec{r}}{\partial\bar{u}^k} = \frac{\partial u^j}{\partial\bar{u}^k}\frac{\partial\vec{r}}{\partial u^j} = \frac{\partial u^j}{\partial\bar{u}^k}\hat{e}_j.$$

Since a physical vector $\vec{v}$ is independent of what particular set of coordinates is used to describe it, it follows that the components $\bar{v}^k$, k = 1,2,3, of $\vec{v}$ in the new coordinate system are given by

$$(14) \qquad \bar{v}^k = \frac{\partial \bar{u}^k}{\partial u^j} v^j ,$$

which expresses the transformation law from unbarred to barred coordinates.

It is clearly possible to define a second set of algebraic quantities, which we call the covariant components of a vector $\vec{v}$, by setting

$$(15) \qquad v_k = \hat{e}_k \cdot \vec{v} = \sum_{j=1}^{3} \hat{e}_k \cdot \hat{e}_j \; v^j = g_{kj} v^j ,$$

where we have set

$$(16) \qquad g_{kj} = \hat{e}_k \cdot \hat{e}_j ,$$

a definition consistent with equation (6) of the previous section.

Introducing a set of <u>reciprocal base vectors</u> by the relations

$$(17) \qquad \hat{e}^i = \frac{1}{\Delta} (\hat{e}_j \times \hat{e}_k), \qquad i,j,k \text{ cyclic},$$

where

$$(18) \qquad \Delta = \hat{e}_1 \cdot (\hat{e}_2 \times \hat{e}_3),$$

we can write

$$(19) \qquad \vec{v} = v^k \hat{e}_k = v_i \hat{e}^i .$$

Note that neither the $\hat{e}^k$, k = 1,2,3, nor the $\hat{e}_k$, k = 1,2,3, need be unit vectors.

It also follows that the scalar product of two vectors can be written either as

(20) $\vec{w}\cdot\vec{v} = (w^k\hat{e}_k)\cdot(v^j\hat{e}_j) = g_{kj}w^kv^j$

or as

(21) $\vec{w}\cdot\vec{v} = (w^k\hat{e}_k)\cdot(v_j\hat{e}^j) = w^kv_k,$

by virtue of the orthogonality $\hat{e}_k\cdot\hat{e}^j = \delta^j_k$ which follows from the definition in equation (17).

(b) *Spherical polar coordinates*

As an example, we consider spherical polar coordinates. Let $u^1 = r$, $u^2 = \theta$, $u^3 = \phi$ be the curvilinear (spherical polar) coordinates defined by

(22a) $x_3 = z = r\cos\theta,$

(22b) $x_2 = y = r\sin\theta\sin\phi,$

(22c) $x_1 = x = r\sin\theta\cos\phi,$

Expressing the components of the position vector in spherical polar coordinates, we have

(23) $$\vec{r} = x\hat{i} + y\hat{j} + z\hat{k}$$

$$= r\sin\theta\cos\phi\ \hat{i} + r\sin\theta\sin\phi\ \hat{j} + r\cos\theta\ \hat{k}.$$

Hence

(24) $$\hat{e}_1 = \frac{\partial\vec{r}}{\partial r} = \sin\theta\cos\phi\ \hat{i} + \sin\theta\sin\phi\ \hat{j} + \cos\theta\ \hat{k} = \hat{r},$$

(25) $$\hat{e}_2 = \frac{\partial\vec{r}}{\partial\theta} = r\cos\theta\cos\phi\ \hat{i} + r\cos\theta\sin\phi\ \hat{j} - r\sin\theta\ \hat{k} = r\hat{\theta},$$

(26) $$\hat{e}_3 = \frac{\partial\vec{r}}{\partial\theta} = -\ r\sin\theta\sin\phi\ \hat{i} + r\sin\theta\cos\phi\ \hat{j} = r\sin\theta\ \hat{\phi},$$

and

(27) $$g_{jk} = \hat{e}_j\cdot\hat{e}_k = \begin{bmatrix} 1 & 0 & 0 \\ 0 & r^2 & 0 \\ 0 & 0 & r^2\sin^2\theta \end{bmatrix}.$$

From equations (17) and (18), we then have

$$(28) \quad \hat{e}^1 = \frac{1}{r^2\sin\theta}(\hat{e}_2 \times \hat{e}_3) = \hat{r},$$

$$(29) \quad \hat{e}^2 = \frac{1}{r^2\sin\theta}(\hat{e}_3 \times \hat{e}_1) = \frac{1}{r^2\sin\theta}[(r\sin\theta)\hat{\phi} \times \hat{r}] = \frac{\hat{\theta}}{r},$$

$$(30) \quad \hat{e}^3 = \frac{\hat{\phi}}{r\sin\theta}.$$

Finally we note that from equation (7)

$$(31) \quad (ds)^2 = (dr)^2 + r^2(d\theta)^2 + r^2\sin^2\theta(d\phi)^2.$$

(c) *Lengths, areas, and volumes*

In curvilinear coordinates distance is expressed through the relation $(ds)^2 = g_{ij}du^idu^j$. If u^2 and u^3 are set equal to constants, an infinitesimal change in position along the line of intersection of these two surfaces, denoted $d\vec{r}$, is given by

$$(32) \quad d\vec{r}_1 = \frac{\partial \vec{r}}{\partial u^1} du^1 = du^1\hat{e}_1.$$

Therefore the distance ds_1 between the points u^1, u^2, u^3, and $u^1 + du^1$, u^2, u^3 measured along the line of intersection of u^2 = constant, u^3 = constant has the form

$$(33) \quad (ds_1) = |d\vec{r}_1| = \sqrt{g_{11}}\, du^1.$$

Similarly $ds_2 = \sqrt{g_{22}}\, du^2$, and $ds_3 = \sqrt{g_{33}}\, du^3$.

Consider u^1 = constant, and the element of area bounded by the lines in this surface defined by

u^2 = constant, $u^2 + du^2$ = constant,

u^3 = constant, $u^3 + du^3$ = constant.

The hatched area shown in Figure 7.2 is denoted da_1, and is given to first order in small quantities by

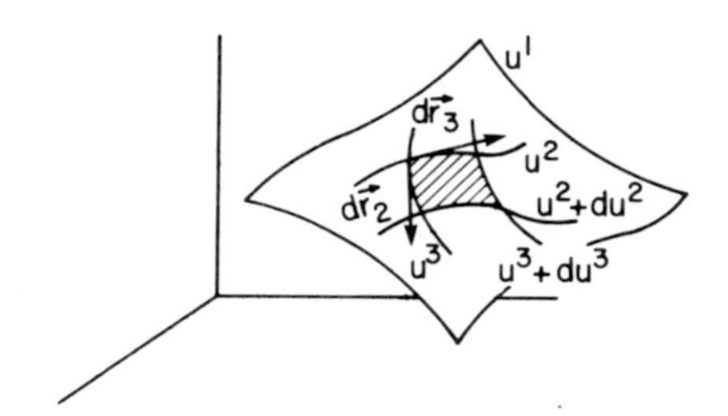

Figure 7.2. Surface area element in curvilinear coordinates.

(34) $$da_1 = |d\vec{r}_2 \times d\vec{r}_3| = |\hat{e}_2 \times \hat{e}_3 \, du^2du^3|$$

$$= \sqrt{(\hat{e}_2 \times \hat{e}_3)\cdot(\hat{e}_2 \times \hat{e}_3)} \, du^2du^3$$

$$= \sqrt{g_{22}g_{33} - (g_{23})^2} \, du^2du^3,$$

where we have used the vector identity

$$(\vec{a} \times \vec{b})\cdot(\vec{c} \times \vec{d}) = (\vec{a}\cdot\vec{c})(\vec{b}\cdot\vec{d}) - (\vec{a}\cdot\vec{d})(\vec{b}\cdot\vec{c}).$$

The three infinitesimal vectors $d\vec{r}_1$, $d\vec{r}_2$, and $d\vec{r}_3$ define a parallelepiped with volume

(35) $$d\tau = |d\vec{r}_1\cdot(d\vec{r}_2 \times d\vec{r}_3)|$$

$$= |\hat{e}_1\cdot(\hat{e}_2 \times \hat{e}_3)|du^1du^2du^3 = |\Delta|du^1du^2du^2$$

where Δ^2 is the determinant of g_{jk}.

In an orthogonal coordinate system $\hat{e}_i\cdot\hat{e}_j = 0$ for $i \neq j$. Thus $g_{ij} = 0$ for $i \neq j$. It is then convenient to define

(36) $$h_1 = \sqrt{g_{11}}, \quad h_2 = \sqrt{g_{22}}, \quad h_3 = \sqrt{g_{33}},$$

so that the expressions for tangents, areas, and volumes become

(37a) $$(ds)_j = h_j du^j \quad \text{(not summed)},$$

(37b) $$(da)_i = h_j h_k du^j du^k \quad \text{(not summed)}, \; i \neq j \neq k,$$

and

(37c) $$d\tau = h_1h_2h_3du^1du^2du^3.$$

For spherical polar coordinates these expressions become

$$ds_1 = dr, \quad ds_2 = rd\theta, \quad ds_3 = r\sin\theta d\phi,$$

$$da_3 = rdrd\theta, \quad da_2 = r\sin\theta drd\phi, \quad da_1 = r^2\sin\theta d\theta d\phi,$$

$$d\tau = r^2\sin\theta drd\theta d\phi.$$

One can express results of the vector calculus also in terms of curvilinear coordinates. It is not our purpose to do so here as the formal derivations are contained in many books.† We simply quote the results for right-handed orthogonal systems as follows. Let V be a scalar field expressed as a function of the curvilinear coordinates u^1, u^2, u^3. Then, using unit vectors,

$$(38) \qquad (\text{grad } V)_j = \frac{1}{h_j}\frac{\partial V}{\partial u^j}, \qquad j = 1,2,3.$$

Similarly, if $\vec{W}$ is a vector field, then

$$(39) \qquad \text{div } \vec{W} = \frac{1}{h_1h_2h_3}\sum_{i=1}^{3}\frac{\partial}{\partial u^i}(h_jh_kW_i)$$

and

$$(40) \qquad (\text{curl } \vec{W})_i = \frac{1}{h_jh_k}\left(\frac{\partial}{\partial u}(h_kW_k) - \frac{\partial}{\partial u^k}(h_jW_j)\right), \qquad i = 1,2,3,$$

where i,j,k are cyclic. Finally, the Laplacian of a scalar field V can be written

$$(41) \qquad \nabla^2 V = \frac{1}{h_1h_2h_3}\left(\sum_{i=1}^{3}\frac{\partial}{\partial u^i}\frac{h_jh_k}{h_i}\frac{\partial V}{\partial u^i}\right), \qquad i,j,k \text{ cyclic.}$$

For the example of spherical polar coordinates, these expressions take on their familiar forms

$$(42) \qquad \vec{\nabla}V = \hat{r}\frac{\partial V}{\partial r} + \frac{\hat{\theta}}{r}\frac{\partial V}{\partial\theta} + \frac{\hat{\phi}}{r\sin\theta}\frac{\partial V}{\partial\phi},$$

---

†For example, T.M. Apostol, Mathematical Analysis (Addison-Wesley, New York, 1964).

$$(43) \quad \vec{\nabla}\cdot\vec{W} = \frac{1}{r^2}\frac{\partial}{\partial r}(r^2 W_r) + \frac{1}{r\sin\theta}\frac{\partial}{\partial\theta}(\sin\theta\ W_\theta) + \frac{1}{r\sin\theta}\frac{\partial W_\phi}{\partial\phi},$$

$$(44) \quad \text{curl}\ \vec{W} = \frac{1}{r^2\sin\theta}\begin{vmatrix} \hat{r} & r\hat{\theta} & r\sin\theta\ \hat{\phi} \\ \frac{\partial}{\partial r} & \frac{\partial}{\partial\theta} & \frac{\partial}{\partial\phi} \\ W_r & rW_\theta & r\sin\theta\ W_\phi \end{vmatrix},$$

and

$$(45) \quad \nabla^2 V = \frac{1}{r^2}\frac{\partial}{\partial r}\left(r^2\frac{\partial V}{\partial r}\right) + \frac{1}{r^2\sin\theta}\frac{\partial}{\partial\theta}\left(\sin\theta\frac{\partial V}{\partial\theta}\right) + \frac{1}{r^2\sin^2\theta}\frac{\partial^2 V}{\partial\phi^2}.$$

7.4. RIEMANNIAN GEOMETRY

There is no reason why we must restrict ourselves to 2-dimensional surfaces embedded in a 3-dimensional Euclidean space. We could start with an $(n + m)$-dimensional Euclidean space and embed an n-dimensional surface in it for any two positive integers n and m. The differential element of arc length in the general case would be

$$(46) \quad (ds)^2 = \sum_{j=1}^{n}\sum_{k=1}^{n} g_{jk}du^j du^k$$

where g_{jk} is given by an equation similar to equation (6) but with ν running from 1 up to $n + m$.

A topological set of objects (see Appendix D) is said to constitute an n-dimensional space (or manifold) if the objects in some <u>neighbourhood</u> of any particular object in the set may be put into a one-to-one correspondence with all of the ordered n-tuples of real numbers $(u^1,\ldots,u^n)$ such that

$$(47) \quad |u^i - a^i| < b^i, \quad i = 1,\ldots,n,$$

where for each neighbourhood put into a correspondence the a^i and b^i are fixed numbers. The a^i define the particular object and the b^i the limits of the neighbourhood.

Any particular association of objects with ordered n-tuples is called a coordinate system or parametrization. The objects are usually called points and the numbers $u^1,\ldots,u^n$ are called the coordinates of the points. The above inequalities give the range of the coordinate system.

An ordered pair (N,C), where N is a neighbourhood and C a coordinate system defined on that neighbourhood, is called a chart. Let (N,C) and $(\bar{N},\bar{C})$ be two charts such that $N \cap \bar{N}$ is non-empty. Clearly the coordinates $u^1,\ldots,u^n$ and $\bar{u}^1,\ldots,\bar{u}^n$, in the coordinate systems C and $\bar{C}$ respectively, of any point $P \in N \cap \bar{N}$ must be related. We assume that this relationship takes the form

$$u^j = u^j(\bar{u}^1,\ldots,\bar{u}^n), \qquad j = 1,\ldots,n,$$

with inverse

$$\bar{u}^j = \bar{u}^j(u^1,\ldots,u^n), \qquad j = 1,\ldots,n,$$

where the above functions are infinitely differentiable for every $P \in N \cap \bar{N}$ (and define a coordinate transformation on $N \cap \bar{N}$). This assumption guarantees the existence of $\partial\bar{u}^j/\partial u^k$, $\partial u^j/\partial\bar{u}^k$, and all higher-order partial derivatives. In particular the identity

$$\delta^h_k = \frac{\partial u^h}{\partial\bar{u}^j}\frac{\partial\bar{u}^j}{\partial u^k}$$

holds, implying via the product rule for determinants that

$$\frac{\partial(u^1,\ldots,u^n)}{\partial(\bar{u}^1,\ldots,\bar{u}^n)}\frac{\partial(\bar{u}^1,\ldots,\bar{u}^n)}{\partial(u^1,\ldots,u^n)} = 1.$$

Thus the Jacobian of the transformation

$$J = \left|\frac{\partial u^k}{\partial\bar{u}^j}\right| = \frac{\partial(u^1,\ldots,u^n)}{\partial(\bar{u}_1,\ldots,\bar{u}^n)}$$

does not vanish on $N \cap \bar{N}$.

Exactly what is meant by a 'neighbourhood' is contained in the topology of the space. A neighbourhood need not be a 'small' collection of points, and in fact in all cases one particular neighbourhood of any point is the whole space. However, there is no guarantee that this neighbourhood may be put in a one-to-one correspondence with a coordinate system. In general, a space will require more than one coordinate system to cover it. The 2-dimensional surface of a sphere, for example, requires a minimum of two local coordinate systems to cover it without singularities.

The rotation group is a 3-dimensional manifold since rotations may be put into a correspondence with the ordered 3-tuples (ϕ,θ,ψ), where ϕ,θ,ψ are the Euler angles. However, this coordinate system (see Figure 3.1) is not suitable for the whole space. If $\theta = 0$, for a given rotation only $\phi + \psi$ is determined, and if $\theta = \pi$, for a given rotation only $\phi - \psi$ is determined. At these singular points of the parametrization (ϕ,θ,ψ) it is no longer possible to associate a unique value of the Euler angles with each rotation. The singularity of the Euler angle parametrization about the origin (i.e. the identity rotation) makes it inappropriate for certain applications.

Consider an n-dimensional space not referred a priori to a higher-dimensional Euclidean space and let there be n^2 functions $g_{jk}(u^1,\ldots,u^n)$, $j,k = 1,2,\ldots,n$, such that the quantity $g_{jk}(u^1,\ldots,u^n)du^j du^k$ is invariant under change of coordinates. The distance between two nearby points $(u^1,\ldots,u^n)$ and $(u^1 + du^1, \ldots, u^n + du^n)$ may then be defined by

$$(48)\quad (ds)^2 = eg_{jk}du^j du^k$$

where e is chosen as +1 or -1 in order that ds^2 is always positive. Hence

$$(49)\quad e = g_{jk}\frac{du^j}{ds}\frac{du^k}{ds}\ .$$

We call an n-dimensional space Riemannian if the distance between nearby points $(u^1,\ldots,u^n)$ and $(u^1 + du^1, \ldots, u^n + du^n)$ is given by equation (48) where $g_{jk} = g_{kj}$ are functions of the $u^1,\ldots,u^n$ only and at every point the determinant of the n x n matrix with components g_{jk} is different from zero, i.e. $|g_{jk}| \neq 0$. (Without loss of generality we may assume that g_{jk} is symmetric on interchange of j with k.)

A Riemannian space of dimension n is called Euclidean if it may be covered by a single coordinate system in which

$$(50) \qquad (ds)^2 = du^j du^j = (du^1)^2 + \ldots + (du^n)^2.$$

An n-dimensional Riemannian space is called flat if it may be covered by a singe coordinate system in which

$$(51) \qquad (ds)^2 = \sum_{\alpha=1}^{n} C_\alpha (du^\alpha)^2$$

where C_α is either +1 or -1. A Euclidean space is flat but the converse is not necessarily true.

Riemannian spaces arise naturally in the description of many physical phenomena. For example, consider a classical system of N particles with position vectors $\vec{r}_i$ and masses M_i, $i = 1,2,\ldots,N$. The position of every particle in the system is specified by 3N parameters. If the motion of the particles is subject to certain constraints the 3N coordinates may not all be independent so that a new parametrization $u^1,\ldots,u^n$, $n < 3N$, may be introduced where the parameters $u^1,\ldots,u^n$ are now all independent and uniquely specify the positions of all the particles in the system, viz.

$$(52) \qquad \vec{r}_i = \vec{r}_i(u^1,\ldots,u^n), \quad i = 1,2,\ldots,N.$$

The $u^1,\ldots,u^n$ are called generalized coordinates.

The positions of the N particles constitute a n-dimensional space (called configuration space) since it may be put into a correspondence with the ordered n-tuples $(u^1,\ldots,u^n)$. The total kinetic energy of the system, denoted T, is given by

$$(53) \qquad T = \sum_{i=1}^{N} \frac{1}{2} M_i \left(\frac{d\vec{r}_i}{dt}\right)\cdot\left(\frac{d\vec{r}_i}{dt}\right) = \sum_{i=1}^{N} \frac{1}{2} M_i \left(\frac{\partial \vec{r}_i}{\partial u^j}\right)\cdot\left(\frac{\partial \vec{r}_i}{du^k}\right)\frac{du^j}{dt}\frac{du^k}{dt} .$$

The total kinetic energy must be independent of the particular set of generalized coordinates used. Hence the expression

$$\sum_{i=1}^{N} \frac{1}{2} M_i \frac{\partial \vec{r}_i}{\partial u^j} \cdot \frac{\partial \vec{r}_i}{\partial u^k} du^j du^k$$

is an invariant with respect to coordinate transformations. The 'distance' between $(u^1,\ldots,u^n)$ and the nearby point $(u^1 + du^1,\ldots,u^n + du^n)$ may be defined as

$$(ds)^2 = g_{jk} du^j du^k$$

where

$$(54) \qquad g_{jk} = \sum_{i=1}^{N} M_i \frac{\partial \vec{r}_i}{\partial u^j} \cdot \frac{\partial \vec{r}_i}{\partial u^k} .$$

With this definition for distance the kinetic energy reduces to the simple expression

$$(55) \qquad T = \frac{1}{2}\left(\frac{ds}{dt}\right)^2 .$$

If the system is conservative (i.e. the energy is constant) it will trace out a curve in configuration space (as it evolves in time) such that the integral quantity

$$(56) \qquad \oint \sqrt{T}\, ds$$

is an extremum.† Hence the geometry of configuration space determines in part the motion of the system. This illustration provides a conceptual forerunner to the theory of general relativity in which Einstein reduced space-time dynamics to 4-dimensional geometry.

7.5. CHANGE OF COORDINATES - GENERALIZED TENSORS

For the remainder of this chapter, unless otherwise stated, we are dealing with a neighbourhood in an n-dimensional Riemannian space which can be covered by a single coordinate system. Let one parametrization of the neighbourhood be the coordinate numbers $u^1,\ldots,u^n$ which we collectively denote by u^k, and let

---

†See H. Goldstein, Classical Mechanics (Addison-Wesley, Reading, Mass., 1950), p. 232.

$\bar{u}^1,\ldots,\bar{u}^n$ be a second parametrization (where the Jacobian of the transformation $|\partial\bar{u}^j/\partial u^k| \neq 0$ for all points in the neighbourhood) and $\bar{u}^k$ again denotes the collective set.

Consider a curve $u^k(s)$ parametrized by its arc length s which passes through a point P_0 with coordinates ${}_0u^k$. Since arc length s is a scalar quantity (i.e. independent of the parametrization of the space) it follows from the chain rule that

$$(57) \qquad {}_0\left(\frac{d\bar{u}^k}{ds}\right) = \sum_{j=1}^{n} {}_0\left(\frac{\partial\bar{u}^k}{\partial u^j}\right) {}_0\left(\frac{du^j}{ds}\right)$$

where the subscript 0 means that the quantity is evaluated at the point P_0. We call the ordered set of n components ${}_0(du^j/ds)$, $j = 1,2,\ldots,n$, the tangent vector to the curve $u^k(s)$ at the point P_0. Any ordered set of n quantities v^k, $k = 1,\ldots,n$, which transform as

$$(58) \qquad \bar{v}^k = {}_0\left(\frac{\partial\bar{u}^k}{\partial u^j}\right)v^j$$

under change of parametrization from the unbarred to barred coordinates, is called a contravariant vector at the point P_0. As in Chapter 2 we denote the contravariant vector with components v^k, $k = 1,\ldots,n$, by the same symbol as its components v^k or alternatively by $\vec{v}$. The tangent vector is a particular example of a contravariant vector.

If at each point u^ℓ in a set of points S in the n-dimensional Riemannian space there is a contravariant vector v^k, then the components are functions of the parameters $(u^1,\ldots,u^n)$, and the ordered set of functions $v^k(u^1,\ldots,u^n)$, $k = 1,\ldots,n$, is called a contravariant vector field on the set S (or simply a contravariant vector field).

A second kind of vector exists. Consider, for example, a scalar field ϕ,

$$(59) \qquad \phi(u^1,u^2,\ldots,u^n) = \bar{\phi}(\bar{u}^1,\ldots,\bar{u}^n),$$

that is, a function which is pointwise invariant under coordinate change. The n quantities

(60) $\quad {}_{o}\phi_{;k} = {}_{o}(\partial\phi/\partial u^{k}), \qquad k = 1,\ldots,n,$

where ;k denotes differentiation with respect to u^k, are connected to the n quantities

(61) $\quad {}_{o}\bar{\phi}_{;k} = {}_{o}(\partial\phi/\partial\bar{u}^{k}), \qquad k = 1,\ldots,n,$

by a different transformation, namely

(62) $$ {}_{o}\bar{\phi}_{ik} = {}_{o}\left(\frac{\partial\bar{\phi}}{\partial\bar{u}^{k}}\right) = {}_{o}\left(\frac{\partial\phi}{\partial u^{j}}\right){}_{o}\left(\frac{\partial u^{j}}{\partial\bar{u}^{k}}\right) = {}_{o}\left(\frac{\partial u^{j}}{\partial\bar{u}^{k}}\right){}_{o}\phi_{;j}. $$

We call the ordered set of n components ${}_{o}\phi_{;k}$ the gradient vector. Any ordered set of n components W_k, $k = 1,\ldots,n$, which transform like the gradient vector, viz.

(63) $$ \bar{w}_{k} = {}_{o}\left(\frac{\partial u^{j}}{\partial\bar{u}^{k}}\right)w_{j}, $$

under change of parametrization from the unbarred to barred coordinates is called a covariant vector at the point P_o. We denote the covariant vector with components w_k, $k = 1,\ldots,n$, by the same symbol as its components w_k or otherwise $\vec{w}^*$. Note that the gradient vector itself is covariant.

Covariant transformation character is denoted by a subscript index, and contravariant transformation character is denoted by a superscript index. Covariant and contravariant vectors are different types of vectors (i.e. elements of different vector spaces). The covariant vectors are most properly viewed as linear maps of contravariant vectors into scalars. If $\vec{a}^*$ is a covariant vector with components a_j, $j = 1,\ldots,n$, and $\vec{b}$ is a contravariant vector with components b^j, $j = 1,\ldots,n$, the action of the linear map $\vec{a}^*$ on $\vec{b}$ is given by

(64) $\quad \vec{a}^*(\vec{b}) = a_j b^j.$

Under change of parametrization from unbarred to barred coordinates the $a_j b^j$ transform as

$$(65)\qquad \bar{a}_k \bar{b}^k = \left(\frac{\partial u^\ell}{\partial \bar{u}^k}\right)_0 \left(\frac{\partial \bar{u}^k}{\partial u^m}\right)_0 a_\ell b^m = \left(\frac{\partial u^\ell}{\partial u^m}\right)_0 a_\ell b^m = \delta^\ell_m a_\ell b^m = a_\ell b^\ell$$

where the symbol δ^ℓ_m has the usual meaning, i.e. it has the value 1 if $\ell = m$ and zero otherwise. Thus $a_j b^j$ is a scalar quantity.

If at each point u^ℓ in a set of points S in the n-dimensional Riemannian space there is a covariant vector v_k, then the ordered set of functions $v_k(u^1,\ldots,u^n)$, $k = 1,\ldots,n$, is called a covariant vector field on the set S (or more simply a covariant vector field).

A second-rank covariant tensor at the point P_0 is defined as an ordered set of n^2 components, say $T_{k\ell}$, $k,\ell = 1,2,\ldots,n$, which transform under change of parametrization from the unbarred to barred coordinates like

$$(66)\qquad \bar{T}_{ji} = \left(\frac{\partial u^\ell}{\partial \bar{u}^j}\right)_0 \left(\frac{\partial u^k}{\partial \bar{u}^i}\right)_0 T_{\ell k}.$$

Multiplying equation (66) by $\left(\frac{\partial \bar{u}^j}{\partial u^m}\right)_0 \left(\frac{\partial \bar{u}^i}{\partial u^n}\right)_0$ and summing over i and j, we get

$$(67)\qquad \left(\frac{\partial \bar{u}^j}{\partial u^m}\right)_0 \left(\frac{\partial \bar{u}^i}{\partial u^n}\right)_0 \bar{T}_{ji} = \left(\frac{\partial u^\ell}{\partial \bar{u}^j}\right)_0 \left(\frac{\partial \bar{u}^j}{\partial u^m}\right)_0 \left(\frac{\partial u^k}{\partial \bar{u}^i}\right)_0 \left(\frac{\partial \bar{u}^i}{\partial u^n}\right)_0 T_{\ell k}$$

$$= \left(\frac{\partial u^\ell}{\partial u^m}\right)_0 \left(\frac{\partial u^k}{\partial u^n}\right)_0 T_{\ell k} = \delta^\ell_m \delta^k_n T_{\ell k} = T_{mn},$$

which is the formula for the inverse transformation (i.e. from barred to unbarred coordinates).

If at each point u^ℓ in a set of points S in the n-dimensional space there is a second-rank covariant tensor T_{jk}, then the ordered set of functions $T_{jk}(u^1,\ldots,u^n)$, $k,j = 1,\ldots,n$, is called a second-rank covariant tensor field on the set S.

We denote the second-rank covariant tensor (or tensor field) with components T_{jk}, $j,k = 1,\dots,n$, by T_{jk} or just by T.

A second-rank contravariant tensor at the point P_o is defined as an ordered set of n^2 components, say $T^{k\ell}$, $k,\ell = 1, \dots, n$, which transform under change of parametrization from the unbarred to barred coordinates like

$$(68) \qquad \bar{T}^{ji} = {}_o\!\left(\frac{\partial \bar{u}^j}{\partial u^\ell}\right) {}_o\!\left(\frac{\partial \bar{u}^i}{\partial u^k}\right) T^{\ell k}.$$

In a fashion similar to that leading to equation (67) the inverse transformation (i.e. from barred to unbarred coordinates) is given by

$$(69) \qquad {}_o\!\left(\frac{\partial u^m}{\partial \bar{u}^j}\right) {}_o\!\left(\frac{\partial u^n}{\partial \bar{u}^i}\right) \bar{T}^{ji} = {}_o\!\left(\frac{\partial u^m}{\partial u^\ell}\right) {}_o\!\left(\frac{\partial u^n}{\partial u^k}\right) T^{\ell k} = \delta^m_\ell \delta^n_k \, T^{\ell k} = T^{mn}.$$

Second-rank contravariant tensor fields are defined in an analogous fashion to second-rank covariant tensor fields.

We denote the second-rank contravariant tensor (or tensor field) with components T^{jk}, $j,k = 1,\dots,n$, by T^{jk} or T.

A second-rank mixed tensor at the point P_o is defined as an ordered set of n^2 components, say T^j_k, $j,k = 1,\dots,n$, which transform under change of parametrization from the unbarred to barred coordinates like

$$(70) \qquad \bar{T}^j_k = {}_o\!\left(\frac{\partial \bar{u}^j}{\partial u^\ell}\right) {}_o\!\left(\frac{\partial u^i}{\partial \bar{u}^k}\right) T^\ell_i.$$

Multiplying equation (70) by

$${}_o\!\left(\frac{\partial u^m}{\partial \bar{u}^j}\right) {}_o\!\left(\frac{\partial \bar{u}^k}{\partial u^n}\right)$$

and summing over j and k, we get the inverse transformation formula

$$(71) \qquad T^m_n = {}_o\!\left(\frac{\partial u^m}{\partial \bar{u}^j}\right) {}_o\!\left(\frac{\partial \bar{u}^k}{\partial u^n}\right) \bar{T}^j_k.$$

Second-rank mixed tensor fields are defined analogously to second-rank covariant tensor fields, and we denote the second-rank mixed tensor (or tensor field) with components T^j_k, $j,k = 1,\dots,n$, by T^j_k or simply T.

Higher-rank tensors are defined and denoted in a similar manner as second-rank tensors.

Examples of tensors and tensor fields are not difficult to find. For example, from the requirement that 'length' be defined as an invariant or scalar we have

$$(72)\qquad (ds)^2 = eg_{jk}du^j du^k = e\bar{g}_{nm}d\bar{u}^m d\bar{u}^n = e\bar{g}_{nm}\frac{\partial \bar{u}^n}{\partial u^j}\frac{\partial \bar{u}^m}{\partial u^k}du^j du^k$$

and hence

$$(73)\qquad g_{jk} = \frac{\partial \bar{u}^n}{\partial u^j}\frac{\partial \bar{u}^m}{\partial u^k}\bar{g}_{nm},$$

which shows that the n^2 functions g_{jk} are the components of a second-rank covariant tensor field.

The n^2 quantities δ^i_j, $j,i = 1,\dots,n$, are the components of a mixed second-rank tensor field (called the Kronecker delta) whose components are <u>independent</u> of the parametrization of the space (i.e. $\bar{\delta}^i_j = \delta^i_j$) and independent of position in the space (i.e. $\delta^i_j(u^1,\dots,u^n) = \delta^i_j$).

As in Chapter 1, where Cartesian tensors were discussed, a vector is called a first-rank tensor and a scalar is called a tensor of rank zero.

7.6. TENSOR ALGEBRA

Let M and N be two tensors at a point P_0 with components

$$M^{j_1\dots j_s}_{k_1\dots k_r} \text{ and } N^{j_1\dots j_s}_{k_1\dots k_r}, \quad j_1,\dots,j_s = 1,\dots,n;\ k_1,\dots,k_r = 1,\dots,n,$$

respectively. Under change of parametrization of n^{r+s} quantities

$$\left[M^{j_1 \ldots j_s}_{k_1 \ldots k_r} + N^{j_1 \ldots j_s}_{k_1 \ldots k_r}\right], \quad j_1, \ldots, j_s = 1, \ldots, n; \; k_1, \ldots, k_r = 1, \ldots, n,$$

transform as the components of a tensor at P_o which is covariant on r indices and contravariant on s indices. We call the tensor with components

$$\left[M^{j_1 \ldots j_s}_{k_1 \ldots k_r} + N^{j_1 \ldots j_s}_{k_1 \ldots k_r}\right], \quad j_1, \ldots, j_s = 1, \ldots, n; \; k_1, \ldots, k_r = 1, \ldots, n,$$

the sum of the tensors M and N and denote it by $M \oplus N$.

One cannot add two tensors which have a different number of covariant or contravariant indices. If N^j and M^j are arbitrary first-rank contravariant tensors at the points P_o and P_1 respectively ($P_o \neq P_1$), then the components $(N^j + M^j)$, $j = 1, \ldots, n$, transform under change of parametrization from unbarred to barred coordinates like

$$(74) \qquad \bar{N}^j + \bar{M}^j = \left(\frac{\partial \bar{u}^j}{\partial u^k}\right)_o N^k + \left(\frac{\partial \bar{u}^j}{\partial u^k}\right)_1 M^k.$$

Therefore the quantities $(N_j + M_j)$, $j = 1, \ldots, n$, are the components of a first-rank tensor only if

$$(75) \qquad \left(\frac{\partial \bar{u}^j}{\partial u^k}\right)_o = \left(\frac{\partial \bar{u}^j}{\partial u^k}\right)_1, \quad k, j = 1, \ldots, n,$$

and addition between tensors at different points is not defined unless the coefficients $\partial \bar{u}^k / \partial u^j$, $k, j = 1, \ldots, n$, are constant. When we dealt with Cartesian tensors the coefficients $\partial \bar{u}^k / \partial u^j$ were the direction cosines of a rotation (which are independent of position in the coordinate system) and so there was no need to associate tensor addition only with a particular point.

The outer product or tensor product of two tensors T and M (at a point P_o) with components $T^{m_1 \ldots m_a}_{n_1 \ldots n_b}$,

$m_1,\ldots,m_a = 1,\ldots,n$, $n_1,\ldots,n_b = 1,\ldots,n$, and $M^{m_1\ldots m_s}_{n_1\ldots n_r}$, $m_1,\ldots,$ $m_s = 1,\ldots,n$, $n_1,\ldots,n_r = 1,\ldots,n$, respectively is the tensor (at point P_o) with components $T^{m_1\ldots m_a}_{n_1\ldots n_b} \cdot M^{m_1\ldots m_s}_{n_1\ldots n_r}$ and is denoted by $T \otimes M$ or by the same symbol as its components. In general the tensor product of two tensors is defined only if the tensors are evaluated at the same point, i.e. tensor sums and products for tensor fields should be regarded as pointwise constructions.

From any tensor at the point P_o with at least one covariant and one contravariant index we can form another tensor at the point P_o whose components are derived from the original tensor's components by summing on a pair of indices one of which is covariant and the other contravariant. This process is called contraction.

If we form the outer product of two tensors and then sum on a pair of indices (one covariant and the other contravariant) one from each tensor we say that we have taken an inner product of the two tensors.

It follows readily from the above definitions (see problem 7) that if the result of taking the product (inner or outer) of a given ordered set of elements with a tensor at the point P_o with arbitrary components is a tensor at P_o, then the ordered set of elements is also a tensor at the point P_o. This result is called the quotient theorem for tensors. The quotient theorem was used when it was proved that the n^2 quantities g_{jk} are the components of a tensor.

The distinction between tensors and tensor fields has been emphasized here since in physics it is common to refer to both as simply tensors. We shall adopt this practice for the remainder of the chapter except for some isolated cases where it could lead to confusion.

Since $|g_{jk}| \neq 0$ there exists a unique ordered set of n^2 elements g^{ik}, $i,k = 1,\ldots,n$, such that

(76) $g^{ik}g_{jk} = \delta^i_j.$

Let A_j be an arbitrary covariant vector. Since $|g_{ij}| \neq 0$ there exists a contravariant vector B^k such that

(77) $g_{kj}B^k = A_j.$

Multiplying the above equation by $g^{\ell j}$ and summing over j gives

(78) $g^{\ell j}g_{kj}B^k = g^{\ell j}A_j$

or

(79) $\delta^\ell_k B^k = B^\ell = g^{\ell j}A_j.$

But A_j was an arbitrary covariant vector; hence by the quotient theorem for tensors the n^2 quantities g^{jk}, $j,k = 1,\ldots,n$, are the components of a second-rank contravariant tensor. The tensors g^{jk} and g_{jk} are called fundamental tensors.

The metric tensor g_{jk} provides a natural isomorphism (i.e. a one-to-one linear map) between contravariant and covariant vectors. Any contravariant vector $\vec{v}$ with components v^j, $j = 1,\ldots,n$, is mapped into the covariant vector $\vec{v}^* = g(\vec{v})$ with components v_j, $j = 1,\ldots,n$, defined by

(80) $v_j = g_{jk}v^k.$

Multiplying by $g^{\ell j}$ and summing over j we get

(81) $g^{\ell j}v_j = v^\ell,$

which is the formula for the inverse transformation $\vec{v} = g^{-1}(\vec{v}^*)$.

In equation (80) we say that the tensor g_{jk} was used to lower an index and in equation (81) we say that g^{jk} was used to raise an index. One can use the fundamental tensors to raise and lower indices of higher-rank tensors also. Consider for example the second-rank covariant tensor A_{ij}. From this tensor we can form the tensors $A^i_j = g^{i\ell}A_{\ell j}$, $A^\ell_j = g^{i\ell}A_{ji}$, and $A^{ij} =$

$g^{i\ell}g^{jk}A_{\ell k}$, all of which can be transformed into A_{ij} by lowering one or two indices. We call two tensors which can be made equal by lowering or raising a number of indices (with the fundamental tensors) associate.

When we dealt with Cartesian tensors we had $g_{jk} = \delta_{jk}$ so that all associate tensors had identical components. Thus there was no need to make the distinction between covariant and contravariant tensors.

7.7. THE SCALAR PRODUCT

Let $d\vec{s}$ be the differential contravariant vector with components du^j, $j = 1,\ldots,n$. From equation (48)

$$(ds)^2 = e d\vec{s}\cdot d\vec{s} = e g_{jk} du^j du^k.$$

In general the scalar product between two contravariant vectors $\vec{v}$ and $\vec{u}$ (at the same point) with components v^j and u^j, $j = 1,\ldots,n$, respectively, is denoted $\vec{v}\cdot\vec{u}$ and defined by†

$$(82) \qquad \vec{v}\cdot\vec{u} = g_{jk} v^j u^k.$$

The magnitude of a contravariant vector $\vec{v}$, denoted either by $||\vec{v}||$ or simply by v, is defined by

$$(83) \qquad ||\vec{v}||^2 = e\vec{v}\cdot\vec{v}, \qquad ||\vec{v}|| \geq 0,$$

where e is +1 or -1 so that $||\vec{v}||^2 \geq 0$.

A contravariant vector $\vec{v}$ is called null if $||\vec{v}|| = 0$. If $\vec{v}$ and $\vec{u}$ are two non-null contravariant vectors, then the angle θ between them is defined by

$$(84) \qquad \cos\theta = \frac{\vec{v}\cdot\vec{u}}{||\vec{v}||\ ||\vec{u}||}, \qquad \theta \in (0,2\pi).$$

†The metric tensor g_{jk} is evaluated at the same point as the vectors $\vec{v}$ and $\vec{u}$ in equation (82).

Two contravariant vectors $\vec{u}$ and $\vec{v}$ are called orthogonal if $\vec{v}\cdot\vec{u} = 0$, and if they are non-null this means the angle between them is $\pi/2$.

The isomorphism g induces in a natural fashion a scalar product between covariant vectors. We define the scalar product between two covariant vectors $\vec{v}^*$ and $\vec{u}^*$ with components v_j and u_j, $j = 1,\ldots,n$, respectively by

$$\vec{v}^*\cdot\vec{u}^* = (g^{-1}(\vec{v}^*))\cdot(g^{-1}(\vec{u}^*)) = g_{jk}(g^{-1}(\vec{v}^*))^j(g^{-1}(\vec{u}^*))^k$$
$$= g_{jk}g^{jn}v_n g^{km}u_m = g^{nm}v_n u_m.$$

The norm of a covariant vector $\vec{v}^*$ and the angle θ between two non-null covariant vectors $\vec{v}^*$ and $\vec{u}^*$ (a covariant vector $\vec{v}^*$ is null if $||\vec{v}^*|| = 0$) are also given by equations (83) and (84) respectively with * placed in the appropriate places. Two covariant vectors $\vec{v}^*$ and $\vec{u}^*$ are called orthogonal if $\vec{v}^*\cdot\vec{u}^* = 0$.

Let $\vec{v}$ and $\vec{u}$ be any two contravariant vectors and λ an arbitrary real number. Consider the quadratic function of λ

$$f(\lambda) = (\vec{v} + \lambda\vec{u})\cdot(\vec{v} + \lambda\vec{u}) = (\vec{v}\cdot\vec{v}) + 2\lambda(\vec{v}\cdot\vec{u}) + \lambda^2(\vec{u}\cdot\vec{u}). \tag{85}$$

If the metric is positive definite (i.e. for any contravariant vector $\vec{\omega}$, $\vec{\omega}\cdot\vec{\omega} = g_{jk}\omega^j\omega^k \geq 0$ with equality if and only if $\vec{\omega} = 0$) then $f(\lambda)$ has at most one real root. It follows that its discriminant is negative or zero,

$$4(\vec{v}\cdot\vec{u})^2 - 4(\vec{v}\cdot\vec{v})(\vec{u}\cdot\vec{u}) \leq 0,$$

so that

$$|(\vec{v}\cdot\vec{u})| \leq ||\vec{v}||\ ||\vec{u}||. \tag{86}$$

Equation (86) is known as the Cauchy-Schwartz inequality.

Conversely, if the metric is indefinite (i.e. the quadratic form $\vec{\omega}\cdot\vec{\omega} = g_{jk}\omega^j\omega^k$ can take on both positive and negative values) the Cauchy-Schwartz inequality no longer holds. Consider $\vec{v}$, $\vec{u}$ in equation (85) such that the span of the set

$\{\vec{u},\vec{v}\}$ contains vectors $\vec{\omega}^1$ and $\vec{\omega}^2$ such that $\vec{\omega}^1 \cdot \vec{\omega}^1 > 0$ and $\vec{\omega}^2 \cdot \vec{\omega}^2 < 0$. Then the function $f(\lambda)$ takes on both positive and negative values and hence has two distinct real roots. Thus the discriminant is positive, implying that†

$$|(\vec{u}\cdot\vec{v})| > ||\vec{u}|| \; ||\vec{v}||.$$

It follows that the notion of the angle between two vectors introduced in equation (84) is well defined only for positive definite metrics.

7.8. TENSORS AS MULTILINEAR MAPS

Covariant vectors were defined as linear maps of contravariant vectors into scalars. Higher-rank tensors can also be looked upon as operators. For example, consider a second-rank mixed tensor Q (at some point) with components Q^i_j, $i,j = 1,2,\ldots,n$. Let $\vec{u}^*$ and $\vec{v}$ be respectively covariant and contravariant vectors (at the same point as Q) with components u_i, $i = 1,\ldots,n$, and v^j, $j = 1,\ldots,n$. The mixed tensor Q is a bilinear operator (or map) which maps two vectors, one contravariant and the other covariant, into a scalar (at the same point as Q) in the following fashion:

$$Q(\vec{u}^*,\vec{v}) = Q^i_j u_i v^j. \tag{87}$$

The operator Q has two arguments and Q is linear in each argument (i.e. $Q(\vec{u}^*, a\vec{v} + b\vec{w}) = aQ(\vec{u}^*,\vec{v}) + bQ(\vec{u}^*,\vec{w})$ and $Q(a\vec{u}^* + b\vec{y}^*, \vec{v}) = aQ(\vec{u}^*,\vec{v}) + bQ(\vec{y}^*,\vec{v})$, where a, b are scalars at the same point as Q and $\vec{y}^*$ and $\vec{v}$ are respectively covariant and contravariant vectors at the same point as Q). The first argument of Q only accepts covariant vectors and the second only contravariant vectors. The mixed tensor Q also acts as a linear map, operating on a covariant vector to produce another covariant vector (i.e. $Q(\vec{u}^*,\_)$ is a covariant vector with components $Q^i_j u_i$, $j = 1,\ldots,n$) or operating on a contravariant vector to

†We have assumed that $(\vec{v}\cdot\vec{v})(\vec{u}\cdot\vec{u}) > 0$.

produce a contravariant vector (i.e. $Q(\_,\vec{v})$ is a contravariant vector with components $Q^i_j v^j$, $i = 1,\ldots,n$).

If R is a second-rank contravariant tensor with components R^{ij}, $i,j = 1,\ldots,n$, it is also a bilinear map; however, in this case it operates on two covariant vectors to produce a scalar. That is

$$(88) \quad R(\vec{u}^*,\vec{y}^*) = R^{ij}u_i y_j$$

where $\vec{u}^*$ and $\vec{y}^*$ are covariant vectors (at the same point as R) with components u_i, $i = 1,\ldots,n$, and y_j, $j = 1,\ldots,n$, respectively. The second-rank contravariant tensor R also acts as a linear map, operating on a covariant vector and producing a contravariant vector (i.e. $R(\vec{u}^*,\_)$ is a contravariant vector with components $R^{ij}u_i$, $j = 1,\ldots,n$, and $R(\_,\vec{u}^*)$ is also a contravariant vector but with components $R^{ij}u_j$, $i = 1,\ldots,n$).

In a similar manner a second-rank covariant tensor may be looked upon as a bilinear map, operating on two contravariant vectors to produce a scalar, or as a linear map, mapping a contravariant vector into a covariant vector. Higher-rank tensors act as multilinear maps or operators in an analogous fashion.

In Chapter 2 Cartesian tensors were discussed, and it was shown that they may be pictured as multilinear maps. Since for Cartesian tensors associate tensors have identical components there was no need to distinguish there between tensors operating on covariant vectors and tensors acting on contravariant vectors. This additional complexity of a Riemannian space is the price one must pay for considering phenomena operating or taking place in a curved space. In general relativity, for example, this geometric complexity is more than offset by the simplicity gained in describing the system dynamics as corresponding to geodesics in the appropriate Riemannian space.

REFERENCES

1 H.S.M. Coxeter, Introduction to Geometry. John Wiley and Sons Inc., New York, 1969
2 L.P. Eisenhart, Riemannian Geometry. Princeton University Press, Princeton, 1960
3 H. Goldstein, Classical Mechanics. Addison-Wesley, Reading, Mass., 1950
4 Erwin Kreyszig, Differential Geometry and Riemannian Geometry. University of Toronto Press, Toronto, 1968
5 D. Lawden, Tensor Calculus and Relativity. Methuen and Co., London, 1967
6 D. Lovelock and H. Rund, Tensors, Differential Forms, and Variational Principles. John Wiley and Sons, New York, 1975
7 C.W. Misner, K.S. Thorne, and J.A. Wheeler, Gravitation. W.H. Freeman and Company, San Francisco, 1973
8 E. Schroedinger, Space-Time Structure. Cambridge at the University Press, 1950
9 B. Spain, Tensor Calculus. Oliver and Boyd, New York, 1965
10 Michael Spivak, Calculus on Manifolds. W.A. Benjamin Inc., New York, 1965
11 I.S. Sokolnikoff, Tensor Analysis and Applications to Geometry and Mechanics of Continua. John Wiley and Sons, New York, 1964
12 A. Trautman, F.A.E. Pirani, and H. Bondi, Brandeis Summer Institute in Theoretical Physics, Volume One: Lectures on General Relativity. Prentice-Hall Inc., Englewood Cliffs, N.J., 1964
13 C.E. Weatherburn, Riemannian Geometry and the Tensor Calculus. Cambridge at the University Press, 1950

PROBLEMS

1 Cylindrical coordinates ρ, θ, z are defined by the equations

$$x = \rho\cos\theta, \quad y = \rho\sin\theta, \quad z = z.$$

(i) Find the basis vectors $\hat{e}_k$ and the metric tensor g_{jk}.
(ii) From the metric tensor find an expression for an element of volume in cylindrical coordinates.
(iii) Is the two-dimensional subspace on the surface of a cylinder a Euclidean or non-Euclidean world? Explain.

2 Parabolic coordinates u,v,w are defined by the equations

$$x = \sqrt{uv}\cos w, \quad y = \sqrt{uv}\sin w, \quad z = (u - v)/2$$

where $0 \leq w < 2\pi$, $u,v \geq 0$.

(i) Draw diagrams to illustrate surfaces of constant u,v, and w.
(ii) Calculate the basis vectors $\hat{e}_j$ and the metric tensor g_{jk}.
(iii) Find expressions for the elements of length, surface area, and volume in terms of u, v, and w.

3 Suppose ξ,η are curvilinear coordinates for 2-dimensional Euclidean space and that

$$x = x(\xi,\eta), \quad y = y(\xi,\eta)$$

are such that $z = x + iy$ is an analytic function of $\xi + i\eta$. Show that

$$\frac{\partial(x,y)}{\partial(\xi,\eta)}\left(\frac{\partial^2}{\partial x^2} + \frac{\partial^2}{\partial y^2}\right) = \frac{\partial^2}{\partial \xi^2} + \frac{\partial^2}{\partial \eta^2}$$

where $\partial(x,y)/\partial(\xi,\eta)$ is the Jacobian of the transformation.

4 Oblate spheroid coordinates ξ,η,θ are defined by

$$z = \sinh\xi\cos\eta, \quad \rho = \cosh\xi\sin\eta, \quad \theta = \theta$$

where $x = \rho\cos\theta$, $y = \rho\sin\theta$, and $0 \leq \xi < \infty$, $0 \leq \eta \leq \pi$.

An oblate spheroid is the surface defined by $\xi = \xi_o$ (η and θ arbitrary).

(i) Show that as $\xi_o \to \infty$ the oblate spheroid $\xi = \xi_o$ becomes a large sphere.
(ii) Show that as $\xi_o \to 0$ the oblate spheroid $\xi = \xi_o$ becomes a flat circular plate of unit radius.
(iii) In electrostatic theory the potential V satisfies the differential equation $\nabla^2 V = 0$ in regions where there is no charge. If there is no charge at infinity then $V(\vec{r}) \to 0$ as $|\vec{r}| \to \infty$. Show that the potential V for the region surrounding the oblate spheroid $\xi = \xi_o$ charged to a unit potential satisfies

$$\frac{d^2V}{d\xi^2} + \tanh\xi\,\frac{dV}{d\xi} = 0$$

where $V = 1$ on $\xi = \xi_o$ and $V \to 0$ as $\xi \to \infty$, and that the solution to this boundary-value problem is

$$V = \frac{\tan^{-1}(e^{-\xi})}{\tan^{-1}(e^{-\xi_o})} .$$

(Hint: Use the symmetry with respect to θ and η to conclude that V is independent of these variables).

(iv) Calculate the total charge on the oblate spheroid.

(v) Show that the potential in the region surrounding a circular plate (of unit radius) charged to a potential of one is given by

$$V = \frac{4}{\pi} \tan^{-1}(e^{-\xi}).$$

5 Explain why in a small neighbourhood of any point (of a Riemannian space with a continuous metric tensor) the space is approximately flat.

6 Let A_{ij} be a second-rank covariant tensor and B^i_j be a second-rank mixed tensor.

(i) Show that if $A_{ij} = A_{ji}$ (or $A_{ij} = -A_{ji}$) in a coordinate system C then $\bar{A}_{ij} = \bar{A}_{ji}$ (or $\bar{A}_{ij} = -\bar{A}_{ji}$) in every other coordinate system $\bar{C}$.

(ii) Show that if $B^i_j = B^j_i$ (or $B^i_j = -B^j_i$) in system C then it is not necessarily true that $\bar{B}^i_j = \bar{B}^j_i$ (or $\bar{B}^i_j = -\bar{B}^j_i$) in every other coordinate system $\bar{C}$.

7 (i) Let T be an arbitrary symmetric second-rank tensor with components T^{jk}, $j,k = 1,2,\ldots,n$. If the n^2 quantities a_{jk}, $j,k = 1,2,\ldots,n$, satisfy the equation

$$a_{jk}T^{jk} = V$$

where the V is a scalar, is it true that the n^2 quantities a_{jk}, $j,k = 1,2,\ldots,n$, are the components of a covariant second-rank tensor?

(ii) Prove the quotient theorem for tensors.

8 We define a relative tensor of weight W at a point P_o as an ordered set of components $T^{i_1\ldots i_m}_{j_1\ldots j_p}$, $i_1\ldots i_m, j_1\ldots j_p =$ $1,2,\ldots,n$, which transforms like

$$\bar{T}^{i_1\ldots i_m}_{j_1\ldots j_p} = J^W_o \left(\frac{\partial \bar{u}^{i_1}}{\partial u^{\ell_1}}\right)_o \cdots \left(\frac{\partial \bar{u}^{i_m}}{\partial u^{\ell_m}}\right)_o \left(\frac{\partial u^{k_1}}{\partial \bar{u}^{j_1}}\right)_o \cdots \left(\frac{\partial u^{k_p}}{\partial \bar{u}^{j_1}}\right)_o T^{\ell_1\ldots\ell_m}_{k_1\ldots k_p}$$

under change of parametrization from unbarred to barred coordinates, where J_o is the Jacobian of the transformation (i.e. $|\partial u^i/\partial \bar{u}^j|$ evaluated at point P_o.

(i) Let $\varepsilon^{i_1\ldots i_n}$ denote a contravariant tensor density in an n-dimensional Riemannian space. If $\varepsilon^{i_1\ldots i_n}$ is anti-symmetric on interchange of any pair of indices show that:

$\varepsilon^{i_1\ldots i_n} = 0$ if two indices are the same,

$= \varepsilon^{12\ldots n}$ if $i_1\ldots i_n$ is an even permutation of $1,2,\ldots,n$,

$= -\varepsilon^{12\ldots n}$ if $i_1\ldots i_n$ is an odd permutation of $1,2,\ldots,n$.

(ii) Prove if in one coordinate system $\varepsilon^{12\ldots n} = 1$ (where as before $\varepsilon^{i_1\ldots i_n}$ is completely antisymmetric) then under change of parametrization

$$\bar{\varepsilon}^{12\ldots n} = JE = 1$$

where $E = |\partial \bar{u}^i/\partial u^j|$, and thus the tensor density $\varepsilon^{j_1\ldots j_n}$ $(\varepsilon^{12\ldots n} = 1)$ possesses components $-1,0,1$.

(iii) Prove that if A_{ij} is a covariant tensor of rank two then the determinant $|A_{ij}|$ is a relative tensor (of rank zero) of weight two.

(iv) If A^{ij} is a contravariant tensor of rank two show that the determinant $|A^{ij}|$ is then a relative tensor (of rank zero) of weight -2.

(v) Explain how pseudovectors arise as a special case of tensor densities.

8
Special relativity

8.1. INTRODUCTION

A central purpose of this book has been to explore the extent to which linear theories successfully describe natural phenomena. The reader is reminded that many physical constructs or 'objects' are describable as elements of a suitably chosen linear vector space. The content of the physical theory is then to show how these constructs are related to one another. In a linear theory the relationships are linear. For example, one has a linear relationship between the angular momentum and the angular velocity of a rigid body rotating with one point fixed. Angular momentum arises as a dynamically conserved quantity whereas angular velocity describes the geometric orientation of the body as a function of time. Another example is afforded by the time-development operator acting linearly on quantum states in wave mechanics.

In the present chapter we explore yet another important area of mathematical physics which exploits linearity successfully, namely Einstein's theory of special relativity. The familiar transformations of Lorentz and Einstein,

$$(1a) \quad x' = \gamma(x - \beta ct), \quad y' = y, \quad z' = z,$$

$$(1b) \quad t' = \gamma(t - \beta x/c), \quad \beta = v/c, \quad \gamma = (1 - \beta^2)^{-\frac{1}{2}},$$

represent a comparison of 'world events' as measured in two inertial frames separating uniformly along the x,x' axes with velocity v. Clearly, these equations represent a linear transformation from one set of space-time variables $(\vec{r},t)$ to another set $(\vec{r}',t')$.

In Chapter 2 a coordinate system was described as a 3-dimensional net for identifying points in space relative to a given point. By asking what seemed like a trivial question, namely how rotated observers would compare measurements on the same physical phenomena, we were led to the concept of tensors and to the elegant mathematical theory associated with transformation groups. We can now ask the more extensive question: how do inertial observers who are rotated and uniformly moving apart compare measurements? Since it turns out that space and time variables can be accommodated in a 4-dimensional world and both space rotations and Lorentz transformations appear as hyper-rotations in that world, we can anticipate that our more extensive question will again lead to an extended concept of tensors (so-called 4-tensors) and a more general set of transformation groups. This is the primary goal of the present chapter.

Our purpose is not to derive the Lorentz transformations from a set of axioms, or to show how they follow from experimental observations. It is assumed that the reader has already been down this road. Rather, we are interested in examining the conceptual framework and in exploring the consequences of the theory, bearing in mind analogues and relationships already developed in previous chapters in this book. Moreover, we shall be concerned in Chapter 9 with an examination of the way in which relativity influences the formulation of quantum mechanics and some consequences of this influence relating to the classification of elementary particles.

8.2. PRELIMINARIES

Before proceeding in depth with our considerations of special relativity, there are several concepts associated with measurement theory and coordinate systems which need clarification. We again remind the reader that the test of whether a concept is acceptable or not is really a test as to whether it is convenient and also self-consistent with other constructs in the theory.

By an observer we shall mean a person capable of moving about in a 3-dimensional world with a set of recording instruments, a set of synchronized clocks, and some standardized metre rods. For any 'event' which occurs, the observer must be able to identify its nature and record both its spatial and time coordinates in his 'frame of reference.' Such events are called world events and can be regarded as occurring at world points in a space-time coordinate system.

It is well to emphasize that we are dealing in special relativity with inertial systems, i.e. with systems of coordinates in which the laws of physics appear to take on their simplest form (in the sense that the 'apparent forces' of pure kinematical origin vanish). It is a matter of experience that coordinate systems at rest or moving uniformly with respect to the 'fixed' stars are inertial to a sufficiently good approximation for normal practice.

It is also a matter of experience that uniformly separating observers always agree on the velocity (direction and speed) of their relative motion. This has the significance that the velocity parameter in the Lorentz transformation is a single quantity common to both observers.

Finally, there are two related questions which should be addressed: what constitutes a uniformly ticking clock and how are different clocks synchronized?

We can take as a uniformly ticking clock one which records equal intervals between the vibrations of a simple harmonic oscillator, such as a vibrating spring, an oscillating pendulum, or an ammonium molecule undergoing inversion vibrations.† The problem of synchronization is essentially a trivial one for clocks at rest relative to one another, since experience shows that all such clocks which are identical in their construction run at the same rate independently of their locations in space

†The nitrogen atom has difficulty knowing whether it should stay symmetrically placed on the right or left side of the plane containing the three hydrogen atoms. As a result it vibrates between these positions.

(i.e. in field-free space; strong gravitation does in fact affect the rate of a clock). A practical method for checking the synchronization of two clocks, A and B, which depends on the axiom that light travels with the same speed in all directions (i.e. on the isotropy of space),† is as follows. Let a pulse of light be emitted from A, reflected at B, and received again at A. If t_A and t'_A are respectively the times on clock A when the pulse was emitted and received and if the time on clock B when the pulse was reflected is given by

$$t_B = (t_A + t'_A)/2,$$

then we say the clocks at A and B are synchronous.

Synchronism of 'stationary' clocks running at uniform rates has the significance in analysis that the 'stationary' observer can take as much time as he wishes placing the clocks, synchronizing them, and going back after an event to record the readings on the clocks. For relatively moving observers this is not so; synchronism only has meaning when two relatively moving clocks are momentarily coincident in space.

8.3. GALILEAN RELATIVITY

In the era previous to Einstein's relativity, a kind of classical relativity held sway, which seemed so obvious that it was rarely stated if at all. We now refer to this relativity of classical mechanics as Galilean relativity. It was based on notions of absolute space and time consistent with Newton's formulation of mechanics. Galilean relativity shared with special relativity the concept of inertial reference frames, that is preferred coordinate frames in 3-dimensions where the laws of mechanics appear to take on their simplest forms. In Galilean relativity mechanical laws are invariant with respect

---

†We assume in this chapter that space is isotropic and homogeneous.

to the transformation group†

(2a) $\bar{x}^j = a_{jk}x^k + v^jt + d^j, \quad j,k = 1,2,3,$

(2b) $\bar{t} = t + e,$

where the barred and unbarred indices refer to two different Galilean observers. The quantities a_{jk} constitute the orthogonal rotation matrix discussed in Chapter 2, which permits the two observers to use different orientations for their Cartesian coordinate frames. Upper and lower indices are used in anticipation of the distinction to be made explicit later between contravariant and covariant vectors in a Lorentz space-time metric. The vector v^j gives the separation velocity of the barred and unbarred observers, and the constants e and d^j allow the observers the freedom of using different initial clock settings and different origins of coordinates at that initial time.

From equations (2a) and (2b), it follows that two Galilean observers agree upon accelerations:

(3) $$\frac{d^2\bar{x}^j}{d\bar{t}^2} = a_{jk}\frac{d^2x^k}{dt^2}$$

except for a rotation a_{jk} which lines up the two coordinate frames. If we assume that the 'real' forces causing accelerations are also Galilean invariants (in the sense of equation (3)), Newton's laws are preserved in form and we say that they are covariant (form invariant) with respect to Galilean transformations.

Unfortunately for Galilean invariance Newton's laws do not hold without modification when v is large. It is not our intention here to review the historic events leading to Einstein's modification of the Newtonian laws and his replacement of

---

†In this chapter small Roman letters take on the values 1,2,3 and Greek letters include reference to the time coordinate and take on the values 0,1,2,3 or 1,2,3,4 depending on whether a Lorentz or Minowski metric is being used.

Galilean relativity by special relativity. Suffice it to say that Newton's laws break down even classically (apart from the need for quantum modifications at the microscopic level) when observing physical objects moving at a high speed. Since light quanta, according to relativity, are always moving at velocity $c \simeq 3 \times 10^8$ m/sec for all inertial observers, it is not surprising that Galilean relativity failed entirely for electromagnetic phenomena. It is remarkable that Maxwell's theory already incorporated the new relativity before it was invented by Lorentz and Einstein.

8.4. POSTULATES OF SPECIAL RELATIVITY

Our intention is not to establish a minimum axiomatic basis for special relativity; rather it is to provide a sufficient foundation for exposition and understanding of the theory. The postulates set out below are designed to provide this foundation. The particular set chosen is by no means unique. Some of the background considerations leading to this set have already been discussed in previous sections.

POSTULATE 1: <u>There exist inertial frames of reference in which the laws of physics take on a particularly simple form. Reference frames stationary with respect to the 'fixed' stars are inertial, as are all reference frames moving at constant velocity with respect to them</u>.

POSTULATE 2: <u>A frame of reference can be chosen as a set of three Cartesian coordinate axes together with the necessary metre sticks, synchronous clocks, and detection apparatus</u>.

POSTULATE 3: <u>Communication between observers is carried out by means of electromagnetic (light) rays; such rays travel necessarily with the same velocity</u> ($\simeq 3 \times 10^8$ m/sec) <u>for all inertial observers</u>.

POSTULATE 4: <u>The Poincaré transformations</u>

$$(4) \qquad \bar{x}^\nu = L^\nu_\mu \, x^\mu + d^\nu$$

relate the space-time measurements x^μ of a world event (a physical phenomenon localized in space and time) as seen by observer O in an 'unbarred' inertial frame to the space-time measurements $\bar{x}^\mu$ as seen by the observer $\bar{O}$ in a 'barred' inertial frame. Here x^ν, $\nu = 0,1,2,3$, stands for ct,x,y,z and is called a 4-vector. The four quantities d^ν are independent of x^ν and can be made identically zero by an appropriate choice of coordinate origins and initial settings of clocks. The fourth rank (matrix) L^ν_μ is also independent of x^ν, but depends on six parameters, the three Euler angles orienting the barred frame with respect to the unbarred, and the three components of the relative velocity with which the two frames separate. The explicit form of L^ν_μ is determined by the requirement of invariant 4-length between two neighbouring world events,

$$\eta_{\mu\nu} d\bar{x}^\mu d\bar{x}^\nu = \eta_{\mu\nu} dx^\mu dx^\nu, \tag{5}$$

where the Lorentz metric $\eta_{\mu\nu}$ is given by

$$\eta_{\mu\nu} = \begin{bmatrix} -1 & 0 & 0 & 0 \\ 0 & 1 & 0 & 0 \\ 0 & 0 & 1 & 0 \\ 0 & 0 & 0 & 1 \end{bmatrix} = \eta^{\mu\nu}. \tag{6}$$

POSTULATE 5: In the Poincaré transformation space-time coupling occurs only in the direction of relative motion and not perpendicular to it.

Remarks: It is not immediately evident that the condition (5) guarantees the correct form for the L^μ_ν coefficients in eq. (4). We shall show that (5) implies a set of constraints on the L^μ_ν analogous to the orthogonality conditions which we studied in Chapter 3 pertaining to 3-dimensional rotation matrices. In fact, the conditions we shall derive are just sufficient in number to guarantee that only 6 of the 16 components of the L^μ_ν tensor are independent (representing the 6 parameters referred to above).

First we observe that the differential form of (4) is

$$d\bar{x}^\nu = L^\nu_\mu \, dx^\mu = \frac{\partial \bar{x}^\nu}{\partial x^\mu} \, dx^\mu . \tag{7}$$

Substituting expressions of the form (7) into (5) then yields

$$\eta_{\sigma\lambda} dx^\sigma dx^\lambda = \eta_{\mu\nu} d\bar{x}^\mu d\bar{x}^\nu = \eta_{\mu\nu} L^\mu_\sigma L^\nu_\lambda dx^\sigma dx^\lambda \tag{8}$$

where we have used judicious relabelling of dummy variables. Since this equation must be satisfied for arbitrary values of dx^σ and dx^λ, we must have

$$\eta_{\sigma\lambda} = \eta_{\mu\nu} L^\mu_\sigma L^\nu_\lambda . \tag{9}$$

This is just the set of pseudo-orthogonal conditions to which we referred above. They constitute 10 independent equations, which reduce the number of free parameters from 16 to 6. Equation (9) is the necessary and sufficient conditions for

$$\bar{x}^\nu = L^\nu_\mu \, x^\mu + d^\nu \tag{4}$$

to be a Poincaré transformation.

If we restrict ourselves to a Poincaré transformation in which the two frames are at rest but rotated so that $t = \bar{t}$, it is simple to check that equations (9) reduce to the statement that orthogonal matrices (i.e. the 3 x 3 space part of L^μ_ν) obey the equation $L\tilde{L} = I$. The corresponding 3 x 3 orthogonal matrix is just the rotation matrix relating rotated observers, as discussed in Chapter 3 (see problem 2).

We wish now to check that equation (4) also yields the Lorentz transformation for uniformly separating observers. Again, without essential loss in generality, we take the special case where the x,y,z directions coincide with their barred counterparts and where, moreover, the relative velocity v is along the $x = x^1$ direction: Postulate 5 (and the isotropy of space) then guarantees that

$$L^2_2 = L^3_3 = 1,$$

$$L^2_0 = L^0_2 = L^3_0 = L^0_3 = 0,$$

and $$L^i_j = 0 \quad \text{for } i \neq j, \qquad i,j = 1,2,3.$$

Hence L^μ_ν takes on the form

$$L^\mu_\nu = \begin{bmatrix} \gamma & \delta & 0 & 0 \\ \alpha & \beta & 0 & 0 \\ 0 & 0 & 1 & 0 \\ 0 & 0 & 0 & 1 \end{bmatrix} . \tag{10}$$

It is then straightforward, if somewhat tedious, to show that the conditions (9) reduce (10) to the form†

$$L^\mu_\nu = \begin{bmatrix} \gamma & -\sqrt{\gamma^2-1} & 0 & 0 \\ -\sqrt{\gamma^2-1} & \gamma & 0 & 0 \\ 0 & 0 & 1 & 0 \\ 0 & 0 & 0 & 1 \end{bmatrix} . \tag{11}$$

Thus

$$d\bar{x} = \gamma dx - \sqrt{\gamma^2 - 1}\, cdt$$

from (7) and (11). To determine γ we observe that in the special case where observer O simply records the position of the origin of $\bar{O}$ we have $d\bar{x} = 0$ and $dx/dt = v$. Hence

$$\gamma = \frac{1}{\sqrt{1 - v^2/c^2}} = \frac{1}{\sqrt{1 - \beta^2}} , \tag{12}$$

and L^μ_ν is seen to give the usual form of the Lorentz transformation.

---

†We have made a choice of signs in (11). Actually four independent choices exist corresponding to whether time and/or space reversal is considered. We return to this question shortly. We have used upper indices to label rows, lower indices to label columns in (11).

8.5. PROPERTIES OF THE POINCARÉ GROUP AND ITS SUBGROUPS

It can be easily shown that the Poincaré transformations form a group (non-compact), but the transformations in equation (4) are non-linear because of the presence of the space-time translation vector d^ν. The Poincaré group is sometimes referred to as the inhomogeneous Lorentz group because of this term. The subgroup with d^ν set equal to zero is referred to as the homogenous Lorentz group, or simply the Lorentz group.

To show that the Poincaré transformations form a group, we must show that an identity exists, that successive transformations compose into a single equivalent transformation and obey associative laws, and finally that each transformation has an inverse.

It is at once evident that the Kronecker delta function δ^μ_ν satisfies (9) so that $L^\mu_\nu = \delta^\mu_\nu$, $d^\nu = 0$ corresponds to the identity transformation wherein observer O compares his space-time measurements x^μ with themselves:

$$\bar{x}^\mu = \delta^\mu_\nu\, x^\nu = x^\mu.$$

To show that successive transformations compose, consider three Poincaré observers, O, $\bar{O}$, and $\bar{\bar{O}}$. We have

$$\bar{x}^\mu = L^\mu_\nu(12)x^\nu + d^\mu(12)$$

and

$$\bar{\bar{x}}^\lambda = L^\lambda_\mu(23)\bar{x}^\mu + d^\lambda(23);$$

hence

$$\text{(13a)}\quad \bar{\bar{x}}^\lambda = L^\lambda_\mu(23)\{L^\mu_\nu(12)x^\nu + d^\mu(12)\} + d^\lambda(23)$$

$$= L^\lambda_\nu(13)x^\nu + d^\lambda(13)$$

where

$$\text{(13b)}\quad L^\lambda_\nu(13) = L^\lambda_\mu(23)L^\mu_\nu(12)$$

and

(13c) $d^\lambda(13) = L^\lambda_\mu(23)d^\mu(12) + d^\lambda(23)$.

Similarly, we can exhibit an inverse for any given transformation; the inverse to the transformation

$$\bar{x}^\mu = L^\mu_\nu(12)x^\nu + d^\mu(12)$$

is just

(14a) $x^\lambda = L^\lambda_\sigma(21)\bar{x}^\sigma + d^\lambda(21)$

and $L^\lambda_\sigma(21)$ satisfies the relationship

(14b) $\delta^\lambda_\nu = L^\lambda_\sigma(12)L^\sigma_\nu(21)$

and $d^\lambda(21)$ satisfies the relationship

(14c) $d^\lambda(12) = -L^\lambda_\sigma(12)d^\sigma(21)$.

Equation (14b) is satisfied, according to (9), by the choice

(15) $L^\lambda_\sigma(21) = \eta^{\lambda\mu}\; L^\varepsilon_\mu(12)\eta_{\varepsilon\sigma}$.

Equation (14c) is then satisfied in a straightforward manner. Associativity of successive transformations is also easily shown but the proof will be omitted here.

We now restrict ourselves to the Lorentz (homogenous) subgroup of the Poincaré group:

$$\bar{x}^\mu = L^\mu_\nu\; x^\nu.$$

Equation (13b) makes it clear that group composition is isomorphic to matrix multiplication. We take advantage of this by identifying the Lorentz group as the group of matrices L^μ_ν (upper index rows, lower columns).

An important subgroup of the Lorentz group is the _discrete group_ consisting of the identity, time inversion, space inversion, and space-time inversion (so-called inversion subgroup):

$$I = \begin{bmatrix} 1 & 0 & 0 & 0 \\ 0 & 1 & 0 & 0 \\ 0 & 0 & 1 & 0 \\ 0 & 0 & 0 & 1 \end{bmatrix}, \quad P = \begin{bmatrix} 1 & 0 & 0 & 0 \\ 0 & -1 & 0 & 0 \\ 0 & 0 & -1 & 0 \\ 0 & 0 & 0 & -1 \end{bmatrix},$$

(16)

$$T = \begin{bmatrix} -1 & 0 & 0 & 0 \\ 0 & 1 & 0 & 0 \\ 0 & 0 & 1 & 0 \\ 0 & 0 & 0 & 1 \end{bmatrix}, \quad J = \begin{bmatrix} -1 & 0 & 0 & 0 \\ 0 & -1 & 0 & 0 \\ 0 & 0 & -1 & 0 \\ 0 & 0 & 0 & -1 \end{bmatrix},$$

with group table:

.	I	P	T	J
I	I	P	T	J
P	P	I	J	T
T	T	J	I	P
J	J	T	P	I

The symmetry about the main diagonal arises because the inversion groups is Abelian (commutative).

We can distinguish uniquely between the four discrete inversion operators by the values of their determinants $|L^{\mu}_{\nu}|$ and their components L^{o}_{o}:

	I	P	T	J
$\lvert L^{\mu}_{\nu}\rvert$	1	-1	-1	1
L^{o}_{o}	1	1	-1	-1

Interpreting equation (9) as a matrix equation

(9) $\quad \eta = \tilde{L}\eta L$

it becomes clear that for any Lorentz transformation

(17) $\quad |L^{\mu}_{\nu}| = 1 \text{ or } -1.$

Also from equation (9)

(18) $$(L^0_0)^2 = 1 + \sum_{i=1}^{3} (L^i_0)^2 \geq 1$$

so that

(19) $$L^0_0 \geq 1 \quad \text{or} \quad L^0_0 \leq -1.$$

It follows that <u>any Lorentz transformation may be written as a product of a restricted Lorentz transformation</u> (i.e. one with determinant 1 and $L^0_0 \geq 1$) <u>and one of the four inversion group transformations</u>.

Unless otherwise specified we shall restrict our further discussion to the restricted Lorentz group. A further subgroup of this subgroup is the set of special Lorentz transformations corresponding to observers separating with uniform velocity v along the x^1-axis. The appropriate transformations then have the form given in equation (11).

It is clear from equation (11) that we may re-parametrize L^μ_ν as follows:

(20) $$L^\mu_\nu = \begin{bmatrix} \gamma & -\beta\gamma & 0 & 0 \\ -\beta\gamma & \gamma & 0 & 0 \\ 0 & 0 & 1 & 0 \\ 0 & 0 & 0 & 1 \end{bmatrix} = \begin{bmatrix} \cosh\phi & -\sinh\phi & 0 & 0 \\ -\sinh\phi & \cosh\phi & 0 & 0 \\ 0 & 0 & 1 & 0 \\ 0 & 0 & 0 & 1 \end{bmatrix}$$

where

(21) $$\beta = v/c = \tanh\phi.$$

In this parametrization, a Lorentz transformation corresponds to a hyper-rotation in terms of hyperbolic angles.

Lorentz transformations of this kind are often called <u>boosts</u> and denoted $B(\phi)$. From matrix multiplication, we can easily verify that successive boosts satisfy

(22) $$B(\phi_1)B(\phi_2) = B(\phi_1 + \phi_2)$$

which, from (21), implies the famous velocity addition formula

$$v_{13} = \frac{v_{12} + v_{23}}{1 + \frac{v_{12} \cdot v_{23}}{c^2}} \tag{23}$$

corresponding to the transformation from observer O_1 to O_3 following from the successive transformations $O_1 \to O_2$ and $O_2 \to O_3$.

We may observe that since ordinary rotations in 3-space are a subgroup of the Lorentz group, a boost in an arbitrary direction can be accomplished by first aligning the coordinate axes, then boosting along the x^1-direction, and finally re-aligning the axes to their original position. Mathematically this can be expressed as

$$\tilde{A}\; B(v)\; A \tag{24}$$

where A is an appropriate rotation matrix. In fact every restricted Lorentz transformation can be written as the product of a boost with a rotation (see problem 2).

Finally we wish to note that the apparatus of tensor analysis introduced in Chapter 7 applies in the present case. Limiting ourselves to changes of coordinates which are restricted Lorentz transformations we may apply the definition of tensors given in the previous chapter. Then it is clear that x^μ as well as dx^μ transform as contravariant vectors. It follows from

$$x_\mu = \eta_{\mu\nu} x^\nu$$

that the covariant components of a world event are $-ct,x,y,z$ and hence the two fundamental tensors are identical:

$$\eta_{\mu\nu} = \eta^{\mu\nu}.$$

Multiplying equation (9) by $\eta^{\varepsilon\sigma}$ and summing on σ then yields

$$\text{(25a)} \quad \eta^{\varepsilon\sigma}\eta_{\sigma\lambda} = \delta^{\varepsilon}_{\lambda} = \eta^{\varepsilon\sigma} L^{\mu}_{\sigma} \eta_{\mu\nu} L^{\nu}_{\lambda} = L^{\varepsilon\mu} L_{\mu\lambda}.$$

Thus

$$\text{(25b)} \quad L^{\varepsilon\mu}L_{\mu\lambda} = \delta^{\varepsilon}_{\lambda}$$

where $\delta^{\varepsilon}_{\lambda}$ is the Kronecker delta.

It is a straightforward matter to show that the 4-gradient $\partial/\partial x^{\nu}$ transforms as

$$\text{(26)} \quad \frac{\partial}{\partial x^{\alpha}} = \frac{\partial \bar{x}^{\beta}}{\partial x^{\alpha}} \frac{\partial}{\partial \bar{x}^{\beta}} = L^{\beta}_{\alpha} \frac{\partial}{\partial \bar{x}^{\beta}} .$$

We can also calculate the transformation properties of $\partial/\partial x_{\alpha}$ as follows:

$$\text{(27)} \quad \bar{x}_{\sigma} = \eta_{\sigma\mu} \bar{x}^{\mu} = \eta_{\sigma\mu}L^{\mu}_{\nu} x^{\nu} = \eta_{\sigma\mu} L^{\mu}_{\nu} \lambda^{\nu\lambda} x_{\lambda}.$$

Hence

$$\text{(28)} \quad \frac{\partial}{\partial x_{\alpha}} = \frac{\partial \bar{x}_{\sigma}}{\partial x_{\alpha}} \frac{\partial}{\partial \bar{x}_{\sigma}} = \eta_{\sigma\mu} L^{\mu}_{\nu} \eta^{\nu\alpha} \frac{\partial}{\partial \bar{x}_{\sigma}} = L^{\alpha}_{\sigma} \frac{\partial}{\partial \bar{x}_{\sigma}} .$$

Hence,† from (26) and (28),

$$\text{(29)} \quad \frac{\partial^2}{\partial x^{\alpha} \partial x_{\alpha}} = L^{\beta}_{\alpha} \eta_{\sigma\mu} L^{\mu}_{\nu} \eta^{\nu\alpha} \frac{\partial^2}{\partial \bar{x}^{\beta} \partial \bar{x}_{\sigma}}$$

$$= L_{\sigma\nu} L^{\nu\beta} \frac{\partial^2}{\partial \bar{x}^{\beta} \partial \bar{x}_{\sigma}} , \quad \text{from equation (25b),}$$

$$= \delta^{\beta}_{\sigma} \frac{\partial^2}{\partial \bar{x}^{\beta} \partial \bar{x}_{\sigma}} = \frac{\partial^2}{\partial \bar{x}^{\beta} \partial \bar{x}_{\beta}} .$$

Hence we see that the 4-dimensional Laplacian

$$\text{(30)} \quad - \frac{\partial^2}{\partial x^{\alpha} \partial x_{\alpha}} = \frac{1}{c^2} \frac{\partial^2}{\partial t^2} - \nabla^2$$

---

†Note that there is an ambiguity in writing $L^{\alpha}_{\sigma} = \eta_{\sigma\mu}L^{\mu}_{\nu}\eta^{\nu\alpha}$. This is usually removed by writing the Lorentz transformations as $x^{\mu} = L^{\mu}{}_{\nu}x^{\nu}$ so that $L_{\sigma}{}^{\alpha} = \eta_{\sigma\mu}L^{\mu}{}_{\nu}\eta^{\nu\alpha}$.

is a scalar operator and takes a scalar field (or a vector field) into a scalar field (or a vector field).

Let $\phi(x^\mu)$ be a scalar field. Then by definition

$$(31) \quad \phi(x^\mu) = \bar{\phi}(\bar{x}^\mu),$$

and the wave equation

$$(32) \quad -\frac{\partial^2\phi}{\partial x^\alpha \partial x_\alpha} = \frac{1}{c^2}\frac{\partial^2\phi}{\partial t^2} - \nabla^2\phi = 0$$

is form-invariant under the Lorentz group.

Finally, we observe that this equation has an invariant solution

$$(33) \quad \phi = \phi_0 \, e^{ik^\alpha x_\alpha}$$

where ϕ_0 is a scalar amplitude and k^α is any 4-vector satisfying the condition

$$(34) \quad k^i k_i + k^0 k_0 = 0.$$

We identify k^i as the wave number 3-vector and k^0 as the frequency divided by c:

$$(35) \quad \vec{k}\cdot\vec{k} = \omega^2/c^2 = (2\pi/\lambda)^2$$

where λ is the wavelength. In quantum mechanics, this equation describes the propagation of light quanta with energy $E = \hbar\omega$ and momentum $\vec{p} = \hbar\vec{k}$, except that the scalar fields are replaced by vector fields.

8.6. MINOWSKI SPACE

An alternative formulation of special relativity can be achieved using the unit tensor as a metric tensor provided the 'time coordinate' is treated as imaginary. It follows that there is no distinction between covariant and contravariant vectors. With the choice $x^1 = x_1 = x$, $x^2 = x_2 = y$, $x^3 = x_3 = z$, and $x^4 = x_4 = ict$, the condition that the 'distance' between neighbouring world events is an invariant becomes

$$\delta_{\mu\nu}dx_\mu dx_\nu = dx_\nu dx_\nu = (dx)^2 + (dy)^2 + (dz)^2 - c^2(dt)^2$$
$$= (d\bar{x})^2 + (d\bar{y})^2 + (d\bar{z})^2 - c^2(d\bar{t})^2 = \delta_{\mu\nu}d\bar{x}_\mu d\bar{x}_\nu = d\bar{x}_\mu d\bar{x}_\mu . \tag{36}$$

Since no signal can travel faster than light, this expression is negative or zero for connected events. In any case one can define 'distance' as

$$e dx_\mu dx_\mu$$

where e is $\mp$ as $dx_\mu dx_\mu$ is $\mp$ so that distance is always positive.

With this parametrization, the restricted Lorentz transformation (we use Λ for L in Minkowski metric)

$$\bar{x}_\mu = \Lambda_{\mu\nu}x_\nu \tag{37}$$

satisfies formally an orthogonality condition

$$\Lambda_{\nu\lambda}\Lambda_{\nu\mu} = \delta_{\lambda\mu} \tag{38}$$

which is sometimes referred to as pseudo-orthogonality because of the imaginary fourth coordinate. For a boost along the 3-direction, $\Lambda_{\mu\lambda}$ clearly has the form

$$\Lambda_{\nu\lambda} = \begin{bmatrix} 1 & 0 & 0 & 0 \\ 0 & 1 & 0 & 0 \\ 0 & 0 & \gamma & i\beta\gamma \\ 0 & 0 & -i\beta\gamma & \gamma \end{bmatrix} . \tag{39}$$

In Minkowski space, we can also introduce the 4-dimensional Laplacian as a scalar operator (D'Alembertian):

$$= \frac{\partial^2}{\partial x_\alpha \partial x_\alpha} = \frac{\partial^2}{\partial \bar{x}_\alpha \partial \bar{x}_\alpha} . \tag{40}$$

Correspondingly for a scalar field ϕ, we can construct a form-invariant scalar wave equation

$$\Box\phi = \left(\nabla^2 - \frac{1}{c^2}\frac{\partial^2}{\partial t^2}\right)\phi = 0 \tag{41}$$

with invariant solutions of the form

$$\phi = \phi_0\, e^{ik_\mu x_\mu}. \tag{42}$$

8.7. MAXWELL'S EQUATIONS IN COVARIANT FORM

We have previously noted that although Newton's equations are covariant under Galilean transformations Maxwell's equations are not. However, Maxwell's equations are covariant under Lorentz transformations. In the present section we shall transcribe Maxwell's equations into 4-tensor form, using the Minkowski metric, so that their invariance becomes manifest.

Maxwell's equations in standard form (c.g.s. units) are:

$$\begin{aligned}
&\vec{\nabla}\cdot\vec{E} = 4\pi\rho, && \text{I}\\
&\vec{\nabla}\times\vec{B} - \frac{1}{c}\frac{\partial\vec{E}}{\partial t} = \frac{4\pi}{c}\vec{J}, && \text{II}\\
&\vec{\nabla}\cdot\vec{B} = 0, && \text{III}\\
&\vec{\nabla}\times\vec{E} + \frac{1}{c}\frac{\partial\vec{B}}{\partial t} = 0. && \text{IV}
\end{aligned} \tag{43}$$

The first is an expression of Coulomb's law and the existence of electric charge monopoles. The third equation denies the existence of magnetic monopoles. Equation II expresses Ampere's law as generalized by Maxwell, and equation IV is essentially Faraday's law of induction.

In 3-dimensional tensor form Maxwell's equations are:

$$\begin{aligned}
&\frac{\partial E_k}{\partial x_k} = 4\pi\rho, && \text{I}\\
&\varepsilon_{ijk}\frac{\partial B_k}{\partial x_j} - \frac{1}{c}\frac{\partial E_i}{\partial t} = \frac{4\pi}{c}J_i, && \text{II}\\
&\frac{\partial B_k}{\partial x_k} = 0, && \text{III}
\end{aligned}$$

and

$$\varepsilon_{ijk} \frac{\partial E_k}{\partial x_j} + \frac{1}{c} \frac{\partial B_i}{\partial t} = 0. \qquad \text{IV}$$

To cast these equations in covariant form (with respect to restricted Lorentz transformations) we introduce a 4-vector current density

$$(44) \qquad J_\mu = (\vec{J}, ic\rho)$$

and a second-rank electromagnetic field strength 4-tensor

$$(45) \qquad F_{\mu\nu} = \begin{bmatrix} 0 & B_3 & -B_2 & -iE_1 \\ -B_3 & 0 & B_1 & -iE_2 \\ B_2 & -B_1 & 0 & -iE_3 \\ iE_1 & iE_2 & iE_3 & 0 \end{bmatrix}.$$

We shall not give here the proof (originally due to Lorentz and Poincaré and predating Einstein's special relativity) that, in fact, these quantities have the appropriate tensor properties.† Suffice it to say that experiment bears out this hypothesis and that the equations

$$(46) \qquad \bar{F}_{\mu\nu} = \Lambda_{\mu\tau}\Lambda_{\nu\lambda}F_{\tau\lambda}$$

give the correct prescription for the transformation properties of $\vec{E}$ and $\vec{B}$.

In terms of the 4-vector current density J_μ and the electromagnetic field strength 4-tensor $F_{\mu\nu}$, the eight Maxwell equations can be cast into two sets of four equations each:

$$(47) \qquad \frac{\partial F_{\mu\nu}}{\partial x_\nu} = \frac{4\pi}{c} J_\mu,$$

$$(48) \qquad \frac{\partial F_{\mu\nu}}{\partial x_\lambda} + \frac{\partial F_{\lambda\mu}}{\partial x_\nu} + \frac{\partial F_{\nu\lambda}}{\partial x_\mu} = 0$$

†See for example, J. Aharoni, The Special Theory of Relativity (Clarendon Press, Oxford, 1959).

which are manifestly covariant (with respect to restricted Lorentz transformations).

Since div$\vec{B}$ = 0 we may introduce a vector potential $\vec{A}$ in the usual fashion, such that

(49) $\vec{B} = \text{curl } \vec{A}.$

Substituting into Faraday's law gives

(50) $\text{curl}[\vec{E} + (1/c)\partial\vec{A}/\partial t] = 0.$

Hence we may introduce a scalar potential Φ, such that

(51) $\vec{E} = -\nabla\Phi - \frac{1}{c}\frac{\partial\vec{A}}{\partial t}.$

Using these definitions and the relations (in Cartesian coordinates)

(52a) $(\vec{\nabla}\text{x})(\vec{\nabla}\text{x}) = \text{curl curl} = \vec{\nabla}(\vec{\nabla}\cdot\ ) - \nabla^2$

and

(52b) $(\vec{\nabla}\text{x})\vec{\nabla} = \text{curl grad} \equiv 0$

we can put Maxwell's equations into the form

(53) $\nabla^2\Phi + \frac{1}{c}\frac{\partial}{\partial t}(\vec{\nabla}\cdot\vec{A}) = -4\pi\rho$

and

(54) $\nabla^2\vec{A} - \frac{1}{c^2}\frac{\partial^2\vec{A}}{\partial t^2} - \vec{\nabla}\left(\vec{\nabla}\cdot\vec{A} + \frac{1}{c}\frac{\partial\Phi}{\partial t}\right) = -\frac{4\pi}{c}\vec{J}.$

The potentials $\vec{A}$ and Φ are not determined uniquely by equations (53) and (54) for if we substitute $\vec{A}'$ for $\vec{A}$ where

(55) $\vec{A}' = \vec{A} + \vec{\nabla}\Omega$

and Ω is any scalar function (called a gauge function) then equation (49) is unchanged (i.e. $\vec{B} = \text{curl } \vec{A}'$). In order that equation (51) also remains unchanged (i.e. $\vec{E} = -\vec{\nabla}\Phi' - (\frac{1}{c}\frac{\partial \vec{A}'}{\partial t})$) we must substitute Φ' for Φ where

$$(56) \qquad \Phi' = \Phi - \frac{1}{c}\frac{\partial \Omega}{\partial t} .$$

Equations (55) and (56) imply that we may restrict our attention to those potentials which satisfy the Lorentz gauge condition

$$(57) \qquad \vec{\nabla}\cdot\vec{A} + \frac{1}{c}\frac{\partial \Phi}{\partial t} = 0,$$

for if $\vec{A}$ and Φ do not satisfy this condition we simply introduce a gauge function Ω satisfying

$$(58) \qquad \nabla^2\Omega - \frac{1}{c^2}\frac{\partial^2\Omega}{\partial t^2} = -\left(\vec{\nabla}\cdot\vec{A} + \frac{1}{c}\frac{\partial \Phi}{\partial t}\right)$$

whereupon

$$(59) \qquad \vec{\nabla}\cdot\vec{A}' + \frac{1}{c}\frac{\partial \Phi'}{\partial t} = \vec{\nabla}\cdot\vec{A} + \frac{1}{c}\frac{\partial \Phi}{\partial t} + \nabla^2\Omega - \frac{1}{c^2}\frac{\partial^2\Omega}{\partial t^2} = 0.$$

The choice (57) has the advantage of casting Maxwell's equations (53) and (54) in a particularly simple and manifestly covariant form. The simplicity is already apparent from (53) and (54):

$$(53a) \qquad \nabla^2\Phi - \frac{1}{c^2}\frac{\partial^2\Phi}{\partial t^2} = -4\pi\rho$$

and

$$(54a) \qquad \nabla^2\vec{A} - \frac{1}{c^2}\frac{\partial^2\vec{A}}{\partial t^2} = -\frac{4\pi}{c}\vec{J}.$$

These four equations are wave equations with inhomogeneous source terms. Their solution gives the potentials $\vec{A}$ and Φ (and through them the fields $\vec{E}$ and $\vec{B}$) in terms of prescribed charge and current density distributions. Manifest covariance follows from the introduction of a 4-vector potential

(60) $A_\mu = (\vec{A}, i\Phi)$.

The fields $\vec{E}$ and $\vec{B}$ are defined in terms of the potentials $\vec{A}$, Φ via the antisymmetric second-rank 4-tensor

(61) $$F_{\mu\nu} = \frac{\partial A_\nu}{\partial x_\mu} - \frac{\partial A_\mu}{\partial x_\nu} .$$

The Lorentz condition becomes the 4-divergence of the 4-potential

(62) $$\frac{\partial A_\mu}{\partial x_\mu} = 0,$$

and the four wave equations (53a) and (54a) relating the potentials to their charge-current source distributions are given compactly by the covariant relations

(63) $$\frac{\partial^2 A_\nu}{\partial x_\lambda \partial x_\lambda} = \Box\ A_\nu = \frac{4\pi}{c} J_\nu .$$

8.8. TWO RELATIVISTIC INVARIANTS

Because of the orthogonality relations given by equation (38) the 256 quantities $i\varepsilon_{\mu\nu\kappa\lambda}$, with $\mu,\nu,\kappa,\lambda = 1,2,3,4$, defined by

$$\varepsilon_{\mu\nu\kappa\lambda} = \begin{cases} 1 \text{ if } (\mu,\nu,\kappa,\lambda) \text{ is an even permutation of } (1,2,3,4), \\ -1 \text{ if } (\mu,\nu,\kappa,\lambda) \text{ is an odd permutation of } (1,2,3,4), \\ 0 \text{ if } (\mu,\nu,\kappa,\lambda) \text{ is not a permutation of } (1,2,3,4), \end{cases}$$

transform as the components of a constant fourth-rank 4-tensor. It can easily be checked that the antisymmetric second-rank 4-tensor $G_{\mu\nu}$ with components

(64) $$G_{\mu\nu} = \begin{bmatrix} 0 & -E_3 & E_2 & -iB_1 \\ E_3 & 0 & -E_1 & -iB_2 \\ -E_2 & E_1 & 0 & -iB_3 \\ iB_1 & iB_2 & iB_3 & 0 \end{bmatrix}$$

is related to the electromagnetic field strength 4-tensor by the equation

(65) $$G_{\mu\nu} = (1/2i)\varepsilon_{\mu\nu\kappa\lambda}F_{\kappa\lambda}.$$

In terms of the 4-tensor $G_{\mu\nu}$ equations (48) become

(66) $$\frac{\partial G_{\mu\nu}}{\partial x_\nu} = 0.$$

Contracting the tensor (or outer) product of the tensors $G_{\kappa\lambda}$ and $F_{\mu\nu}$ with respect to the indices κ and μ as well as λ and ν we get the 4-scalar invariant quantity

(67) $$G_{\kappa\mu}F_{\kappa\mu} = -4(\vec{E}\cdot\vec{B}).$$

Hence the scalar product (in 3-dimensional Euclidean space) of the vectors $\vec{E}$ and $\vec{B}$ is a relativistic invariant.

Forming the tensor product of the electromagnetic field strength 4-tensor with itself we get the fourth-rank 4-tensor $F_{\kappa\lambda}F_{\mu\nu}$. Contracting this fourth-rank 4-tensor with respect to κ and μ as well as λ and ν yields the scalar

(68) $$F_{\kappa\nu}F_{\kappa\nu} = 2(|\vec{B}|^2 - |\vec{E}|^2).$$

Hence the difference between the density of magnetic and electric field energies $(|\vec{B}|^2 - |\vec{E}|^2)/8\pi$ is also a relativistic invariant.

It can be shown that $\vec{E}\cdot\vec{B}$ and $|\vec{E}|^2 - |\vec{B}|^2$ are the only two independent relativistic invariants which can be formed from the electromagnetic field strength 4-tensor. In particular the electromagnetic field energy density $U = (|\vec{E}|^2 + |\vec{B}|^2)/8\pi$ is not a relativistic invariant.

8.9. COVARIANCE OF THE LORENTZ FORCE EQUATION - CONSERVATION LAWS

The Lorentz force density on the charge-current density distribution due to the action of an impressed field $\vec{E},\vec{B}$ is given by

(69) $$\vec{f} = \rho\vec{E} + \frac{1}{c}(\vec{J} \times \vec{B})$$

or alternatively, in 4-vector notation,

(70) $$f_\mu = \frac{1}{c} F_{\mu\nu} J_\nu$$

where

(71) $$f_4 = \frac{if_0}{c} = \frac{i}{c}(\vec{E}\cdot\vec{J}),$$

implying that f_0 is the rate at which the electromagnetic field does work on the charge-current distribution (per unit volume). Substituting for J_ν from Maxwell's equations we obtain

(72) $$f_\mu = \frac{1}{4\pi} F_{\mu\nu} \frac{\partial F_{\nu\lambda}}{\partial x_\lambda}.$$

The electromagnetic energy-momentum stress 4-tensor, denoted $S_{\mu\nu}$, is defined by

(73) $$S_{\mu\nu} = -\frac{1}{4\pi}(F_{\mu\lambda}F_{\lambda\nu} + \frac{1}{4}\delta_{\mu\nu}F_{\lambda\tau}F_{\lambda\tau}).$$

Taking the 4-gradient, we have

$$\frac{\partial S_{\mu\nu}}{\partial x_\nu} = -\frac{1}{4\pi}\left[F_{\mu\lambda}\left(\frac{\partial F_{\lambda\nu}}{\partial x_\nu}\right)+\left(\frac{\partial F_{\mu\lambda}}{\partial x_\nu}\right)F_{\lambda\nu} + \frac{1}{2}\delta_{\mu\nu}F_{\lambda\tau}\left(\frac{\partial F_{\lambda\tau}}{\partial x_\nu}\right)\right].$$

If we rename some indices this relation can be written

$$\frac{\partial S_{\mu\nu}}{\partial x_\nu} = -\frac{1}{4\pi}\left[F_{\mu\lambda}\left(\frac{\partial F_{\lambda\nu}}{\partial x_\nu}\right)+ \frac{1}{2}F_{\lambda\tau}\left(\frac{\partial F_{\lambda\tau}}{\partial x_\mu}\right)+\left(\frac{\partial F_{\mu\lambda}}{\partial x_\tau}\right)F_{\lambda\tau}\right]$$

which can be rewritten, using the antisymmetry of $F_{\mu\nu}$, as

$$\frac{\partial S_{\mu\nu}}{\partial x_\nu} = -\frac{1}{4\pi}\left(F_{\mu\lambda}\frac{\partial F_{\lambda\nu}}{\partial x_\nu}\right) - \frac{1}{8\pi}F_{\lambda\tau}\left(\frac{\partial F_{\lambda\tau}}{\partial x_\mu} + \frac{\partial F_{\mu\lambda}}{\partial x_\tau} + \frac{\partial F_{\tau\mu}}{\partial x_\lambda}\right).$$

Using Maxwell's equations (48) this expression finally reduces to

(74) $$\frac{\partial S_{\mu\nu}}{\partial x_\nu} = -\frac{1}{4\pi}\left(F_{\mu\lambda}\frac{\partial F_{\lambda\nu}}{\partial x_\nu}\right) = -f_\mu.$$

The components of $S_{\mu\nu}$ may be written as

$$(75) \qquad S_{\mu\nu} = \begin{bmatrix} S_{11} & S_{12} & S_{13} & icg_1 \\ S_{21} & S_{22} & S_{23} & icg_1 \\ S_{31} & S_{32} & S_{33} & icg_3 \\ icg_1 & icg_2 & icg_3 & -U \end{bmatrix}$$

where the vector in 3-dimensional Euclidean space with components g_j, $j = 1,2,3$, denoted $\vec{g}$, is given by the electromagnetic momentum density and is defined by

$$(76) \qquad \vec{g} = \frac{1}{4\pi c} (\vec{E} \times \vec{B}) = \frac{1}{c^2} \vec{S},$$

and U is the energy density

$$(77) \qquad U = \frac{1}{8\pi} (|\vec{E}|^2 + |\vec{B}|^2).$$

The 3 x 3 electromagnetic stress tensor S_{ij} has components

$$(78) \qquad S_{ij} = -\frac{1}{4\pi}\Big(E_iE_j + B_iB_j - \frac{1}{2}\delta_{ij}(|\vec{E}|^2 + |\vec{B}|^2)\Big), \qquad i,j = 1,2,3.$$

In terms of these quantities (74) can be rewritten as

$$(79) \qquad f_0 = \frac{c}{i} f_4 = (\vec{E}\cdot\vec{J}) = -\frac{c}{i}\frac{\partial S_{kj}}{\partial x_j} - \frac{c}{i}\frac{\partial S_{44}}{\partial x_4} = -\vec{\nabla}\cdot\vec{S} - \frac{\partial U}{\partial t},$$

$$(80) \qquad f_k = -\frac{\partial S_{k\nu}}{\partial x_\nu} = -\frac{\partial S_{kj}}{\partial x_j} - \frac{\partial S_{k4}}{\partial x_4} = -\frac{\partial S_{kj}}{\partial x_j} - \frac{\partial g_k}{\partial t}.$$

Identifying the volume integral of f_0 with the rate of change of the total mechanical energy T we have upon integration of equation (79)

$$(81) \qquad \frac{\partial}{\partial t}(T + U) = -\int_V \vec{\nabla}\cdot\vec{S}\, d^3x = -\int_S \vec{S}\cdot d\vec{a}$$

where $\mathcal{U}$ is the total electromagnetic energy in the volume V bounded by the surface S. The vector $\vec{S}$ representing energy flow per unit area is called Poynting's vector. Equation (81) expresses the law of conservation of energy.

Identifying the volume integral of f_k as the kth component of the rate of change of the total mechanical momentum $\vec{P}$ and the volume integral of g_k as the kth component of the total electromagnetic momentum $\vec{G}$, equation (80) becomes upon integration

$$(82) \qquad \frac{\partial}{\partial t}(P_k + G_k) = -\int_V \frac{\partial S_{kj}}{\partial x_j}\, d^3x = -\int_S n_j S_{kj}\, da$$

where n_1, n_2, n_3 are the components of a unit vector normal to the surface S (bounding the volume V) with area element da. Evidently the momentum flow per unit area, which equals the force transmitted, has components $n_j S_{ij}$. Equation (82) expresses the law of conservation of momentum.

8.10. RELATIVISTIC MECHANICS

Maxwell's equations, which predated Einstein's development of special relativity, are nonetheless relativistically covariant as we have seen in Section 9. However, Newtonian mechanics had to be altered to bring them into conformity with the propositions of special relativity. This is an interesting bit of history, but is in fact not surprising, since particles usually travel with velocities much less than that of light, whereas electromagnetic quanta (photons) are always moving at the speed of light and hence the equations that describe them must necessarily be covariant.

In the present section we develop a relativistic formulation of mechanics using heuristic arguments rather than rigorous proofs. We first introduce an invariant for a moving particle called the proper time. Instantaneously, in any inertial reference frame, a particle has a well-defined velocity $\vec{v}$ of magnitude less than c, the velocity of light. Even if the particle is accelerating, its velocity is momentarily well defined and one can find some Lorentz frame which is co-moving with the particle at that instant. In such a frame (barred frame) the particle is instantaneously at rest. Hence, the invariant 4-distance corresponding to an interval of time dt can be written

$$dx_\nu dx_\nu = d\bar{x}_\nu \, d\bar{x}_\nu = -c^2(d\bar{t})^2 = -c^2(d\tau)^2$$

$$= (dx)^2 + (dy)^2 + (dz)^2 - c^2dt^2$$

$$= -c^2(dt)^2(1 - \beta^2)$$

where $\beta = v/c$. Hence $d\tau$, which is called the proper time, is given by

(83) $$d\tau = (1/\gamma)dt,$$

the famous Einstein time-dilation formula.

Since dx_ν is a 4-vector and $d\tau$ is a scalar, it follows that the 4-velocity u_ν defined by

(84) $$u_\nu = dx_\nu/d\tau$$

is a 4-vector. If we take the rest mass of the particle M_0 also to be a scalar, we have a 4-vector momentum p_ν defined by

(85) $$p_\nu = M_0 u_\nu$$

and we can write a generalization of Newton's laws by introducing a 4-vector force K_ν defined by

(86) $$K_\nu = d(M_0 u_\nu)/d\tau.$$

It is important to realize that the components u_1, u_2, u_3 are not the usual three velocity components, since they involve the usual space intervals in the inertial frame, but a time interval as measured by an observer moving instantaneously with the particle. Indeed we can write u_ν as

$$(87)\quad u_k = \frac{dx_k}{d\tau} = \gamma\frac{dx_k}{dt} = \gamma v_k, \quad k = 1,2,3,$$

$$(88)\quad u_4 = ic\,\frac{dt}{d\tau} = ic\gamma.$$

As $|\vec{v}| \to 0$, $p_\nu \to (\vec{p}, iM_0c)$ where $\vec{p}$ is the ordinary momentum used in Newtonian mechanics. Clearly in the same limit $K_j \to F_j$, $j = 1,2,3$, where F_j is the jth component of Newton's force $\vec{F}$, and the fourth component has the following significance:

$$(89)\quad K_4 = \frac{d}{d\tau}(M_0u_4) = iM_0c\,\frac{d\gamma}{d\tau} = iM_0\,\gamma\,\frac{d}{dt}\left(1 - \frac{v_kv_k}{c^2}\right)^{-\frac{1}{2}}$$

$$= iM_0c\gamma\left(1 - \frac{v_kv_k}{2}\right)^{-\frac{3}{2}}\frac{v_j}{c^2}\,\frac{dv_j}{dt} = \frac{iM_0}{c}\,\gamma^4\,\frac{dv_k}{dt}\,v_k$$

$$\to \frac{i}{c}\,\vec{F}\cdot\vec{v} \text{ as } |\vec{v}| \to 0.$$

We see that the fourth component of a 4-force K_ν is proportional in the low-velocity limit to the 'rate of working' of the force in 3-dimensional Euclidean space with components K_j. Finally, if we interpret generally that

$$(90)\quad K_4 = \frac{i}{c}\,(\text{rate of working by external forces})$$

then

$$(91)\quad K_4 d\tau = \frac{i}{c}\,(\text{increase in energy in the time } d\tau)$$

$$= d(M_0u_4) = dp_4 = icM_0d(\gamma).$$

Einstein interpreted the total energy of the particle E as given by

$$(92)\quad E = -icp_4 = M_0c^2\gamma.$$

Since, as $\gamma \to 1$, $E \to E_0 = M_0c^2$ one has the possible interpretation of a rest energy M_0c^2 and a kinetic energy

(93) $T = M_0c^2\gamma - M_0c^2 = M_0c^2(\gamma - 1)$.

One can also introduce a dynamic mass $M = \gamma M_0$ so that

(94) $E = Mc^2$ and $T = (M - M_0)c^2$.

We also observe that the scalar product $p_\nu p_\nu$ may now be written

(95) $p_\nu p_\nu = (p_1^2 + p_2^2 + p_3^2) - E^2/c^2$.

In a coordinate system in which the particle is instantaneously at rest, clearly we have $\bar{p}_\nu \bar{p}_\nu = M_0c^2$ so that generally one can write

(96) $E^2 = c^2p^2 + M_0^2c^4$

where $p = (p_1^2 + p_2^2 + p_3^2)^{\frac{1}{2}}$.

8.11. APPLICATIONS OF RELATIVISTIC MECHANICS

We now consider briefly three applications of relativistic mechanics.

(a) *Doppler effect*

For a light quantum one has momentum proportional to energy through the factor c,

$$E = cp.$$

Consider two frames O and O' which coincide at time $t = 0$ but with O' moving to the right (z-axis) with velocity v (see Figure 8.1).

Let a source of light on O' emit photons of energy $E' = \hbar\omega'$ at an angle θ' with respect to the z'-axis. The photons are observed by O to come from a source of frequency ω and at angle θ with respect to the z-axis. The question is: how do the two observers compare results according to special relativity?

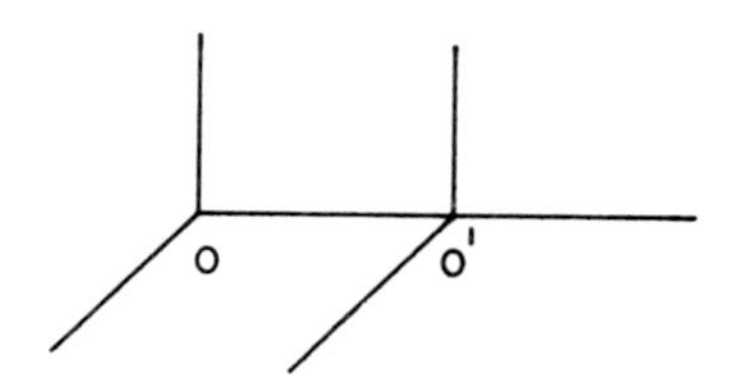

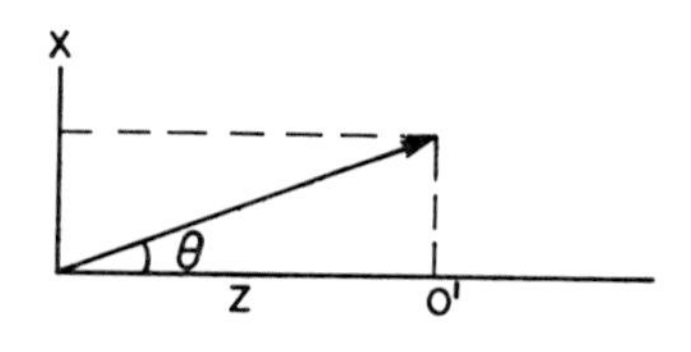

Figure 8.1. Illustration of frames O and O'.

From the 4-vector nature of energy and momentum, we have

$$p_1 = p_1', \qquad p_2 = p_2',$$

$$p_3 = \gamma(p_3' - i\beta p_4), \qquad p_4 = \gamma(p_4' + i\beta p_3').$$

But $$p_1' = \frac{\hbar\omega'}{c}\sin\theta',$$

$$p_2' = 0,$$

$$p_3' = \frac{\hbar\omega'}{c}\cos\theta',$$

$$p_4' = \frac{i\hbar\omega'}{c}.$$

Hence $$p_1 = \frac{\hbar\omega}{c}\sin\theta = \frac{\hbar\omega'}{c}\sin\theta',$$

$$p_2 = 0 = p_2',$$

$$p_3 = \frac{\hbar\omega}{c}\cos\theta = \frac{\gamma\hbar\omega'}{c}(\cos\theta' + \beta),$$

$$p_4 = \frac{i\hbar\omega}{c} = \frac{\gamma i\hbar\omega'}{c}(1 + \beta\cos\theta').$$

From the last equation

$$\omega = \gamma\omega'(1 + \beta\cos\theta')$$

so that the second last equation becomes

$$\cos\theta = \frac{\cos\theta' + \beta}{1 + \beta\cos\theta'};$$

therefore

$$\omega = \gamma\omega'(1 + \beta\cos\theta') = (\omega'/\gamma)(1 - \beta\cos\theta).$$

Hence the O observer sees both a shifted frequency and an angle abberation. Note that for $\theta' = 0$, i.e. forward emission,

$$\cos\theta = \frac{1 + \beta}{1 + \beta} = 1$$

so that

$$\theta = 0$$

and O also detects forward emission.

For backward emission, $\theta' = 180°$, $\cos\theta' = -1$, and we get

$$\cos\theta = \frac{-1 + \beta}{1 - \beta}$$

so that

$$\theta = 180°$$

and O also detects backward emission. At all other angles, the two observers disagree.

For forward emission,

$$\omega = \gamma(1 + \beta)\omega' = \frac{\omega'}{\gamma(1 - \beta)},$$

which is just the non-relativistic Doppler effect modified by the factor γ.

For backward emission, again

$$\omega = \gamma\omega'(1 - \beta) = \frac{\omega'}{\gamma(1 + \beta)},$$

which is again the usual Doppler effect modified by the factor γ.

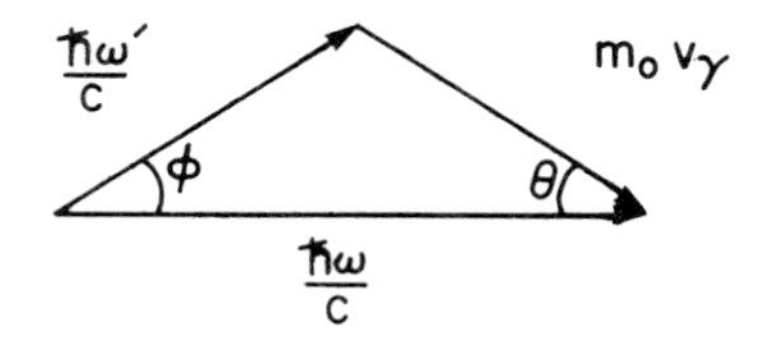

Figure 8.2. Momentum conservation.

(b) *Compton effect*

The Compton effect provides yet another example of relativistic dynamics. Consider the scattering of a gamma ray photon (Figure 8.2) of energy $\hbar\omega$ by a 'free' electron (valence electrons bound in atoms are almost 'free' at the energies where the Compton effect is large). Energy and momentum conservation yield the equations

$$\frac{\hbar\omega'}{c}\sin\phi = m_o v\gamma \sin\theta,$$

$$\frac{\hbar\omega'}{c}\cos\phi + m_o v\gamma\cos\theta = \frac{\hbar\omega}{c},$$

$$\hbar\omega = \hbar\omega' + (\gamma - 1)m_o c^2.$$

In these equations $\hbar\omega'$ is the energy of the scattered photon, v is the velocity of the ejected electron, and θ, ϕ are the angles of the ejected electron and scattered photon, respectively. Eliminating θ from the first two equations, we get

$$m_o^2 v^2\gamma^2 = \left(\frac{\hbar(\lambda\!\!\!^{-} - \lambda\!\!\!^{-}{}')}{\lambda\!\!\!^{-}{}'\lambda\!\!\!^{-}}\right)^2 + \frac{2\hbar^2}{\lambda\!\!\!^{-}\lambda\!\!\!^{-}{}'}(1 - \cos\phi),$$

where $\lambda\!\!\!^{-} = \lambda/2\pi$. Using the third equation to eliminate $\lambda\!\!\!^{-}\lambda\!\!\!^{-}{}'$ then yields the famous relation

$$\Delta\lambda = \lambda' - \lambda = \frac{\hbar}{m_o c}(1 - \cos\phi)$$

which shows that the shift in wavelength is independent of the energy of the incident gamma ray and depends only on the scattering angle. The quantity $\hbar/m_o c$ is called the Compton wavelength and is a measure of the distance at which quantum effects become important.

(c) *Threshold of π_0 production*

Consider a proton beam scattering on a hydrogen gas target at high energies

$$p + p \to p + p + \pi_0.$$

Let v be the incident velocity and V_0 the corresponding centre-of-mass velocity.† The corresponding quantities γ and γ_0 are the usual relativistic factors

$$\gamma = \frac{1}{\sqrt{1 - v^2/c^2}}, \qquad \gamma_0 = \frac{1}{\sqrt{1 - V_0^2/c^2}}.$$

At threshold the final state consists of a cluster of three particles moving with the centre of mass but with no relative motion. Hence the energy and momentum relationships are

$$M_0c^2(1 + \gamma) = (2M_0 + m_0)c^2\gamma_0,$$

$$M_0v\gamma = (2M_0 + m_0)V_0\gamma_0,$$

where m_0 is the meson mass and M_0 the nucleon mass. Eliminating $2M_0 + m_0$ by taking ratios, we get

$$v = \frac{\gamma + 1}{\gamma} V_0$$

which, combined with definitions of γ and γ_0, yields

$$\gamma_0^{-2} = 2/(1 + \gamma).$$

Substituting back into the above energy relationship in the lab before and after collision, we get after squaring

---

†For a relativistic system, the concept of centre of mass loses in part its classical meaning. One can still define, however, a Lorentz frame in which the total relativistic 3-momentum is zero. The relative velocity associated with this frame we call V_0.

$$\gamma = \frac{1}{2}\left(2 + \frac{m_o}{M_o}\right)^2 - 1.$$

Hence the threshold kinetic energy is

$$T = M_o c^2(\gamma - 1) = \frac{1}{2} m_o c^2\left(4 + \frac{m_o}{M_o}\right).$$

Putting in $m_o = 264$ electron masses and $M_o = 1836$ electron masses, we obtain

$$T = 273 \text{ MeV},$$

which is approximately twice the meson rest energy.

8.12. A DOUBLE-VALUED REPRESENTATION OF THE RESTRICTED LORENTZ GROUP

In Chapter 3 we studied a double-valued representation of the rotation group in terms of 2 x 2 unitary matrices with unit determinant. Such matrices are specified by three independent parameters which can be chosen as the three Euler angles characterizing the rotation. A 3-vector was represented by a traceless Hermitian matrix and the transformation itself was bilinear. The associated unilinear operation corresponded to transformation on a two-component <u>spinor</u>. In the present section and the following, we show how 4-vectors behave as second-rank spinors under restricted Lorentz transformations. Two observations may help to motivate our development. The first observation is that a 2 x 2 Hermitian matrix without the traceless condition requires four parameters for its specification, and can be used to represent a 4-vector as a second-rank spinor. The second observation is that a 2 x 2 matrix with complex entries and determinant 1 requires six parameters for its specification. As we shall see shortly these six parameters can be used to represent the three Euler angles and the three components of the separation velocity for observers O and $\bar{O}$.

All that is required for our development is to show that 2 x 2 Hermitian matrices X with complex entries preserve their hermiticity under bilinear transformations of the type

$$(97) \qquad \bar{X} = QXQ^{+}$$

where Q is a 2 x 2 matrix with unit determinant. That is to say, if X has the form

$$(98) \qquad X = \begin{bmatrix} c + d & a - ib \\ a + ib & d - c \end{bmatrix}$$

then $\bar{X}$ also has this form:

$$(99) \qquad \bar{X} = \begin{bmatrix} \bar{c} + \bar{d} & \bar{a} - i\bar{b} \\ \bar{a} + i\bar{b} & \bar{d} - \bar{c} \end{bmatrix}$$

where a,b,c,d and $\bar{a},\bar{b},\bar{c},\bar{d}$ are real parameters. Since Q has det 1, det $\bar{X}$ = det X. If we choose $d = x^0$, $c = x^3$, $a = x^1$, and $b = x^2$, this determinant invariance becomes

$$(100) \qquad (x^0)^2 - (x^1)^2 - (x^2)^2 - (x^3)^2 = (\bar{x}^0)^2 - (\bar{x}^1)^2 - (\bar{x}^2)^2 - (\bar{x}^3)^2,$$

which is just the invariance associated with restricted Lorentz transformations. Thus equation (97) is the bilinear equivalent of the linear transformations

$$\bar{x}^{\mu} = L^{\mu}_{\nu}\, x^{\nu}$$

which leave the quadratic form (100) invariant.

From our previous work in Chapter 3 we can also see that successive restricted Lorentz transformations L(12)L(23) = L(13) have a bilinear analogue in Q(13) = ±(Q(12)Q(23)) where the ± sign is admissible because Q(13) occurs twice in the bilinear transformation, hence the double-valued nature of the representation. It is easy to show that the choice of matrix

$$(101) \qquad Q = \begin{bmatrix} e^{i(\psi+\phi)/2}\cos\theta/2 & ie^{i(\psi+\phi)/2}\sin\theta/2 \\ ie^{-i(\psi+\phi)/2}\sin\theta/2 & e^{-i(\psi+\phi)/2}\cos\theta/2 \end{bmatrix}$$

corresponds to a spatial rotation through angles ϕ,θ,ψ, while leaving $x^0 = ct$ fixed. Similarly it is easy to show that the choice

$$(102)\quad Q = \begin{bmatrix} e^{-\alpha/2} & 0 \\ 0 & e^{\alpha/2} \end{bmatrix}$$

implies a boost in the x^3 direction. That is, if $\bar{X} = QXQ^+$, then

$$\bar{x}^1 = x^1, \quad \bar{x}^2 = x^2,$$

$$\bar{x}^3 = \cosh\alpha \cdot x^3 - \sinh\alpha \cdot x^0,$$

$$\bar{x}^0 = \cosh\alpha \cdot x^0 - \sinh\alpha \cdot x^3.$$

If boosting in an arbitrary direction occurs, one can compose the appropriate Q-matrix by rotating first until the boosting can be carried out along the x^3-axis and then rotating back again to the original orientation.

Given a restricted Lorentz transformation L we can find two 2 x 2 matrices differing in sign which give an equivalent transformation. These Q-matrices are denoted $\pm D^{\frac{1}{2}0}(L)$. The set of all matrices $\pm D^{\frac{1}{2}0}(L)$ forms a double-valued representation of the restricted Lorentz group which we denote by $\mathcal{D}^{\frac{1}{2}0}$. It can be easily seen that their complex conjugates also form a double-valued representation, which we denote by $\mathcal{D}^{0\frac{1}{2}}$.

Not every 2 x 2 matrix Q of unit determinant, however, is an element of $\mathcal{D}^{\frac{1}{2}0}$. For example, the matrix

$$(103)\quad Q = \begin{bmatrix} 0 & 1 \\ -1 & 0 \end{bmatrix}$$

does not correspond to a restricted Lorentz transformation since

$$(104)\quad \begin{bmatrix} 0 & 1 \\ -1 & 0 \end{bmatrix} \begin{bmatrix} x^0 + x^3 & x^1 - ix^2 \\ x^1 + ix^2 & x^0 - x^3 \end{bmatrix} \begin{bmatrix} 0 & 1 \\ -1 & 0 \end{bmatrix} = \begin{bmatrix} \bar{x}^0 + \bar{x}^3 & \bar{x}^1 - i\bar{x}^2 \\ \bar{x}^1 + i\bar{x}^2 & \bar{x}^0 - \bar{x}^3 \end{bmatrix}$$

implies that $\bar{x}^1 = x^1$, $\bar{x}^2 = -x^2$, $\bar{x}^3 = x^3$, $\bar{x}^0 = -x^0$.

The representation $\mathcal{D}^{\frac{1}{2}0}$ cannot be extended to a representation of the full Lorentz group in the obvious manner since, for example, there does not exist a 2 x 2 matrix of determinant 1 which corresponds to the spatial reflection P (see problem 6). However, we can form a double-valued representation of the full Lorentz group by associating with each restricted Lorentz transformation L two 4 x 4 matrices $\pm D(L)$, where

$$(105)\quad D(L) = \begin{bmatrix} D^{\frac{1}{2}0}(L) & 0 \\ 0 & \varepsilon[D^{\frac{1}{2}0}(L)]^*\varepsilon^{-1} \end{bmatrix} = \begin{bmatrix} D^{\frac{1}{2}0}(L) & 0 \\ 0 & \varepsilon D^{0\frac{1}{2}}(L)\varepsilon^{-1} \end{bmatrix},$$

$$(106)\quad \varepsilon = \begin{bmatrix} 0 & 1 \\ -1 & 0 \end{bmatrix},$$

and the inversion matrices P, T, J are given by $\pm D(P)$, $\pm D(T)$, $\pm D(J)$, respectively, where

$$(107)\quad D(P) = \begin{bmatrix} 0 & iI \\ iI & 0 \end{bmatrix}, \quad D(T) = \begin{bmatrix} 0 & iI \\ -iI & 0 \end{bmatrix}, \quad D(J) = \begin{bmatrix} I & 0 \\ 0 & -I \end{bmatrix}.$$

Here I is the 2 x 2 identity matrix.

8.13. SPINORS

An ordered pair ξ^1, ξ^2 which transform under rotations like

$$(108)\quad \begin{bmatrix} \bar{\xi}^1 \\ \bar{\xi}^2 \end{bmatrix} = \begin{bmatrix} e^{i(\psi+\phi)/2}\cos\theta/2 & ie^{i(\psi-\phi)/2}\sin\theta/2 \\ ie^{-i(\psi-\phi)/2}\sin\theta/2 & e^{-i(\psi+\phi)/2}\cos\theta/2 \end{bmatrix} \begin{bmatrix} \xi^1 \\ \xi^2 \end{bmatrix}$$

where (ϕ,θ,ψ) are the Euler angles of the rotation, and which transform under boosts in the $x^3 = z$ direction like

$$(109)\quad \begin{bmatrix} \bar{\xi}^1 \\ \bar{\xi}^2 \end{bmatrix} = \begin{bmatrix} e^{-\frac{1}{2}\alpha} & 0 \\ 0 & e^{\frac{1}{2}\alpha} \end{bmatrix} \begin{bmatrix} \xi^1 \\ \xi^2 \end{bmatrix}$$

where $v = c \tanh \alpha$ is the velocity associated with the boost, is called a contravariant first-rank spinor. (A contravariant first-rank spinor with components ξ^A, $A = 1,2$, is denoted here by the same symbol as its general component ξ^A.) In this chapter capital Roman letter indices are used to denote two-component quantities, and when two such indices are repeated summation is to occur over the values 1,2 of those indices.

The double-valued representation of the restricted Lorentz group $\mathcal{D}^{\frac{1}{2}0}$ acts on the space of spinors ξ^A. The double-valuedness of the representation is due to the fact that a rotation through 2π radians, as far as first-ranked spinors are concerned, is not equivalent to a rotation through zero radians (see Chapter 3 for more details). We write

$$\bar{\xi}^B = \alpha^B{}_A \, \xi^A \tag{110}$$

where $\bar{\xi}^B$ is the result of transforming ξ^A by an arbitrary restricted Lorentz transformation. The components $\alpha^B{}_A$, $B,A = 1,2$, are determined by equations (108) and (109); however, in most cases all one needs to know is that the $\alpha^B{}_A$ are the components of a 2 x 2 matrix with determinant 1; that is,

$$\alpha^1{}_1\alpha^2{}_2 - \alpha^2{}_1\alpha^1{}_2 = 1. \tag{111}$$

If χ^A is a first-rank contravariant spinor with components χ^1, χ^2 then the complex conjugate of χ^A is denoted by $\chi^{\dot{A}}$ and has components $\chi^{\dot{1}} = (\chi^1)^*$ and $\chi^{\dot{2}} = (\chi^2)^*$. Under restricted Lorentz transformations the $\chi^{\dot{A}}$, $\dot{A} = \dot{1},\dot{2}$, transform like

$$\bar{\chi}^{\dot{B}} = (\alpha^B{}_A)^* \chi^{\dot{A}} = \alpha^{\dot{B}}{}_{\dot{A}} \, \chi^{\dot{A}}. \tag{112}$$

An ordered set of two components $\chi^{\dot{A}}$ which transform under restricted Lorentz transformations as shown in equation (112) is also called a first-rank contravariant spinor. The double-valued representation of the restricted Lorentz group $\mathcal{D}^{0\frac{1}{2}}$ acts on the 2-dimensional space of spinors $\chi^{\dot{A}}$.

Since the matrices $\alpha^B{}_A$ have determinant 1, if ξ^A and χ^B are two first-rank contravariant spinors then

$$\xi^1\chi^2 - \xi^2\chi^1 = \varepsilon_{AB}\xi^A\chi^B = \xi^A\chi_A,$$

where

$$\begin{bmatrix} \varepsilon_{11} & \varepsilon_{12} \\ \varepsilon_{21} & \varepsilon_{22} \end{bmatrix} = \begin{bmatrix} 0 & 1 \\ -1 & 0 \end{bmatrix} \quad \text{and} \quad \chi_A = \varepsilon_{AB}\chi^B,$$

is invariant under restricted Lorentz transformations. We call $\varepsilon_{AB}\xi^A\chi^B$ the inner or scalar product of ξ^A with χ^B. The quantities χ_B, B = 1,2, are the components of a first-rank covariant spinor and transform under restricted Lorentz transformations like

$$\bar{\chi}_A = \varepsilon_{AB}\bar{\chi}^B = \varepsilon_{AB}\alpha^B{}_R\chi^R = (\varepsilon_{AB}\alpha^B{}_R\varepsilon^{RD})\chi_D = \alpha_A{}^D\chi_D,$$

where

$$\begin{bmatrix} \varepsilon^{11} & \varepsilon^{12} \\ \varepsilon^{21} & \varepsilon^{22} \end{bmatrix} = \begin{bmatrix} 0 & 1 \\ -1 & 0 \end{bmatrix} \quad \text{so that } \varepsilon_{AB}\varepsilon^{RB} = \delta^R_A.$$

It follows from the definition of the scalar product that

$$\xi_A\chi^A = -\xi^A\chi_A \quad \text{where } \xi_A = \varepsilon_{AB}\xi^B$$

and hence $\xi_A\xi^A = 0$. Of course, similar statements hold for spinors $\xi^{\dot{A}}$ and $\chi^{\dot{B}}$. That is, the expression

$$\xi^{\dot{1}}\chi^{\dot{2}} - \xi^{\dot{2}}\chi^{\dot{1}} = \varepsilon_{\dot{A}\dot{B}}\xi^{\dot{A}}\chi^{\dot{B}} = \xi^{\dot{A}}\chi_{\dot{A}},$$

where

$$\begin{bmatrix} \varepsilon_{\dot{1}\dot{1}} & \varepsilon_{\dot{1}\dot{2}} \\ \varepsilon_{\dot{2}\dot{1}} & \varepsilon_{\dot{2}\dot{2}} \end{bmatrix} = \begin{bmatrix} \varepsilon^{\dot{1}\dot{1}} & \varepsilon^{\dot{1}\dot{2}} \\ \varepsilon^{\dot{2}\dot{1}} & \varepsilon^{\dot{2}\dot{2}} \end{bmatrix} = \begin{bmatrix} 0 & 1 \\ -1 & 0 \end{bmatrix}$$

and $\chi_{\dot{A}} = \varepsilon_{\dot{A}\dot{B}}\, \chi^{\dot{B}}$, is invariant under restricted Lorentz transformations and is called the inner product or scalar product of $\xi^{\dot{A}}$ and $\chi^{\dot{B}}$. The quantities $\chi_{\dot{B}}$, $\dot{B} = \dot{1},\dot{2}$, are also the components of a first-rank covariant spinor and transform under restricted Lorentz transformations like

$$\bar{\chi}_{\dot{A}} = \varepsilon_{\dot{A}\dot{B}}\bar{\chi}^{\dot{B}} = \varepsilon_{\dot{A}\dot{B}}\alpha^{\dot{B}}{}_{\dot{R}}\chi^{\dot{R}} = (\varepsilon_{\dot{A}\dot{B}}\alpha^{\dot{B}}{}_{\dot{R}}\varepsilon^{\dot{R}\dot{D}})\chi_{\dot{D}} = \alpha_{\dot{A}}{}^{\dot{D}}\chi_{\dot{D}}\,.$$

It is evident from the transformation properties of spinors ϕ^{A} and $\chi_{\dot{C}}$ that the double-valued representation of the full Lorentz group mentioned previously acts on the space of bispinors $(\phi^{A},\chi_{\dot{C}})$ provided the spinors $\phi^{A},\chi_{\dot{C}}$ have the correct reflection properties.

From the four types of first-rank spinors we can form quantities which transform as higher-ranked spinors by taking outer products. For example, the outer product of $\xi^{\dot{A}}$ with χ_{B} is the mixed second-rank spinor $\psi^{\dot{A}}_{B} = \xi^{\dot{A}}\chi_{B}$, which transforms under restricted Lorentz transformations like

$$\bar{\psi}^{\dot{A}}_{B} = \alpha^{\dot{A}}{}_{\dot{C}}\;\alpha_{B}{}^{D}\psi^{\dot{C}}_{D}\,.$$

Other types of second- and higher-ranked tensors have analogous transformations properties.

We have already encountered one second-rank spinor $x^{A\dot{C}}$ whose components are given by

$$\begin{bmatrix} x^{1\dot{1}} & x^{1\dot{2}} \\ x^{2\dot{1}} & x^{2\dot{2}} \end{bmatrix} = \begin{bmatrix} x^{0} + x^{3} & x^{1} - ix^{2} \\ x^{1} + ix^{2} & x^{0} - x^{3} \end{bmatrix}. \tag{113}$$

That is,

$$x^{A\dot{C}} = x^{\mu}\sigma_{\mu}^{A\dot{C}} \tag{114}$$

where

$$(115)\quad \begin{bmatrix} \sigma_\mu^{1\dot{1}} & \sigma_\mu^{1\dot{2}} \\ \sigma_\mu^{2\dot{1}} & \sigma_\mu^{2\dot{2}} \end{bmatrix} = \begin{cases} \begin{bmatrix} 1 & 0 \\ 0 & 1 \end{bmatrix}, & \mu = 0, \\ \begin{bmatrix} 0 & 1 \\ 1 & 0 \end{bmatrix}, & \mu = 1, \\ \begin{bmatrix} 0 & -i \\ i & 0 \end{bmatrix}, & \mu = 2, \\ \begin{bmatrix} 1 & 0 \\ 0 & -1 \end{bmatrix}, & \mu = 3. \end{cases}$$

These four matrices are just the unit matrix and the Pauli spin matrices. Of course, we can also form the second-rank spinor

$$(116)\quad x_{A\dot{C}} = \varepsilon_{AB}\varepsilon_{\dot{C}\dot{D}}\, x^{B\dot{D}}$$

which is related to the covariant components of an event in space-time by

$$(117)\quad x_{A\dot{C}} = x_\mu \sigma^\mu_{A\dot{C}}$$

where

$$(118)\quad \sigma^\mu_{A\dot{C}} = \eta^{\mu\nu}\sigma_\nu^{B\dot{V}}\varepsilon_{BA}\varepsilon_{\dot{V}\dot{C}}$$

or more explicitly

$$(119)\quad \begin{bmatrix} \sigma^\mu_{1\dot{1}} & \sigma^\mu_{1\dot{2}} \\ \sigma^\mu_{2\dot{1}} & \sigma^\mu_{2\dot{2}} \end{bmatrix} = \begin{cases} -\begin{bmatrix} 1 & 0 \\ 0 & 1 \end{bmatrix}, & \text{for } \mu = 0, \\ -\begin{bmatrix} 0 & 1 \\ 1 & 0 \end{bmatrix}, & \text{for } \mu = 1, \\ \begin{bmatrix} 0 & -i \\ i & 0 \end{bmatrix}, & \text{for } \mu = 2, \\ -\begin{bmatrix} 1 & 0 \\ 0 & -1 \end{bmatrix}, & \text{for } \mu = 3. \end{cases}$$

The orthogonality and normalization relations

(120a) $\sigma_{\mu}^{A\dot{C}}\ \sigma^{\mu}_{B\dot{D}} = -2\delta^{A}_{B}\ \delta^{\dot{C}}_{\dot{D}}$,

(120b) $\sigma_{\mu}^{A\dot{V}}\ \sigma^{\nu}_{A\dot{V}} = -2\delta^{\nu}_{\mu}$

permit us to invert equations (114) and (117) to obtain

(121) $x^{\nu} = -\frac{1}{2}\ \sigma^{\nu}_{A\dot{C}}\ x^{A\dot{C}}, \quad \nu = 0,1,2,3,$

(122) $x_{\nu} = -\frac{1}{2}\ \sigma_{\nu}^{A\dot{C}} x_{A\dot{C}}$.

The components of an event form a vector and equations (114), (117) and (121), (122) express the relationship between this vector and a second-rank spinor. In general with a tensor $T^{\nu_1 \ldots \nu_p}_{\mu_1 \ldots \mu_q}$ of rank (p + q) we can associate a spinor $T^{A_1\dot{B}_1 \ldots A_p\dot{B}_p}_{C_1\dot{D}_1 \ldots C_q\dot{D}_q}$ of rank 2(p + q) via the equations

(123) $T^{A_1\dot{B}_1 \ldots A_p\dot{B}_p}_{C_1\dot{D}_1 \ldots C_q\dot{D}_q} = \sigma^{\nu_1}_{C_1\dot{D}_1} \ldots \sigma^{\nu_q}_{C_q\dot{D}_q}\ \sigma^{A_1\dot{B}_1}_{\mu_1} \ldots \sigma^{A_p\dot{B}_p}_{\mu_p}\ T^{\mu_1 \ldots \mu_p}_{\nu_1 \ldots \nu_q}$

and conversely

(124) $T^{\mu_1 \ldots \mu_p}_{\nu_1 \ldots \nu_q} = -\frac{1}{2}^{\,p+q}\ \sigma^{C_1\dot{D}_1}_{\nu_1} \ldots \sigma^{C_q\dot{D}_q}_{\nu_q}\ \sigma^{\mu_1}_{A_1\dot{B}_1} \ldots \sigma^{\mu_p}_{A_p\dot{B}_p}\ T^{A_1\dot{B}_1 \ldots A_p\dot{B}_p}_{C_1\dot{D}_1 \ldots C_q\dot{D}_q}$.

As well as giving the relationship between 4-tensors and spinors equations (123) and (124) also may be used to relate 4-tensor and spinor operators. For example, the operator ∂_{μ} with components

(125) $\partial_{\mu} = \dfrac{\partial}{\partial x^{\mu}}, \quad \mu = 0,1,2,3,$

transforms under restricted Lorentz transformations like a covariant vector

$$(126)\quad \bar{\partial}_\mu = \partial/\partial x^\mu = L^\nu_\mu\ \partial/\partial x^\nu = L^\nu_\mu \partial_\nu$$

and hence is called a 4-vector operator. With this operator we associate the spinor operator $\partial_{A\dot{B}}$ defined by

$$(127)\quad \partial_{A\dot{B}} = \sigma^\mu_{A\dot{B}}\ \partial_\mu ,$$

which yields

$$(128)\quad \begin{bmatrix} \partial_{1\dot{1}} & \partial_{1\dot{2}} \\ \partial_{2\dot{1}} & \partial_{2\dot{2}} \end{bmatrix} = -\begin{bmatrix} \partial_0 + \partial_3 & \partial_1 - i\partial_2 \\ \partial_1 + i\partial_2 & \partial_0 - \partial_3 \end{bmatrix} .$$

From the covariant spinor operator $\partial_{D\dot{B}}$ we can form the contravariant spinor operator $\partial^{A\dot{C}} = \varepsilon^{AD}\varepsilon^{\dot{C}\dot{B}}\partial_{D\dot{B}}$

$$\begin{bmatrix} \partial^{1\dot{1}} & \partial^{1\dot{2}} \\ \partial^{2\dot{1}} & \partial^{2\dot{2}} \end{bmatrix} = -\begin{bmatrix} \partial_0 - \partial_3 & -\partial_1 + i\partial_2 \\ -\partial_1 - i\partial_2 & \partial_0 + \partial_3 \end{bmatrix} .$$

Spinor analysis, more or less in the form used here, was first developed by van der Waerden† but is essentially already contained in a book by Weyl.††

†B.L. van der Waerden, Gottinger Nachrichten (1929), p. 100.

††H. Weyl, Gruppentheorie und Quantenmechaniks (Hirzel, Leipzig, 1928). For a clear and concise account of the general principles of spinor analysis, including the spinor formulation of the Dirac and Maxwell equations, see O. Laporte and G.E. Uhlenbeck, Phys. Rev. 37:1380 (1931).

REFERENCES

1 J. Aharoni, The Special Theory of Relativity. Oxford University Press, London, 1959
2 H. Arzeliès, Relativistic Kinematics. Pergamon Press, London, 1966
3 D. Bohm, The Special Theory of Relativity. W.A. Benjamin Inc., New York, 1965
4 J. Jackson, Classical Electrodynamics. John Wiley and Sons, Inc., New York, 1962
5 D. Lawden, Tensor Calculus and Relativity. Methuen and Co. Ltd., London, 1967
6 H.A. Lorentz, A. Einstein, H. Minkowski, and H. Weyl, The Principle of Relativity. Dover Publications, New York, 1952
7 C. Misner, K. Thorne, and J. Wheeler, Gravitation. W.H. Freeman and Co., San Francisco, 1976
8 E.R. Paërl, Representations of the Lorentz Group and Projective Geometry. Mathematical Centre, Amsterdam, 1969
9 R.K. Pathria, The Theory of Relativity. Pergamon Press, London, 1974
10 H.M. Schwartz, Introduction to Special Relativity. McGraw-Hill, New York, 1968
11 E. Taylor and J. Wheeler, Spacetime Physics. W.H. Freeman and Company, San Francisco, 1966

PROBLEMS

1 Prove that the Poincaré transformations are the unique transformations

$$\bar{x}^\mu = \bar{x}^\mu(x^\nu)$$

which satisfy equation (5):

$$\eta_{\mu\nu} d\bar{x}^\mu d\bar{x}^\nu = \eta_{\mu\nu} dx^\mu dx^\nu.$$

2 (i) Show that any restricted Lorentz transformation leaving time invariant (i.e. $\bar{t} = t$) has the form

$$L = \begin{bmatrix} 1 & 0 & 0 & 0 \\ 0 & a_{11} & a_{12} & a_{13} \\ 0 & a_{21} & a_{22} & a_{23} \\ 0 & a_{31} & a_{32} & a_{33} \end{bmatrix}$$

where the a_{jk}, j,k = 1,2,3, are the components of a 3-dimensional rotation matrix.

(ii) Prove that every restricted Lorentz transformation may be written as the product of a boost with a rotation.

3 (i) If k_μ is a 4-vector: $k_\mu = (\vec{k}, i\omega_{\vec{k}})$, $\omega_{\vec{k}} > 0$, with magnitude

$$k_\mu k_\mu = \vec{k}^2 - \omega_{\vec{k}}^2 = -m^2$$

and $x_\mu = (\vec{x}, ix_o)$, $y_\nu = (\vec{y}, iy_o)$ are two space-time points, show that

$$\Delta(x_\nu - y_\nu) = \frac{1}{(2\pi)^3} \int_{R^3} \frac{d^3\vec{k}}{\omega_{\vec{k}}} e^{i\vec{k}\cdot(\vec{x}-\vec{y})} \sin\omega_{\vec{k}}(x_o - y_o)$$

is a scalar function

$$\Delta(x_\nu - y_\nu) = \bar{\Delta}(\bar{x}_\nu - \bar{y}_\nu)$$

with $\bar{\Delta} = \Delta$ (i.e. an invariant function).

(ii) Show that $\Delta(x_\nu = y_\nu) = 0$ for all $(x_\nu - y_\nu)^2 \geq 0$.

4 Using the knowledge that $A_\mu = (\vec{A}, i\phi)$ transforms as a 4-vector and the fields of a stationary electron are

$$\vec{A} = 0, \quad \phi = e/r,$$

find the fields $\vec{A}'$, ϕ' for a relativistic observer moving with velocity $\vec{v} = v\hat{z}$. Hence find the fields $\vec{E}$ and $\vec{B}$ associated with a charge moving at constant velocity.

5 A π° meson decays in flight into two gamma rays. The meson has a kinetic energy of 2000 MeV and a rest energy of 135 MeV.

(i) If one gamma ray has an energy of 1400 MeV find the energy of the second gamma ray and the angle at which each gamma ray is emitted.

(ii) Find the energies and angles of each gamma ray in the centre-of-mass frame of reference.

6 Prove that there does not exist a 2 x 2 matrix Q, of determinant 1, such that the transformation $\bar{X} = QXQ^+$,

$$X = \begin{bmatrix} x^0 + x^3 & x^1 - ix^2 \\ x^1 + ix^2 & x^3 - x^0 \end{bmatrix},$$

corresponds to a spatial reflection.

9
Classification of elementary particles

9.1. INTRODUCTION

One of the most extensive and interesting applications of the theory of groups and group representations to the description of physical phenomena is the classification problem for the eigenstates of quantum systems. The classification problem, which is basic and common to all quantum systems whether atomic, molecular, or nuclear, has two important aspects. The first is the purely bookkeeping aspect of counting and labelling states in order to distinguish among them and to gauge their statistical significance; the second is the structural problem of relating the underlying dynamics of the system to the nature of the states which are observed to occur. A familiar example is the spectroscopy of the hydrogen atom, idealized to neglect nuclear recoil and spin effects, in which the bound states of the system are uniquely classified in terms of a principal quantum number n ($n = 1,2,3,\ldots$), an angular momentum quantum number ℓ ($\ell = 0,1,2,\ldots,n-1$), and a magnetic quantum number m ($m = -\ell,-\ell+1,\ldots,\ell$). The quantum number n determines the energy of the bound state

$$E_n = \frac{-m_o e^4}{2n^2\hbar^2}, \qquad m_o = \text{electron mass},\ e = \text{electron charge},$$

and ℓ determines the magnitude $\sqrt{\ell(\ell + 1)}\hbar$ of the angular momentum; for given ℓ there are $2\ell + 1$ degenerate states which can be chosen so as to have an angular momentum projected on the z-axis (axis of quantization) of magnitude $m\hbar$. Since an orbiting charge develops a magnetic moment, the m-degeneracy in the energies of these states can be removed by applying a magnetic field (Zeeman effect).

Even for such a simple system, the question of relating the detailed spectroscopy to the symmetry group of the Hamiltonian is a non-trivial problem. When nuclear recoil, spin, and relativistic effects are included, this problem becomes exceedingly difficult, but particularly interesting and significant.[†]

Closely related to the questions of classification of levels and structural properties of quantum systems is the question of finding a classification scheme for elementary particles and fields. It has become traditional in physics to seek understanding of physical systems in terms of an underlying substructure. This 'reductionist' approach has been enormously successful and one has come to refer to the substructural units as 'elementary particles.' As the number of elementary particles[††] increased dramatically in recent years, a search was begun to find classification schemes which would bring some kind of order out of chaos. The present success of the SU(3)-quark model attests to the value of this kind of phenomenology; much in the way that the Rydberg-Ritz formula provided a useful guide at the end of the last century to physicists seeking to understand the dynamical structure of the hydrogen atom, one hopes that the quark model will provide guidance for discovering the detailed dynamical substructure of the universe of elemental systems.

There are risks involved in this kind of thinking, and it is entirely possible that an absolute faith in reductionist philosophy is not warranted. These questions will be examined in more detail in the final chapter of this book. In the present chapter we accept the central dogma of reductionism and examine

---

[†]See, for example, B.G. Wybourne, Classical Groups for Physicists (Wiley, New York, 1974), Chapter 21.

[††]Elementary particles are more properly referred to as 'resonances' which occur in scattering and reaction processes; most of the so-called elementary particles are unstable against decay, but manifest themselves as sharp enhancements in scattering and reaction cross-sections as the energy is varied. According to Heisenberg's uncertainty principle $\Delta E \Delta t \gtrsim \hbar$, a narrow peak of width ΔE corresponds to a long lifetime Δt for the composite system.

in some detail the promising aspects of group-theoretical classification schemes for systematizing elementary particles.

We have, of course, no direct or 'human' experience of elementary particles; to us elementary particles exist only by inference, and it is indeed remarkable that we have such strong convictions when the inference is so indirect, involuted, and even tenuous. Again we set these basic, almost metaphysical, questions aside and accept the modern inference that elementary structures exist and are distinguished by such structural properties as rest mass, spin, magnetic moment, and some other more esoteric properties such as strangeness, isospin, and baryon number.

When we speak of elementary particles as though they were individual entities, an immediate complication arises by virtue of the deep connection between spin and statistics, first investigated by Pauli,[†] whereby integer spin particles obey Bose-Einstein statistics (the 'social' particles) and half-odd integer spin particles obey the Fermi-Dirac statistics (the 'anti-social' particles). As a result of these considerations an essential complication arises, for even when elementary particles of like kind are separated beyond the range of inter-particle forces, their motions are still not independent, but are statistically correlated by the overall requirement that the wave function must be symmetric (antisymmetric) in its arguments with respect to interchanges of boson (fermion) coordinates.[††] As a consequence, even when like particles are not interacting dynamically, their motions and general behaviour are nonetheless correlated through the statistical requirements. Historically, the upshot of this was the development of quantum field theory (so-called second quantization) in which it is explicitly recognized that the particles have lost their individualities to some degree, and that one must in the purist sense treat all like particles in the universe simultaneously.

[†] Pauli based his arguments on such general principles as Lorentz invariance, microscopic causality, and unitarity. See W. Pauli, Phys. Rev. 58: 716 (1940).

[††] Recall, for example, the discussion in Section 6.9.

On the one hand the consequences are difficult and dramatic in that, in principle, the appearance of a single additional particle (particle creation) or the disappearance of a single additional particle (particle annihilation) means that every other like particle in the universe must adjust its circumstances since the motions of all particles are statistically correlated. On the other hand, if the particles are 'distant' in space or time or both from one another the correlation effects are small and the particles behave more or less independently. These remarks may help to explain why it still makes sense to talk about an 'individual particle' and its structure even though, strictly speaking, one should be talking about fields which describe the whole universe of particles simultaneously.

Regarding the kinematics associated with the internal structure of elementary particles, little progress has been made aside from interesting conjectures and speculations of a primitive kind.† Our considerations are, therefore, directed mainly toward the phenomenological side of the classification problem. To the extent that a Schroedinger-like theory is adequate, the phenomenology of fitting particles into group-theoretical classification schemes reflects on the symmetry properties of the Hamiltonian and hence on the system dynamics, as has been remarked upon in Chapter 5.

In Section 6.6 we discussed briefly the question of inner structure from a non-relativistic point of view. In the non-relativistic theory, one can draw upon an analogy with the centre-of-mass theorem familiar in classical mechanics to separate the particle wave function into a product of two factors, one representing the motion of the centre-of-mass of the system and the other representing the wave function describing the internal dynamics. For relativistic motions of composite systems, the classical centre-of-mass loses its unique

†For example, it is not certain whether one should regard particles as having size and structure on a flat space-time manifold, or whether particles are singularities in a more complicated geometry, or whether either or neither view is valid.

significance and the whole question of separating out internal structure from gross motion of the composite system is fraught with difficulties. For example, in a theorem due to O'Raifeartaigh† it is shown that any attempt to combine internal and Poincaré symmetries non-trivially into a finite-rank Lie group must necessarily fail. Although some progress has been made by going outside the structure of Lie groups, most of the phenomenology of SU(3) and SU(6) has been developed within the viewpoint of global Lie group structures and hence non-relativistically.

In our brief presentation, we shall avoid this basic problem and limit our discussion to the group theoretical classifications of internal structure developed phenomenologically in the literature, except to the extent that the intrinsic spin itself can be regarded as internal structure and represented by the transformation laws of multicomponent, relativistic Schroedinger state vectors (wave functions).

In Section 2, we discuss briefly the historical problem of finding suitable relativistic wave equations for a Schroedinger theory. Particular attention is paid to the Dirac equation and its possible generalizations. Dirac's explicit demonstration of spinor geometry is treated briefly in Section 3; the 'spinor spanner' demonstrates, in effect, the double connectivity of the rotation group R(3). Section 4 deals briefly with the more general question of symmetries of the Hamiltonian and invariance groups, in a non-relativistic context. Isospin is given as an explicit example in Section 5. Extension to SU(3) symmetry is made in Section 6. The quark model and other recent developments are discussed briefly in Section 7.

9.2. RELATIVISTIC WAVE EQUATIONS

We pay particular attention in this section to the structure and form of the Dirac relativistic wave equation, since it represents a first and fundamental step towards understanding the 'internal'

---

†L. O'Raifeartaigh, Phys. Rev. 139: B1052 (1965).

structure of elementary particles. Basically the device used by Dirac was to introduce a 4-component wave function, so that the state of the electron (or any other spin 1/2 particle with non-zero rest mass) requires simultaneous specification of four coupled wave fields defined over the manifold of 4-dimensional space-time.

De Broglie had, in fact, explicitly used relativistic considerations in proposing the form of the free electron wave function

$$\Psi = Ae^{i\vec{k}\cdot\vec{r}-i\omega t} = Ae^{ik_\nu x_\nu} \tag{1}$$

where $\nu = 1,2,3,4$, $E = \hbar\omega$, $\vec{p} = \hbar\vec{k}$, and the Einstein dispersion equation $E^2 = c^2p^2 + m_o^2c^4$ is satisfied. It seemed obvious to Schroedinger that the propagation of de Broglie waves must be governed by the relativistic equation† (Minkowski metric)

$$\left(\frac{\partial^2}{\partial x_\nu \partial x_\nu} - \kappa^2\right)\Psi = 0 \tag{2a}$$

or, equivalently,

$$\left(\nabla^2 - \frac{1}{c^2}\frac{\partial^2}{\partial t^2}\right)\Psi = \left(\frac{m_o c}{\hbar}\right)^2 \Psi. \tag{2b}$$

Here $\kappa = m_o c/\hbar$ is the inverse Compton wavelength, a universal scalar. It is evident†† that the de Broglie waves form a complete set of relativistically invariant wave solutions to the manifestly covariant equation (2a) since $k_\mu x_\mu$ is a 4-scalar and $\partial^2/\partial x_\mu \partial x_\mu$ is a 4-scalar operator.

The problem with equation (2a) is that it is second order in the time, which means that specification of an initial state of the system would require a knowledge of both Ψ and $\partial\Psi/\partial t$ at t_o. On the contrary, Nature seems to opt for the much more

†Equation (2) is usually referred to as the Klein-Gordon equation in meson field theory.

††This follows from the fact that any finite, relativistically invariant wave packet can be expressed as a sum (integral) over suitably weighted de Broglie waves.

simple requirement that only Ψ itself need be specified. Dirac solved the problem by finding a first-order alternative to Schroedinger's relativistic equation, which is however consistent with it. Dirac searched for a Hamiltonian operator which would correspond to the linearized energy dispersion law

$$(3) \qquad E = \pm\sqrt{c^2p^2 + m_o^2c^4}$$

and still be consistent with the definitions

$$(4) \qquad E_{op} \equiv i\hbar\partial/\partial t \quad \text{and} \quad \vec{p}_{op} = -i\hbar\vec{\nabla}.$$

Dirac solved this problem by introducing[†] a 4-component wave function (Dirac spinor or 'bispinor'):

$$(5) \qquad \Psi = \begin{bmatrix} \Psi_1 \\ \Psi_2 \\ \Psi_3 \\ \Psi_4 \end{bmatrix}$$

and a 4 x 4 Hermitian matrix Hamiltonian operator

$$(6) \qquad H = c\vec{\alpha}\cdot\vec{p} + \beta m_o c^2$$

and subjecting this system to the requirements that the Schroedinger equation

$$(7) \qquad H\Psi = i\hbar\partial\Psi/\partial t$$

be satisfied. The $\vec{\alpha}$ and β matrices are then determined up to a unitary equivalence[††] by the requirement that the quadratic

[†]P.A.M. Dirac, Proc. Roy. Soc. A117: 610 (1928). It can easily be shown that at least 4 components are needed and that 4 is a sufficient number.

[††]Let S be any 4 x 4 unitary transformation matrix: $S\alpha_k S^{-1} = \alpha_k'$ $(k = 1,2,3)$ and $S\beta S^{-1} = \beta'$. Then α_k' and β' satisfy the same relations (equations (8)) as α_k and β. In terms of α_k' and β' the Dirac equation becomes $i\hbar\partial\psi'/\partial t = (c\vec{\alpha}'\cdot\vec{p} + \beta' m_o c^2)\psi'$, where $\psi' = S\psi$.

form of Dirac's equation be consistent with the Klein-Gordon equation, i.e.

$$\left(i\hbar\frac{\partial}{\partial t} + c\vec{\alpha}\cdot\vec{p} + \beta m_o c^2\right)\left(i\hbar\frac{\partial}{\partial t} - c\vec{\alpha}\cdot\vec{p} - \beta m_o c^2\right)\psi = 0;$$

thus

$$\left(-\hbar^2\frac{\partial^2}{\partial t^2} - c^2(\alpha_1^2 p_1^2 + \ldots) - \beta^2 m_o^2 c^4 - c^2(\alpha_1 p_1 \alpha_2 p_2 + \alpha_2 p_2 \alpha_1 p_1 + \ldots)\right)\psi = 0$$

and upon rearrangement,

$$\left[-\hbar^2\frac{\partial^2}{\partial t^2} + \hbar^2 c^2\left(\alpha_1^2\frac{\partial^2}{\partial x_1^2} + \ldots\right) - \beta^2 m_o^2 c^4 + \hbar^2 c^2 \quad \alpha_1\alpha_2\frac{\partial^2}{\partial x_1 \partial x_2} + \alpha_2\alpha_1\frac{\partial^2}{\partial x_2 \partial x_1} + \ldots\right]\psi = 0,$$

which reduces to the Klein-Gordon equation on each component of ψ provided that

(8a) $\alpha_1^2 = \alpha_2^2 = \alpha_3^2 = \beta^2 = 1,$

(8b) $\alpha_j\alpha_k + \alpha_k\alpha_j = 0, \quad \alpha_j\beta + \beta\alpha_j = 0 \quad (j \neq k;\ j,k = 1,2,3).$

The self-adjointness of the Hamiltonian yields the further requirement that the $\vec{\alpha}$ and β matrices themselves be self-adjoint (Hermitian):

(8c) $\alpha_j^+ = \alpha_j, \quad \beta^+ = \beta, \quad j = 1,2,3.$

A common choice for $\vec{\alpha}$ and β is

(9) $$\vec{\alpha} = \begin{bmatrix} 0 & \vec{\sigma} \\ \vec{\sigma} & 0 \end{bmatrix}, \quad \beta = \begin{bmatrix} I & 0 \\ O & -I \end{bmatrix}.$$

It is not sufficient, however, for the Dirac theory to be relativistically invariant in quadratic form; the theory must also be relativistically invariant as a Schroedinger equation (i.e. in the form given in equations (6) and (7)). We can cast the Dirac equation into a more compact form, and one which lends itself to interpretation as a covariant equation under Lorentz transformations, by multiplying (6) from the left by the matrix β, dividing out by $\hbar c$, and rearranging terms. The result is

$$\text{(10a)} \quad \left(\gamma_\mu \frac{\partial}{\partial x_\mu} + \kappa\right)\Psi(x_\mu) = 0$$

where

$$\text{(10b)} \quad \begin{aligned} \gamma_k &= -i\beta\alpha_k, & k &= 1,2,3, \\ \gamma_4 &= \beta, & \kappa &= m_o c/\hbar, \end{aligned}$$

and the anticommutation relations (8) can be re-expressed as

$$\text{(11)} \quad \gamma_\mu\gamma_\nu + \gamma_\nu\gamma_\mu = 2\delta_{\mu\nu}, \quad \mu,\nu = 1,2,3,4.$$

The reader can easily verify that the matrices α_x, α_y, α_z, and β in equation (9) are Hermitian, and hence so are the γ_μ matrices.

If we take the Hermitian conjugate (i.e. complex conjugate transposed) of (10a) and multiply through by γ_4 on the right, we obtain the adjoint equation

$$\text{(12)} \quad \frac{\partial}{\partial x_\mu} \bar{\Psi}\gamma_\mu - \kappa\bar{\Psi} = 0$$

where

$$\text{(13)} \quad \bar{\Psi} = \Psi^+\gamma_4 = \tilde{\Psi}^*\gamma_4 = (\psi_1^*, \psi_2^*, -\psi_3^*, -\psi_4^*).$$

Equation (12) is called the adjoint to equation (10a). Multiplying equation (10a) on the left by $\bar{\psi}$ gives

$$\bar{\psi}\gamma_\mu \frac{\partial}{\partial x_\mu} \psi + \kappa\bar{\psi}\psi = 0,$$

and multiplying equation (12) on the right by ψ yields

$$(\partial\bar{\psi}/\partial x_\mu)\gamma_\mu\psi - \kappa\bar{\psi}\psi = 0.$$

The addition of these two equations shows that the Dirac 4-component wave function satisfies a continuity equation

(14) $$\partial(\bar{\Psi}\gamma_\mu\Psi)/\partial x_\mu = 0.$$

Equation (14) is consistent with the interpretation of

$$j_\mu = ic\bar{\Psi}\gamma_\mu\Psi$$

as a 4-current. The probability density is then

(15) $$\rho = -ij_4/c = \Psi_1^*\Psi_1 + \Psi_2^*\Psi_2 + \Psi_3^*\Psi_3 + \Psi_4^*\Psi_4,$$

which is the linear addition of the separate densities relating to each component. By considering the action of the electromagnetic field on the electron, Dirac was led to interpret the four fields ψ_1, ψ_2, ψ_3, and ψ_4 as probability amplitudes for the electron spin-up, the electron spin-down, the positron spin-up, and the positron spin-down. Details of this treatment can be found in many places.†

By virtue of the γ_μ matrices, (10a) represents a set of four coupled equations for the electron and positron amplitudes. By substituting a plane wave solution

(16) $$\Psi = \Psi_o e^{i\vec{k}\cdot\vec{r}-i\omega t}, \quad E = \hbar\omega, \quad \text{and } \vec{p} = \hbar\vec{k}$$

in (10a), one finds that the two components Ψ_1 and Ψ_2 obey the usual energy-momentum dispersion law for positive energies

(17) $$E = \sqrt{c^2p^2 + m_o^2c^4},$$

whereas Ψ_3 and Ψ_4 obey the anomalous dispersion for negative energies

†For example, E. Corinaldesi and F. Strocchi, Relativistic Wave Mechanics (North-Holland, Amsterdam, 1963).

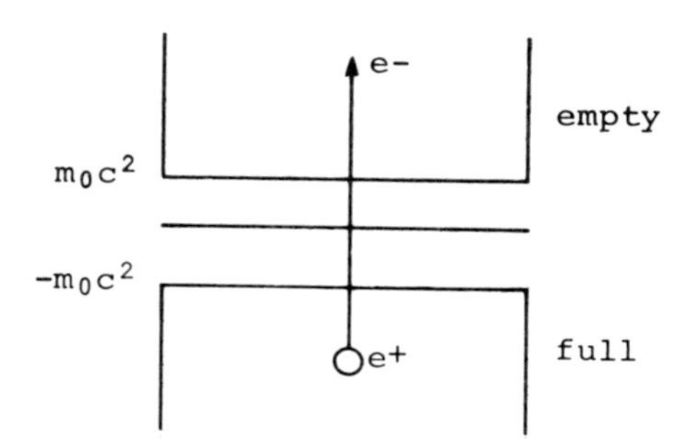

Figure 9.1. Free electron spectrum. The negative continuum of states is normally occupied (vacuum) while the positive energy states are empty. The diagram shows a transition across the $2m_oc^2$ gap corresponding to e^+e^- pair production.

$$E = -\sqrt{c^2p^2 + m_o^2c^4}. \tag{18}$$

Dirac suppressed the disastrous effects† of the presence of negative energy states by supposing that all the negative energy states are filled in the vacuum situation. If an energy in excess of $2m_oc^2$ is supplied to a negative energy electron, it can be promoted across the gap into the positive energy states as an observable electron. The hole left behind in the negative sea then behaves as a positively charged electron (positron) and one has pair production (see Figure 9.1). The reverse process is pair annihilation, which in free space occurs with the emission of two characteristic gamma rays, in order that energy and momentum be conserved.

From equation (37), Chapter 8, we have the space-time transformation law

$$x'_\mu = \Lambda_{\mu\nu}x_\nu \tag{19}$$

for a Lorentz transformation from unprimed to primed coordinates. Referring to equation (10a) this tells us how $\partial/\partial x_\mu$ transforms

†Disastrous because ordinary electrons could fall deeper and deeper into the negative energy states, radiating electromagnetic waves in the process.

and also how the argument of Ψ (namely x_μ) transforms. What is still not clear is how a Lorentz transformation mixes the components of Ψ. For covariance one must have (if we regard the γ_μ as constant matrices)

$$\left(\gamma_\mu \frac{\partial}{\partial x'_\mu} + \kappa\right)\Psi'(x'_\mu) = 0 \tag{20}$$

in the primed system (κ is a universal scalar). If the state vector $\Psi'(x'_\mu)$ is equal to $L\Psi(x_\mu)$, where L is a 4 x 4 transformation matrix, then the adjoint state vector transforms† like $\bar{\Psi}'(x'_\mu) = \bar{\Psi}(x_\mu)L^{-1}$. Equation (20), when transformed, reads

$$\gamma_\mu \Lambda_{\mu\nu} \frac{\partial}{\partial x_\nu} L\Psi + \kappa L\Psi = 0. \tag{20a}$$

Multiplying on the left by L^{-1} gives

$$(L^{-1}\gamma_\mu \Lambda_{\mu\nu} L) \frac{\partial}{\partial x_\nu} \Psi + \kappa\Psi = 0. \tag{20b}$$

Comparison with equation (10a) then yields

$$L^{-1}\gamma_\mu \Lambda_{\mu\nu} L = \gamma_\nu. \tag{21a}$$

From the orthogonality conditions satisfied by the Λ matrix, it follows upon multiplying by $\Lambda_{\lambda\nu}$ and summing on ν that

$$L^{-1}\gamma_\lambda L = \Lambda_{\lambda\nu}\gamma_\nu \tag{21b}$$

is the condition for covariance. These equations imply transformation properties for the 4-current vector:

$$j'_\mu(x'_\lambda) = ic\bar{\Psi}'(x'_\lambda)\gamma_\mu\Psi'(x'_\lambda) = ic\bar{\Psi}(x_\lambda)(L^{-1}\gamma_\mu L)\Psi(x_\lambda)$$

$$= \Lambda_{\mu\nu} ic\bar{\Psi}(x_\lambda)\gamma_\nu\Psi(x_\lambda) = \Lambda_{\mu\nu} j_\nu(x_\lambda).$$

---

†See S. Schweber, An Introduction to Relativistic Quantum Field Theory (Harper and Row, New York, 1962), pp. 74-7.

Note that the $\Lambda_{\mu\nu}$ are matrix elements whereas each γ_ν is a 4 x 4 matrix acting on the 4-component wave function Ψ. Thus $j_\mu(x_\lambda)$ does indeed transform as a 4-vector.

One can also display the Dirac equation in manifestly covariant spinor form as a pair of coupled spinor equations:[†]

(22a) $\partial_{A\dot{B}}\,\phi^A = i\kappa\chi_{\dot{B}}$,

(22b) $\partial^{A\dot{B}}\,\chi_{\dot{B}} = i\kappa\phi^A$.

Here $(\phi^A, \chi_{\dot{B}})$ is a bispinor each part of which transforms under Lorentz transformations as discussed in Section 8.13.

With the introduction of multicomponent wave functions in the Dirac theory it becomes apparent how to remedy the difficulty associated with the Klein-Gordon equation for spin zero particles (bosons), namely the fact that it is second order in the time:

$$\left(\nabla^2 - \frac{1}{c^2}\frac{\partial^2}{\partial t^2}\right)\phi = \left(\frac{m_o c}{\hbar}\right)^2 \phi.$$

Here ϕ is a one-component (scalar) function of the coordinates x_μ. Observing that ϕ and $\partial\phi/\partial t$ represent two independent degress of freedom we rewrite the Klein-Gordon equation as a set of two coupled first-order differential equations for $\psi_1 = \phi$ and $\psi_2 = \partial\phi/\partial t$, i.e.

(23a) $i\hbar\,\dfrac{\partial\psi_1}{\partial t} = i\hbar\psi_2$,

(23b) $i\hbar\,\dfrac{\partial\psi_2}{\partial t} = i\hbar c^2\nabla^2\psi_1 - i\left(\dfrac{m_o^2 c^4}{\hbar}\right)\psi_1$.

Equations (23) constitute the Klein-Gordon equation in Hamiltonian form and describe particles of zero spin. The two

---

[†]See, for example, P. Roman, Theory of Elementary Particles (North-Holland, Amsterdam, 1960). Note that equations (22) are in the same form as (7) if in the latter we choose $\vec{\alpha} = \begin{bmatrix} -\vec{\sigma} & 0 \\ 0 & \vec{\sigma} \end{bmatrix}$, $\beta = \begin{bmatrix} 0 & I \\ I & 0 \end{bmatrix}$.

components of the Klein-Gordon wave function can be shown† to be associated with the particle and antiparticle states for spin zero particles. Similar formulations can be made for vector (spin one) mesons.

The importance of the Dirac equation can hardly be overestimated. Dirac was able immediately to account for electron spin and the existence of anti-fermions as natural consequences of the requirements of quantum theory and special relativity. It now appears that the spin 1/2 particles are the basic structural units of matter, as we know it, and have their fundamental description in terms of the Dirac equation or its field theory equivalent. An exceptional case does arise, however, as pointed out by Weyl, which corresponds to spin 1/2 particles of zero rest mass. Using the Weyl equation (2-component spinors) for the neutrino

$$\partial_{A\dot{B}}\psi^{A} = 0, \tag{24}$$

Yang and Lee were able to account for the so-called τ-θ puzzle†† as a consequence of the non-conservation of parity in weak interactions.

We shall not pursue here the whole history of proposals for relativistic equations corresponding to higher spins and special circumstances.††† In practice the emphasis shifted to the theory of quantized fields in an effort to take into account the appropriate statistics for many-particle systems, and to provide a convenient formalism for the creation and annihilation of particles in elementary processes. Suffice it to say that field theories were plagued with unphysical infinities, and that only in the case of the 0, 1/2, and 1 spin fields has it been possible to use so-called renormalization techniques to systematically remove them. Thus it appears that these are

---

†H. Feshbach and F. Villars, Revs. Mod. Phys. 30: 24 (1958).
††T.D. Lee and C.N. Yang, Phys. Rev. 104: 254 (1956).
†††See, however, P. Roman, Theory of Elementary Particles, for many details.

the fundamental fields. The discovery of unitary symmetry by Gell-Mann and Ne'eman shifted the emphasis from dynamical field theories to phenomenological models purporting to describe internal structure. The problem of uniting these phenomenological models with renormalizable field theories is an area of intensive modern investigation.

9.3. SPINOR SPANNERS

The explicit appearance of spinors in the Dirac theory and the importance of this theory in providing insight into elementary structures again raise the question as to just what spinors are. In Chapter 3 they occurred as a natural extension of the tensor concept to vector spaces defined over the complex field; but this 'natural' extension is based on mathematical possibilities rather than on intuitive physical models. In an attempt to provide greater physical insight, Dirac constructed a model, referred to by Bolker[†] as the 'spinor spanner.' Consider any two rigid structures in 3-dimensional space connected together by any number of non-tangled strings (e.g., a globe inside a box with strings running radially from the surface of the globe to the inside walls of the box). If either structure is now held fixed and the second is rotated $2\pi n$ about any fixed axis, where n is an odd integer, the strings become hopelessly tangled; if n is an even integer or zero, however, the strings may appear to be tangled, but topologically they are not and the system may be restored to its initial situation without further rotation.

This remarkable demonstration shows that spinors 'exist,' since for such a system it takes a rotation through 4π for rotational equivalence to no rotation at all. One could naively picture the electron as a 3-dimensional structure with its Faraday lines of force radiating out to all oppositely charged particles in the rest of the universe. Its spinorial character would then

[†]E.D. Bolker, Proc. Math. Assoc. America, Nov. 1973, pp. 977-84. See also C.W. Misner, K.S. Thorne, and J.A. Wheeler, Gravitation (Freeman, San Francisco, 1973).

represent the fact that its lines of force (which cannot pass through each other) become untangled for every double rotation as it 'spins' about an intrinsic axis. No doubt this naive picture is inadequate, but it probably represents a classical limit to the actual behaviour of a spinning quantum electron.

9.4. INVARIANCE PROPERTIES OF THE HAMILTONIAN

Consider a non-relativistic system with a Hamiltonian H. Let O be a time-independent, self-adjoint operator representing some dynamical quantity. If O commutes with H, then

$$\frac{d}{dt}\langle O\rangle = \frac{d}{dt}\int \psi^*(x,t)O\psi(x,t)\,dx \tag{25}$$

$$= \int\left(\frac{d}{dt}\psi^*(x,t)\right)O\psi(x,t)\,dx + \int \psi^*(x,t)O\frac{d\psi(x,t)}{dt}\,dx.$$

Using the Schroedinger equation

$$\frac{d\psi}{dt} = -\frac{i}{\hbar}(H\psi) \tag{26}$$

this becomes

$$\frac{d}{dt}\langle O\rangle = \frac{i}{\hbar}\int (H\psi)^*(x,t)O\psi(x,t)\,dx - \int \psi^*(x,t)OH\psi(x,t)\,dx$$

$$= \frac{i}{\hbar}\int \psi^*(x,t)\,[H,O]\psi(x,t)\,dx$$

$$= 0,$$

where we have used the self-adjointness of H to get the second-last line and the vanishing of the commutator [O,H] to get the last line. Thus in quantum mechanics (even relativistic) if an operator commutes with the Hamiltonian, it represents a conserved quantity for the system in question. An example is given by the total angular momentum (squared) operator

$$L^2 = -\hbar^2\left[\frac{1}{\sin\theta}\frac{\partial}{\partial\theta}\left(\sin\theta\frac{\partial}{\partial\theta}\right) + \frac{1}{\sin^2\theta}\frac{\partial^2}{\partial\phi^2}\right] \tag{27}$$

which commutes with the Hamiltonian for the non-relativistic theory of the hydrogen atom

$$(28) \qquad H = -\frac{\hbar^2}{2m}\nabla^2 - \frac{e^2}{r} .$$

A second example is the relationship between conservation of momentum and translational invariance[†] of the Hamiltonian for two particles interacting under central forces:

$$(29) \qquad H = -\frac{\hbar^2}{2m_1}\nabla_1^2 - \frac{\hbar^2}{2m_2}\nabla_2^2 + V(|\vec{r}_1-\vec{r}_2|).$$

$$(30) \qquad \vec{P} = -i\hbar\vec{\nabla}_1 - i\hbar\vec{\nabla}_2,$$

$$(31) \qquad [H,\vec{P}] = 0.$$

Translational invariance of the Hamiltonian follows from the observation that with $[H,\vec{P}] = 0$, one also has $[H,T(\vec{a})] = 0$, where $T(\vec{a})$ is the translation operator

$$(32) \qquad T(\vec{a}) = e^{i/\hbar\ \vec{a}\cdot\vec{P}}$$

with the property[††]

$$(33) \qquad T(\vec{a})\psi(\vec{r}_1,\vec{r}_2,t) \equiv T(\vec{a})\psi(\vec{r},\vec{R},t) = \psi(\vec{r},\ \vec{R}+\vec{a},\ t),$$

as can easily be checked.

Both of the above examples serve to illustrate a connection between conserved quantities and invariances of the Hamiltonian under transformations of a group (symmetry group). In

---

[†]Let $\{\phi_n(x)\}$, $n = 0,1,2,\ldots$, be an orthonormal basis for the Hilbert space of square-integrable functions and A be a self-adjoin operator representing some dynamical quantity. Suppose that $A\phi_n(x) = \psi_n(x)$, $n = 0,1,2,\ldots$ Let T be an operator which effects a change of basis $T\phi_n(x) = \phi_n'(x)$, $n = 0,1,2,\ldots$, $\{\phi_n'(x)\}$ also being an orthonormal basis for the Hilbert space. The change of basis transforms the operator A into a new operator A' determined by the requirement that $T\psi_n(x) = A'(T\phi_n(x))$, which implies that $T(A\phi_n(x)) = A'(T\phi_n(x))$ so that $A\phi_n(x) = (T^{-1}A'T)\phi_n(x)$. Hence $A' = TAT^{-1}$. If T commutes with A, then $A' = A$ and we say that A is invariant under the transformation T.

[††]Here $\vec{r} = \vec{r}_1-\vec{r}_2$ is the relative coordinate and $\vec{R} = (m_1\vec{r}_1 + m_2\vec{r}_2)/(m_1 + m_2)$ is the centre-of-mass coordinate.

the first example, conservation of angular momentum is related to invariance of the Hamiltonian under the action of the rotation group operators $R(\vec{\alpha}) = \exp(i\vec{\alpha}\cdot\vec{L}/\hbar)$. In the second example, conservation of total linear momentum is related to invariance of the Hamiltonian under translations.[†]

In quantum field theory, this general connection between symmetry groups and invariance properties is called Noether's theorem after E. Noether who first fully recognized the connection in classical Lagrangian theory.[††]

In addition to invariance under continuous symmetry groups such as the rotation and translation groups mentioned above, one can also consider invariance of the Hamiltonian under the so-called discrete symmetry groups such as time reflection and coordinate inversion. Associated with invariance under coordinate inversion is the concept of conservation of parity. (It was precisely the failure of β-decay theory to obey inversion invariance that forced one to give up parity as a conserved quantity for weak interactions. Parity is conserved for strong and electromagnetic interactions.)

If P is the inversion operator[†††]

$$(34) \qquad P\Psi(\vec{r},t) = \Psi(-\vec{r},t)$$

and if P commutes with the Hamiltonian,

$$(35) \qquad [H,P] = 0,$$

then for any eigenfunction Ψ of H,

$$(36) \qquad H(P\Psi) = P(H\Psi) = E(P\Psi),$$

$P\Psi$ is also an eigenfunction and with the same eigenvalue E.

[†]Note that $[H,R(\vec{\alpha})] = 0$ for all $\vec{\alpha}$ implies that $[H,\vec{L}] = 0$, and $[H,T(\vec{a})] = 0$ for all $\vec{a}$ implies that $[H,\vec{P}] = 0$. This is easily proved by considering infinitesimal transformations.

[††]E. Noether, Nachr. d. Kgl. Ges. d. Wiss. Gottingen (1918), p. 235.

[†††]We assume here that the wave function transforms as a proper scalar. For complications which arise when the wave function is a Dirac spinor, see J.D. Bjorken and S.D. Drell, Relativistic Quantum Mechanics (McGraw-Hill, New York, 1964).

Since H and P commute, they can be simultaneously diagonalized Since $P^2 = I$, the parity operator has eigenvalues ±1 and energy eigenstates of the system can be labelled by the parity quantum number, in addition to the other labels constituting a maximal set.

Consider now the operator T of time-reversal applied to both sides of the Schroedinger equation

(37) $$THT^{-1}T\Psi = T\left(i\hbar \frac{\partial\Psi}{\partial t}\right) = -i\hbar \frac{\partial}{\partial t}(T\Psi),$$

which we rewrite in the form

(38) $$H'\Psi' = -i\hbar\partial\Psi'/\partial t$$

where the primes indicate the time-reversed state or operator, respectively. Now† $\Psi'(x,t) = \Psi(x,-t)$. Hence we can write (38) in the form

(39) $$H'\Psi(x,\tau) = i\hbar \frac{\partial}{\partial\tau}\Psi(x,\tau)$$

where $\tau = -t$. If $H' = THT^{-1} = H$, i.e., if H is invariant under time-reversal, then $\Psi(x,\tau) = \Psi(x,-t)$ obeys the same equation as $\Psi(x,t)$ with the same boundary (time-independent!) conditions, and hence the time-reversed state is also a solution if Ψ is. A system for which $H = H'$ is said to be time-reversal invariant.

An exactly similar statement can be made about complex conjugation, which in quantum mechanics is related to reversing the sign of the electric charge on the particle.††

---

†Again we note that in general it is possible to have $\Psi'_\alpha(x,t) = A_{\alpha\beta}\Psi_\beta(x,-t)$ where Ψ_α is an n-component wave function and $A_{\alpha\beta}$ an n x n matrix.

††This can be seen naively to follow from the well-known fact (see H. Goldstein, Classical Mechanics (Addison-Wesley, Cambridge, Mass., 1950)) that for a particle of charge e moving in an external electromagnetic potential field A_μ, if P_μ is the canonical momentum the actual mechanical momentum is $P_\mu - eA_\mu/c$. In quantum mechanics P_μ is given by $i\hbar\vec{\nabla}$ and hence $P_\mu \to -P_\mu$ under complex conjugation; it follows that for consistency e must go into -e.

Thus, if $H' = CHC^{-1} = H$, we speak of charge-conjugation invariance of the Hamiltonian,† with the implication that if ψ is a solution of the Schroedinger equation for a particle of charge q, so is ψ^* for a particle of charge -q.

One can consider combined invariances like CP or CPT. In a general theorem due originally to Pauli, it can be shown on very general grounds (relativistic invariance, unitarity, and causality) that the Hamiltonian of quantum systems is invariant under the combined operation CPT. This is the celebrated CPT theorem. For some quantum systems one has separate invariance under C, P, and T. Other systems, e.g. the weak interactions,†† violate parity conservation but are invariant under time-reversal or combined CP.

Feynman has pointed out that CPT invariance for the electron has the significance that a positron is essentially an electron moving backward in time. Thus the two diagrams in Figure 9.2 are essentially equivalent descriptions of a freely moving electron. Figure 9.3 then illustrates that pair annihilation corresponds to a normal electron being scattered backward in time.

Finally we remark that time-reversal is an anti-unitary operator in quantum mechanics. Anti-unitary operators are products of a complex-conjugation operator K

$$K^2 = I, \quad K\Psi = \Psi^*,$$

and a unitary operator U:

$$T = KU.$$

For example, if ϕ and χ are wave functions and α and β complex numbers,

---

†Strictly speaking charge conjugation also involves changes in other quantum numbers in addition to charge.
††In fact, the weak interactions also violate CP conservation to a small extent. See, e.g., T.D. Lee and C.S. Wu, Annual Rev. Nuc. Sci. 16:471 (1966), and P.K. Kabir, The CP Puzzle (Academic Press, New York, 1968).

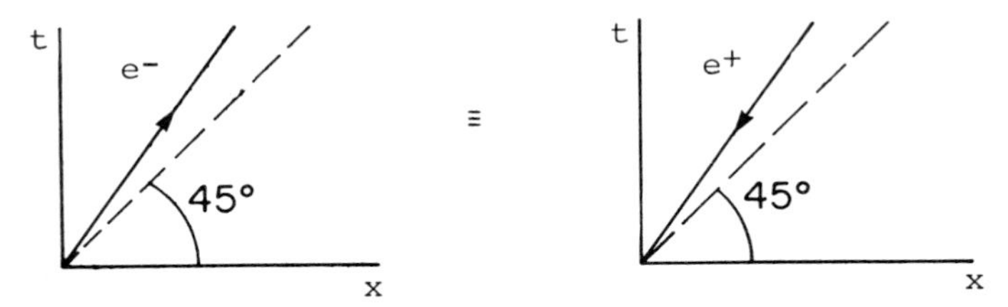

Figure 9.2. The world lines of an electron and a positron moving in reversed space and time directions correspond to physically identical states.

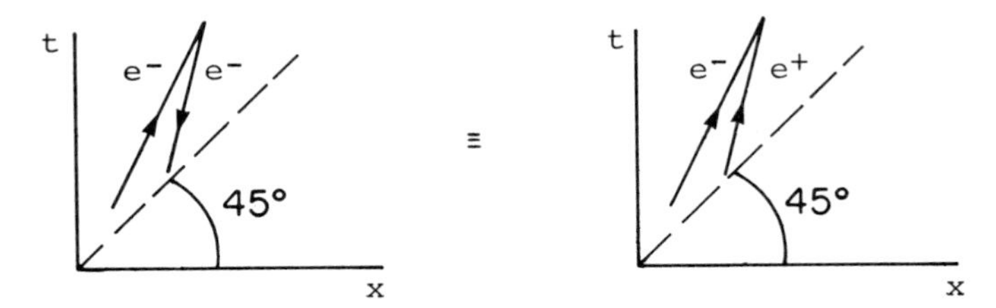

Figure 9.3. The CPT theorem has the consequence that pair annihilation is formally equivalent to a violent scattering in which the electron reverses its path and moves backward in time.

$$T(\alpha\phi + \beta\chi) = \alpha^{*}T\phi + \beta^{*}T\Psi.$$

Since U is unitary, we have

$$\langle\phi|\chi\rangle = \langle U\phi|U\chi\rangle,$$

but

$$\langle T\phi|T\chi\rangle = \langle U\phi|U\chi\rangle^{*} = \langle\phi|\chi\rangle^{*}.$$

However, quantities like transition probabilities involve only $|\langle\phi|\chi\rangle|^2$ and are not affected by time-reversal invariance.

9.5. ISOSPIN AS AN INTERNAL SYMMETRY

Dirac's theory shows how electron spin manifests itself as a part of the internal structure of the electron, and how this structure is representable by the introduction of a multi-component wave function. In the non-relativistic limit where the complication of anti-particle components can be neglected, the Dirac theory reduces to the Pauli theory of electron spin, in which the total wave function is a product of a spatial part representing centre-of-mass motion and a 2-component spin function (spinor) representing internal angular momentum. Spinors were introduced in Chapter 3, and their properties and significance discussed in Chapters 6 and 8. In effect, they form bases for representations of the group SU(2) of unimodular, unitary matrices in 2-dimensions; equivalently they form bases for double-valued representations of the rotation group R(3). The invariance under SU(2) or R(3) of the total spin operator

(40) $$S^2 = \hbar^2(\sigma_x^2 + \sigma_y^2 + \sigma_z^2)/4$$

represents the fact that an arbitrary normalized linear combination of the two basic spin states

(41) $$a\begin{bmatrix}1\\0\end{bmatrix} + b\begin{bmatrix}0\\1\end{bmatrix} = \begin{bmatrix}a\\b\end{bmatrix}$$

with $|a|^2 + |b|^2 = 1$

is a state with the same total angular momentum

(42) $$s(s + 1)\hbar^2 = (1/2)(3/2)\hbar^2 = (3/4)\hbar^2,$$

but with component $\hbar/2$ along some other axis than the z-axis of quantization, i.e. with some other orientation of the spin vector.

In the early years of the nuclear era it was established that nuclear forces were largely charge-independent, i.e., if one could imagine setting aside the electromagnetic field effects, the residual strong and short-range forces between pairs of nucleons seemed independent of whether pairs were nn, np or pp. In the light of this observation, Heisenberg perceived some analogy with the electron spin formalism: just as the $\pm\hbar/2$ states corresponded to different states of the same particle, one could perhaps imagine that (electromagnetic forces aside) the neutron and proton were likewise states of the same particle in some sort of 'charge space'

$$n = \begin{bmatrix} 0 \\ 1 \end{bmatrix} \quad \text{and} \quad p = \begin{bmatrix} 1 \\ 0 \end{bmatrix} \tag{43}$$

so that a general state according to quantum mechanics is represented by an arbitrary, normalized, linear combination

$$ap + bn = a\begin{bmatrix} 1 \\ 0 \end{bmatrix} + b\begin{bmatrix} 0 \\ 1 \end{bmatrix} = \begin{bmatrix} a \\ b \end{bmatrix} \tag{44}$$

with

$$|a|^2 + |b|^2 = 1.$$

From this point of view, the state of the nucleon would be described by a product wave function:

$$\Psi = \psi_{\text{space-spin}}\, \phi_{\text{isospin}} \tag{45}$$

where $\phi_{\text{isospin}} = \begin{bmatrix} a \\ b \end{bmatrix}$ is the internal space of the nucleon. To satisfy the physical observation that mixed states do not in fact occur, but only the 'pure' states $p = \begin{bmatrix} 1 \\ 0 \end{bmatrix}$ and $n = \begin{bmatrix} 0 \\ 1 \end{bmatrix}$ one must assume the operation of a super-selection rule† imposed on Heisenberg's theory of 'isospin.' Just as one can alternatively view the SU(2) transformations in ordinary spin space as (double-valued) representations of the rotation group which preserve the total spin $|\vec{S}|^2 = 3\hbar^2/4$ in R(3), so in isospin theory one can associate with the SU(2) transformations on the 2-dimensional 'charge spinor' space, of the type (44), 'rotations' in a 3-dimensional charge hyperspace where distances along the 3-axis identify the actual electric charge on the particle according to the operator

$$Q = \tfrac{1}{2}(1 + \tau_3), \qquad \tau_3 = \begin{bmatrix} 1 & 0 \\ 0 & -1 \end{bmatrix}, \tag{46}$$

---

†G. Wick, A. Wightman, and E. Wigner, Phys. Rev. 88:101 (1952).

since this operator has eigenvalues 1 and 0 for the proton state $\begin{bmatrix}1\\0\end{bmatrix}$ and the neutron state $\begin{bmatrix}0\\1\end{bmatrix}$, respectively. Here τ_3 is the 3rd component of a Pauli isospin vector $\vec{\tau}$ which is formally identical with the Pauli spin vector $\vec{\sigma}$ but acts in 3-dimensional charge space rather than in ordinary 3-dimensional space.

If one confined oneself only to consideration of single-nucleon wave functions,the isospin formalism would not possess even formal convenience. However, just as with systems of several electrons it is often useful to construct spin-multiplets (i.e., bases for irreducible representations of SU(2) or SO(3)) out of appropriate linear combinations of products of single-particle spin functions, so here one expects that it is useful to construct isospin multiplets for many-nucleon systems. The convenience for two-particle systems (and hence for many-nucleon systems) is twofold: first, one can express charge independence of interactions as invariance of the Hamiltonian under 'rotations' in R(3) charge space† (i.e. under SU(2) on the isospinors); secondly, the isospin formalism lends itself readily to the description of charge-exchange forces since the operator

$$(47) \qquad \tau_1 = \begin{bmatrix}0 & 1\\1 & 0\end{bmatrix}$$

evidently changes neutrons into protons and protons into neutrons. Moreover, it can easily be shown†† that for a many-nucleon system, antisymmetrization of the wave function separately under neutron exchanges and proton exchanges is formally equivalent to simultaneous antisymmetrization with respect to 'nucleon' exchanges in the isospin formalism.

Just as two spin 1/2 particles can couple up to form three triplet states with $s = 1$, $m_s = 1,0,-1$, and one singlet state with $s = 0$, $m_s = 0$, so also can one form 'triplet' and

†Note that this implies that the total isotopic spin also commutes with the Hamiltonian.

††For example, see L. Rosenfeld, Theory of Nuclear Forces (North-Holland, Amsterdam, 1948).

'singlet' states of total isospin T = 1 and T = 0 for the 2-nucleon system:

$$\text{(48)}\quad M_T = \begin{cases} 1 & p(1)p(2) & \\ 0 & \dfrac{p(1)n(2) + n(1)p(2)}{\sqrt{2}} & \dfrac{p(1)n(2) - n(1)p(2)}{\sqrt{2}} \\ -1 & n(1)n(2) & \end{cases}$$

(columns: T = 1 and T = 0)

In these expressions the proton state is the isospin 'up' state and the neutron state is the isospin 'down' state.† Under 'rotations' in isospin space the three triplet states transform among themselves while the singlet goes into itself. Thus the T = 1 states form the basis of a 3-dimensional irreducible representation of the group SO(3), or equivalently SU(2), whereas the T = 0 state is the basis of a one-dimensional irreducible representation. Charge independence thus expresses itself in the statement that isospin is a good quantum number†† for actual 2-nucleon states. For any multiplet, the substates are characterized by the T_3 or M_T value, which is clearly given by

$$\text{(49)}\quad M_T = (Z - N)/2$$

where Z is the proton number and N the neutron number.

To this point the introduction of isospin could be regarded as a mere convenience only but, as often happens in the historical development of physics, concepts introduced as conveniences (e.g., Planck's constant) sometimes turn out to have deep physical significance. As early as 1938, Oppenheimer and Serber found some evidence to suggest that isospin could be

†This is the convention used in elementary particle physics; the opposite convention is used in nuclear physics.

††A quantum number is said to be good if it is the eigenvalue of an operator that commutes with the Hamiltonian. By saying that isospin is a good quantum number we mean that $[\vec{T},H] = 0$ and eigenstates of the Hamiltonian can be chosen to be simultaneously eigenstates of T^2 and T_z.

a good quantum number in nuclei. Following Yukawa's 1937 paper predicting the π-meson, Kemmer† suggested that three kinds of π-mesons, the π° and $\pi^\pm$, forming an isospin triplet, would be useful in formulating the charge-symmetry and charge-independence hypotheses. However, isospin was converted from a convenience into a physically significant concept in strong-interaction physics on the one hand by the discovery of isospin selection rules in nuclear gamma radiations†† and on the other hand by the recognition that isospin is conserved in nuclear reactions.†††

The degeneracy of the two basic isospin states n and p is, of course, not complete. The neutron, in fact, turns into a proton spontaneously with the emission (β-decay) of an electron and a neutrino. Nonetheless it appears that the greater mass of the neutron can be accounted for largely through the effects of the electromagnetic and weak interactions so that the idea of complete degeneracy can be retained if one limits oneself to strong interaction effects only. The validity of this statement is borne out by the observed conservation of isospin T in strong nuclear reactions (characterized by interaction times of the order of 10^{-23} second) but the violation of this conservation law in electromagnetic decays, or decays by weak interactions (characterized by a lifetime $\approx 10^{-10}$ second).

It is clear from the formal analogy with ordinary spin angular momentum that composite systems of strongly interacting particles can be classified according to their isospin multiplet structure. For example, the nucleon T = 1/2 doublet can

---

†See, for example, G. Wentzel, Quantum Theory of Fields (Wiley Interscience, New York, 1949).

††Isospin selection rules for nuclear radiations were first derived by Trainor (Phys. Rev. 83: 895 (1951); 85: 962 (1952)) and later refined by Radicati (Phys. Rev. 87: 521 (1952)). Detailed experimental verification was carried out largely by Wilkinson and collaborators; see D.H. Wilkinson, Isospin in Nuclear Physics (North-Holland, Amsterdam, 1969).

†††R.K. Adair, Phys. Rev. 87: 1041 (1952).

combine with the π-meson T = 1 triplet to give meson-nucleon 2-particle systems with T = 1/2 and 3/2. On the basis of the coupling coefficients involved one can make predictions about the relative intensities of certain nuclear scattering or reaction cross-sections, for example, the equality of the reaction cross-sections

$$(50) \quad \sigma(\pi^{+}n \rightarrow \pi^{\circ}p) = \sigma(\pi^{-}p \rightarrow \pi^{\circ}n),$$

which can be experimentally verified.

9.6. UNITARY SYMMETRY AND SU(3) INVARIANCE

The essence of the isospin formalism is the bringing together of states which are degenerate in energy as a result of a symmetry in the strong-interaction part of the Hamiltonian, namely its invariance under interchange of neutrons and protons (charge independence). An example is the T = 1 isospin multiplet depicted in equation (48) of the previous section. In the absence of Coulomb forces, the three T = 1 states are degenerate in energy: the diproton ($M_T = 1$), the dineutron ($M_T = -1$), and the lowest T = 1 state of the deuteron ($M_T = 0$). Experimentally it is found that the latter state, which has both total angular momentum J = 0 and spin angular momentum S = 0, is not quite bound. It is not surprising then that neither the diproton, which has additional Coulomb energy, nor the dineutron, which has additional electromagnetic mass,† exists as a bound system. The T = 0, $M_T = 0$ isospin state belongs to the ground state of the np system and is bound by about 2 MeV.

Because of the simplicity of the group SU(2) (or equivalently the SO(3) group) as a rank-1 group, the members of an isospin multiplet can be distinguished from one another by a single quantum number T_3 (or M_T as it is sometimes called). It may seem natural to ask whether in the zoo of elementary particles some higher symmetries, such as SU(3) or SU(4), might also

†The neutron is heavier than the proton and in fact goes into the proton by β-decay.

exist. In fact, this seems to be the case although historical developments did not take place in such a smooth manner.

For one thing, the 'zoo' was only populated one or two new particles at a time, so that SU(3) invariance was not arrived at in a very direct manner. Up until 1947 the only strongly interacting particles known were the neutron and proton; in that year the pions, which had been predicted 10 years earlier by Yukawa, were discovered. It then appeared that the nucleons, which are fermions, provide the basic constituents of nuclear matter, while the π-mesons, which are bosons, provide the glue holding nucleons together. However, in 1947 the dramatic discovery of V-shaped tracks in photographic emulsions exposed to cosmic rays indicated the presence of hitherto unknown neutral particles which decayed into protons and charged pions.†

Actually two different neutrals were involved, the Λ and K particles with decay schemes

$$\Lambda^\circ \to p + \pi^-,$$

$$K^\circ \to \pi^+ + \pi^-.$$

Eventually it turned out that there were several kinds of K particles, and some other 'strange' particles as well. What is of interest to us, however, is not the enumeration†† of all newly discovered particles and resonances, but the strange behaviour exhibited by the particles. This behaviour had to do with the long lifetimes which characterized their decay, viz. about 10^{-10} second, which is about one trillion times longer than would have been expected for strongly interacting particles. In fact, such lifetimes are characteristic of the so-called weak interactions such as β-decay.

---

†G.D. Rochester and C.C. Butler, Nature 160: 855 (1947).
††For a detailed summary of particles or resonances and their properties see Revs. Mod. Phys. 48, No. 2, Part II (1976) and periodic updating of the tables given there.

The result of these observations was the introduction of new selection rules which would prevent the strong decay of the Λ and K particles. Presumably, the selection rules would reflect some new conservation laws for strong interactions which were inhibiting strong decay. In line with our previous discussion (Section 4), the new conservation laws would imply additional symmetries in the strong-interaction part of the Hamiltonian, which are not present in the weak-interaction part. An additional aspect was the observation of associated production in which strange particles only appear in certain characteristic pairs.

The experimental observations led Gell-Mann, and Nakato and Nishijima independently, to propose a new quantum number,[†] 'strangeness,' which was conserved only in the strong interactions. Physicists tend now to use a different but related quantum number called hypercharge. It was proposed that, in addition to charge and isospin, hypercharge Y and baryon number[††] B were also conserved in strong interactions. Eventually this orientation of thought led Gell-Mann and Ne'eman[†††] to propose 8-particle multiplets (8-fold way) which were then associated with irreducible representations of the group SU(3).

Actually the first step in the latter development was taken by Sakata[§] when he proposed a scheme for explaining the new particles by mixing together three basic particles, the neutron, the proton, and a new particle, the 'sakaton.' As it turned out, Sakata had the right idea but the wrong particles in his SU(3) triplet. As we shall see, the basic particles in the SU(3) scheme are the quarks of Gell-Mann and Zweig[§§] (see

---

[†]T. Nakato and K. Nishijima, Prog. Theo. Phys. 10:581 (1953); K. Nishijima, Prog. Theo. Phys. 12:107 (1954): M. Gell-Mann, Phys. Rev. 92:833 (1953).

[††]The baryon number is essentially the number of nucleons produced in the decay of an unstable particle. Antiparticles are assigned negative baryon numbers.

[†††]Y. Ne'eman, Nuc. Phys. 26:22 (1961); M. Gell-Mann, Cal. Inst. of Tech. Report CTSL-20 (1961).

[§]S. Sakata, Prog. Theor. Phys. 16:686 (1956).

[§§]M. Gell-Mann, Phys. Letters 8:214 (1964); G. Zweig, CERN Reprint Th. 412 (1964).

Section 7). In fact, quarks have never been observed and may not even exist. Historically it was not the basic 3-multiplet which forms the regular representation of SU(3) which was observed as a multiplet, but an 8-particle multiplet corresponding to an 8-dimensional irreducible representation of SU(3). The inference to the SU(3) scheme was made later by Gell-Mann and Ne'eman.†

In the 8-fold way, the 8 baryons of lowest mass and the 8 mesons of lowest mass separately form multiplets in which members share the same mass, spin, and parity ($J^P = \frac{1}{2}^+$ for the baryons, $J^P = 0^-$ for the mesons). It is presumed that the mass degeneracy is broken by weaker forces, so that the observed masses differ significantly from the basic common mass that all members of a multiplet would have if only the strongest interactions were present. The members of a given multiplet are distinguished from one another by two 'magnetic' quantum numbers,†† which are conveniently chosen to be the charge

(51) $$Q = \frac{1}{2}Y + T_3$$

and the hypercharge Y. The multiplet can be plotted in a (Q,Y) set of axes, with the Q-axis oriented at 30° with respect to the horizontal, and the Y-axis taken vertical. The horizontal axis then gives $\sqrt{3}/2$ times the isospin quantum number T_3.††† The baryon octet is shown in Figure 9.4 and the meson octet in Figure 9.5. Table 9.1 lists the properties of the members of both multiplets.

Once these multiplets were identified with SU(3) symmetry, it was natural to look for possible particle multiplets

---

†Y. Ne'eman, Nuc. Phys. 26:22 (1961); M. Gell-Mann, Cal. Inst. of Tech. Report CTSL-20 (1961).

††There are two quantum numbers because SU(3) is a rank-2 group. We refer to them as 'magnetic' because they are the analogues of the M_J quantum number for the rotation group.

†††Points on the same horizontal (T_3) axis belong to an isospin multiplet. This follows from the detailed structure of SU(3), which contains the subgroup SU(2).

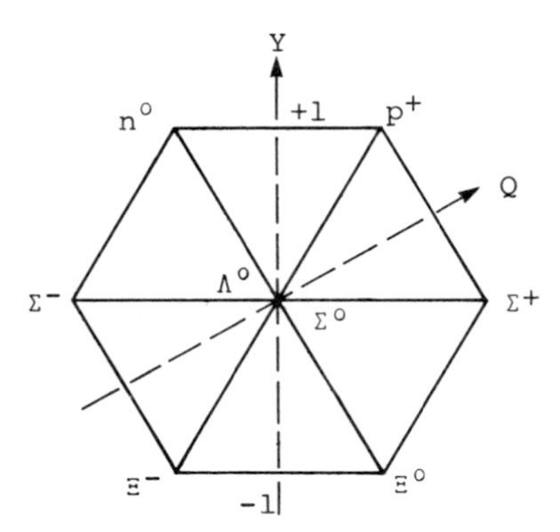

Figure 9.4. The baryon octet in (Y,Q) space. Note the occupancy of the origin by both the $T_z = 0$ member of the Σ isospin triplet and by the isoscalar Λ°.

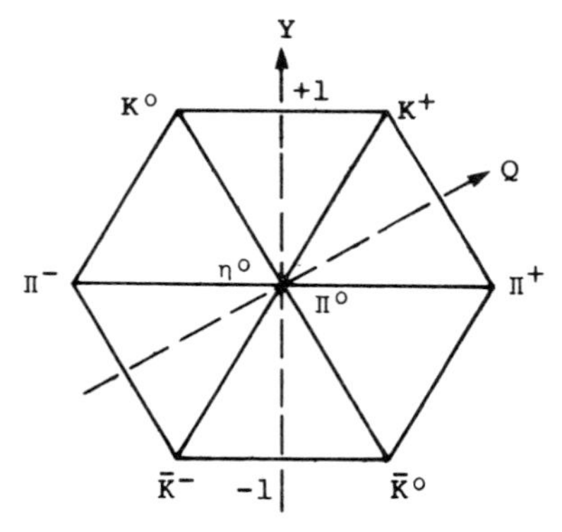

Figure 9.5. The meson octet in (Y,Q) space. Since SU(3) is not a perfect symmetry, the observed particle η° is not really a pure octet state; it is rather a mixture of pure octet state η_8 and a nearby singlet state η_1: $\eta^\circ = \eta_8 \cos\theta + \eta_1 \sin\theta$ where $\theta \simeq -10°$.

corresponding to other irreducible representations of the group SU(3). On this basis, Gell-Mann predicted the Ω^- particle shown at the lower apex of the triangle representing the 10-dimensional multiplet of $J^P = 3/2^+$ baryons shown in Figure 9.6. The experimental identification of the Ω^- left little doubt about the importance of SU(3) as a basic symmetry group for elementary particles.

TABLE 9.1
Properties of the lowest-mass mesons and baryons

	Q	T_3	B	Y
Baryons, $J^P = \frac{1}{2}^+$				
Ξ^o	0	$\frac{1}{2}$	1	-1
Ξ^-	-1	$-\frac{1}{2}$	1	-1
Σ^+	1	1	1	0
Σ^o	0	0	1	0
Σ^-	-1	-1	1	0
Λ^o	0	0	1	0
p	1	$\frac{1}{2}$	1	1
n	0	$-\frac{1}{2}$	1	1
Mesons, $J^P = 0^-$				
$\bar{K}^o$	0	$\frac{1}{2}$	0	-1
K^-	-1	$-\frac{1}{2}$	0	-1
π^+	1	1	0	0
π^o	0	0	0	0
π^-	-1	-1	0	0
η^o	0	0	0	0
K^+	1	$\frac{1}{2}$	0	1
K^o	0	$-\frac{1}{2}$	0	1

We remark finally that the SU(3) phenomenology once again demonstrates the essential correctness of the concept in quantum mechanics that states of a system are represented by elements of a linear vector space which itself may be an outer product of several separate linear vector spaces (e.g. functions

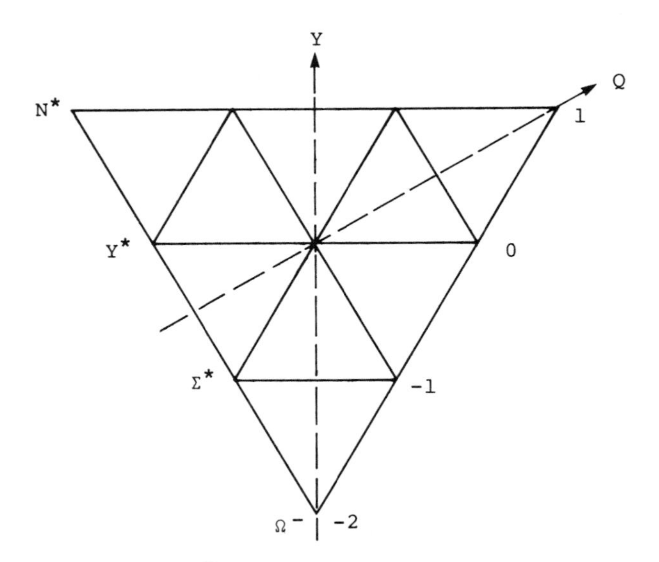

Figure 9.6. The $J^P = 3/2^+$ baryon decimet. The N^* is an isospin quartet, the Y^* an isotriplet, the Σ^* an isodoublet, and the Ω^- an isoscalar. The Q,Y values are set by noting that Ω^- has $Q = -1$, $Y = -2$.

in the space-time continuum, internal spin dynamics, and internal SU(3) symmetry), at least in non-relativistic quantum theory. The symmetries associated with the Hamiltonian are reflected in various invariance groups, such as SU(3), whose irreducible representations correspond to multiplets of basic states which share in common important properties like spin and parity. The additional quantum numbers, such as Q and Y or T_3 and Y, serve to label the members of the multiplet. When the masses of particles in the multiplet are altered by symmetry-breaking forces such as the electromagnetic interaction or weak forces, the mass differences can sometimes be expressed as functions of these additional quantum numbers.

9.7. THE QUARK MODEL

In line with the discussions of Section 6, it is natural to ask what the simplest or most basic multiplets associated with the group SU(3) are and whether there are experimentally observed groups of particles corresponding to these multiplets. It is not our purpose to go into the representation theory of SU(3) in any detail; suffice it to say that the basic representations are 3-dimensional. Other representations can be generated from the basic one by the process of identifying the three underlying vectors in the basic representation, and by taking outer products of such vectors to form symmetry classes of traceless tensors as the bases of more complicated representations. The process is basically the same as that for the general linear and orthogonal groups alluded to in Chapter 5.

From the experimental observations on the 8-fold representation, it is possible to work backwards and to identify the kind of particle (i.e. quantum state) which would necessarily form the three basic states in the 3-dimensional representation. Gell-Mann called these basic particles 'quarks.'[†] They are peculiar 'particles' in that they must possess fractional charges. The commonly accepted notation for the three basic quarks is u, d, and s and their properties are given in Table 9.2. The antiquarks belong to the associated representation denoted $\bar{3}$. The 'weight' diagrams corresponding to the representations 3 and $\bar{3}$ are shown in Figure 9.7.

Using the basic vectors u,d,s and $\bar{u},\bar{d},\bar{s}$ to form products and taking linear combinations of these one can construct traceless symmetry tensors corresponding to higher-dimensional representations of SU(3). The scheme that works and yet fits the phenomenology drawn from experimental observations is one that identifies nucleons and other baryons as being 3-quark systems, and mesons as quark-antiquark systems.

$$qqq \rightarrow \text{baryons}$$

$$q\bar{q} \rightarrow \text{mesons}$$

An arbitrary linear combination of the three quarks

[†]Quarks were first mentioned in James Joyce's Finnegan's Wake (Viking Press, New York, 1939), p. 383.

TABLE 9.2

Properties† of quarks and antiquarks

Quark	Spin	B	Y	Q	T	T_3
u	1/2	1/3	1/3	2/3	1/2	1/2
d	1/2	1/3	1/3	-1/3	1/2	-1/2
s	1/2	1/3	-2/3	-1/3	0	0
$\bar{u}$	1/2	-1/3	-1/3	-2/3	1/2	-1/2
$\bar{d}$	1/2	-1/3	-1/3	1/3	1/2	1/2
$\bar{s}$	1/2	-1/3	2/3	1/3	0	0

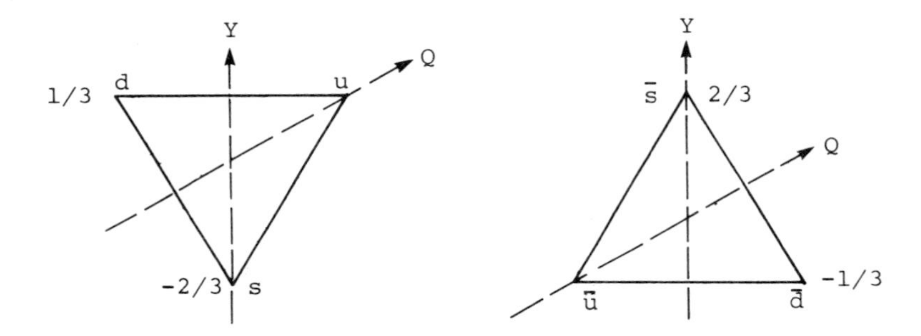

Figure 9.7. The quark and antiquark representation diagrams.

$$(52) \qquad \vec{q} = q_1 u + q_2 d + q_3 s = \begin{bmatrix} q_1 \\ q_2 \\ q_3 \end{bmatrix}$$

transforms under SU(3) like a vector

†In addition to the quantum numbers given in the table, each quark (or antiquark) carries an additional quantum number called colour. The colour quantum number can take on three values which are sometimes denoted red, blue, and yellow.

(53) $q'_\ell = U_{\ell k} q_k$,

the $U_{\ell k}$ being the components of a 3 x 3 unitary matrix. An arbitrary linear combination of antiquarks

(54) $$\vec{\bar{q}} = \bar{q}_{\dot{1}}\bar{u} + \bar{q}_{\dot{2}}\bar{d} + \bar{q}_{\dot{3}}\bar{s} = \begin{bmatrix} \bar{q}_1 \\ \bar{q}_2 \\ \bar{q}_3 \end{bmatrix}$$

transforms under SU(3) as a second type of vector

(55) $$\bar{q}'_{\dot{k}} = U^*_{\dot{k}\dot{\ell}}\ \bar{q}_{\dot{\ell}}$$

where the $U^*_{\dot{k}\dot{\ell}}$, $\dot{k},\dot{\ell} = \dot{1},\dot{2},\dot{3}$, are the components of the complex conjugate of the unitary matrix given in equation (53) and the dotted indices indicate reference to antiquark basis vectors.

Under SU(3) transformations the inner product $q_k\bar{q}_{\dot{k}}$ transforms as a scalar since

(56) $$q'_k\bar{q}'_{\dot{k}} = U_{k\ell}q_\ell U^*_{\dot{k}\dot{m}}\ \bar{q}_{\dot{m}} = (U_{k\ell}U^{\dagger}_{\dot{m}\dot{k}})\ q_\ell\bar{q}_{\dot{m}}$$

$$= \delta_{\ell\dot{m}}\ q_\ell\bar{q}_{\dot{m}} = q_m\bar{q}_{\dot{m}}$$

where we have used the summation convention (i.e. $q_k\bar{q}_{\dot{k}} = q_1\bar{q}_{\dot{1}} + q_2\bar{q}_{\dot{2}} + q_3\bar{q}_{\dot{3}}$). There exist quantities which transform under SU(3) like higher-rank tensors. For example the meson outer product $q_k\bar{q}_{\dot{\ell}}$ transforms as a mixed second-rank tensor:

(57) $$F'_{k\dot{\ell}} = U_{km}U^*_{\dot{\ell}\dot{n}}\ F_{m\dot{n}}\ .$$

Any mixed second-rank tensor $F_{k\dot{\ell}}$ may be decomposed into a sum of irreducible tensors

(58) $$F_{k\dot{\ell}} = \frac{1}{3}\ \delta_{k\dot{\ell}}\ F_{m\dot{m}} + T_{k\dot{\ell}}\ .$$

We have already shown that $F_{m\dot{m}}$ transforms as an irreducible scalar. The tensor

(59) $$T_{k\dot{\ell}} = F_{k\dot{\ell}} - \frac{1}{3}\,\delta_{k\dot{\ell}}\,F_{m\dot{m}}$$

is traceless. Since symmetry between undotted and dotted indices is not preserved under SU(3) (i.e. $T_{k\dot{\ell}} = T_{\dot{\ell}k}$ does not imply $T'_{k\dot{\ell}} = T'_{\dot{\ell}k}$) the Young symmetrizers cannot be used to further reduce $T_{k\dot{\ell}}$. Thus $T_{k\dot{\ell}}$ is an irreducible tensor with eight independent components. It follows that there exists an 8-dimensional irreducible representation of SU(3) which acts on the space of mixed, traceless, second-rank tensors. Table 9.3 gives the mixed, traceless, second-rank tensors which form the meson octet of Figure 9.5.

The baryon outer product of three quarks $q_{k_1}\,q_{k_2}\,q_{k_3}$ transforms as a third-rank tensor $F_{k_1k_2k_3}$ under SU(3):

$$F'_{k_1k_2k_3} = U_{k_1m_1}U_{k_2m_2}\,U_{k_3m_3}\,F_{m_1m_2m_3}.$$

In order to explain the baryon octet and decimet given in Section 6 we consider the irreducible symmetry tensors that may be formed from an arbitrary third-rank tensor $F_{k_1k_2k_3}$ by taking linear combinations. Since no dotted indices appear, the trace no longer serves as a useful method for reduction. The irreducible symmetry tensors are formed by acting on $F_{k_1k_2k_3}$ with the Young operators associated with the standard tableaux

$$\begin{array}{|c|}\hline 1\\\hline 2\\\hline 3\\\hline\end{array}\qquad \begin{array}{|c|c|}\hline 1&2\\\hline 3\\\cline{1-1}\end{array}\qquad \begin{array}{|c|c|}\hline 1&3\\\hline 2\\\cline{1-1}\end{array}\qquad \begin{array}{|c|c|c|}\hline 1&2&3\\\hline\end{array}$$

The first tableau gives an irreducible tensor with one independent component, i.e. a scalar (see problem 4). The next two tableaux give an irreducible tensor with eight independent components. Hence we have the baryon octet of Figure 9.4. The last tableau gives an irreducible tensor with 10 independent components, yielding the decimet of Figure 9.6.

We shall not pursue the quark model further here except to indicate that it has been extended to include a fourth quark c (charm) and perhaps two others b and t (beauty and truth).

TABLE 9.3
The quark content of the meson octet

Particle	Tensor	Non-zero components
π^-	$d\bar{u}$	$F_{2\dot{1}} = 1$
π^+	$u\bar{d}$	$F_{1\dot{2}} = 1$
K^-	$s\bar{u}$	$F_{3\dot{1}} = 1$
K^+	$u\bar{s}$	$F_{1\dot{3}} = 1$
$\bar{K}^\circ$	$s\bar{d}$	$F_{3\dot{2}} = 1$
K°	$d\bar{s}$	$F_{2\dot{3}} = 1$
π°	$1/\sqrt{2}(u\bar{u} - d\bar{d})$	$F_{1\dot{1}} = 1/\sqrt{2}$, $F_{2\dot{2}} = -1/\sqrt{2}$
η	$1/\sqrt{6}(u\bar{u} + d\bar{d} - 2s\bar{s})$	$F_{1\dot{1}} = 1/\sqrt{6}$, $F_{2\dot{2}} = 1/\sqrt{6}$, $F_{3\dot{3}} = -2/\sqrt{6}$

The u,d,s,c,b,t distinctions are sometimes referred to as flavours. In addition to flavour each quark has an additional quantum number called colour as mentioned previously. As a purely phenomenological model explaining unitary symmetry, the quark model has been enormously successful but many questions concerning its dynamical basis remain unanswered at this time.

9.8. GAUGE INVARIANCE

The Schroedinger equation is covariant under the 'gauge' transformation $U = \exp(i\alpha)$, where α is a real number, since

$$(60) \quad H'\psi' = (UHU^{-1})U\psi = UH\psi$$
$$= i\hbar\partial(U\psi)/\partial t = i\hbar \frac{\partial\psi'}{\partial t} .$$

A gauge transformation of this kind is said to be global since the Abelian group of operators $U(\alpha)$ is parametrized by real numbers α which are independent of space and time. If one introduces a space-time dependence into the parameters of some Lie group, the corresponding gauge transformations are said to be local. Local gauge transformations play an important role in our description of elementary particle interactions.[†]

As a simple example consider the Dirac equation

$$(61) \quad (\gamma^\mu \partial_\mu + \kappa)\psi(x_\mu) = 0, \qquad \partial_\mu = \partial/\partial x_\mu,$$

and the Lie group of simple phase transformations U(1):

$$(62) \quad \psi'(x_\mu) = U(\Lambda)\psi(x_\mu)$$

where Λ is a real number independent of space and time, $U(\Lambda) = \exp(iq\Lambda)$, and q is the 'charge' on the particle. Since ψ' also satisfies the Dirac equation

$$(63) \quad (\gamma^\mu \partial_\mu + \kappa)\psi'(x_\mu) = 0$$

we say that the Dirac equation is invariant with respect to the group U(1).

If we allow Λ to have a space-time dependence, however, (63) is no longer valid since it would imply

$$(64) \quad (\gamma^\mu [\partial_\mu + iq\partial_\mu \Lambda] + \kappa)\psi(x_\mu) = 0$$

instead of equation (61). An invariant wave equation can be restored by introducing a 'gauge boson' with a wave function $A_\mu(x_\nu)$ which is a four vector under Lorentz transformations but which transforms under local gauge transformations as

$$(65) \quad A_\mu(x_\nu) \rightarrow A'_\mu(x_\nu) = A_\mu + \partial_\mu \Lambda.$$

The modified Dirac equation

$$(66) \quad (\gamma^\mu [\partial_\mu - iqA_\mu] + \kappa)\psi(x_\nu) = 0$$

is now invariant under U(1).

[†]E.S. Abers and B.W. Lee, Physics Reports 9:1 (1973).

Actually this case was already considered in part in Chapter 8 where it was shown that local gauge tranformations of the type (65) leave Maxwell's equations invariant. In fact, with the identification $q = e/\hbar c$ and A_μ the electromagnetic four potential, equation (66) is just the Dirac equation for the motion of a charged particle in an electromagnetic field. The photon is thus the gauge boson which makes the Dirac equation locally gauge invariant. In some sense, one can say that the requirement of local gauge invariance implies the existence of the electromagnetic field.

The concept of local gauge invariance has been extended by Weinberg and Salam† to the theory of weak and electromagnetic interactions. The appropriate gauge group is the direct product group $SU(2) \otimes U(1)$ which leads to three 'massive' gauge bosons (the so-called $W^+, W^-, Z°$ - two charged and one neutral boson) and the massless photon. The massive bosons mediate interactions between weakly interacting particles much as the photon mediates the electromagnetic interaction between charged particles. Since the exchange of a massive boson requires borrowing its rest energy, according to the uncertainty principle $\Delta E \Delta t \geq \hbar$ the transit time must be correspondingly short. Hence, the weak interactions are effective only at small distances (short ranged) in contrast to the electromagnetic interaction which is still effective at large distances. The direct product group $SU(2) \otimes U(1)$ has a subgroup $SU(2)$ which is a 3-parameter group giving rise to W^+, W^-, and $Z°$; the $U(1)$ subgroup is a one-parameter group and gives rise to the photon.

Similar ideas also apply in the theory of strong interactions and give rise to the theory of quantum chromodynamics (QCD) in rough analogy to quantum electrodynamics (QED). In QCD colour plays a role analogous to electric charge, except there are three colours, red, blue, and yellow. The gauge bosons which mediate the interaction between coloured quarks are called gluons, but unlike photons which do not carry charge, the gluons can

---

†S. Weinberg, Phys. Rev. Letters 19:1264 (1967); and A. Salam in: Elementary Particle Theory, ed. by N. Svartholm (Almqvist and Wiksell, Stockholm, 1968), p. 367.

carry colour (however, like charge, colour is conserved). This implies that gluons can interact with themselves, giving rise to a non-linear theory. Since SU(3) is an 8-parameter Lie group there are eight different kinds of coloured gluons. The theory of coloured quarks and gluons (QCD) has been very successful in systematizing information obtained in high energy collisions producing strongly interacting particles. For example, since gluons couple to colour (and <u>not</u> flavour) we expect an SU(3) flavour symmetry to be apparent whenever mass differences between the u,d, and s quarks are relatively unimportant. Also, it has been shown that QCD is an asymptotically free theory[†] (i.e. at short distances the quarks behave essentially as free particles). This is in agreement with experimental data which exhibit such "scaling" phenomena.[††]

REFERENCES

1 E. Corinaldesi and F. Strocchi, Relativistic Wave Mechanics. North-Holland, Amsterdam, 1963

2 W. Frazer, Elementary Particles. Prentice-Hall Inc., Englewood Cliffs, N.J., 1966

3 J. Kokkedee, The Quark Model. W.A. Benjamin Inc., New York, 1969

4 M. Leon, Particle Physics: An Introduction. Academic Press, New York, 1973

5 H. Lipkin, Lie Groups for Pedestrians. North-Holland, Amsterdam, 1965

PROBLEMS

1 Show that the Dirac free electron Hamiltonian

$$H = c\vec{\alpha}\cdot\vec{p} + \beta m_0 c^2$$

commutes with the component of spin in the direction of motion:

$$\vec{s}\cdot\vec{p} = \frac{\hbar}{2}\,\vec{\sigma}'\cdot\vec{p}.$$

[†]H.D. Politzer, Phys. Reports 14C: 130 (1974).

[††]For example, the ratio $R = \sigma\,(e^+e^- \to \text{hadrons})/\sigma(e^+e^- \to \mu^+\mu^-)$ is constant at high energies above the appearance of resonances.

2 Suppose $\phi(x_\nu)$ satisfies the Klein-Gordon equation

$$\left[\Box - \left(\frac{m_o c}{\hbar}\right)^2\right]\phi(x_\nu) = 0.$$

Defining

$$\psi_1 = \frac{1}{\sqrt{2}}\left(\phi + \frac{i}{\kappa}\frac{\partial\phi}{\partial t}\right), \quad \psi_2 = \frac{1}{\sqrt{2}}\left(\phi - \frac{i}{\kappa}\frac{\partial\phi}{\partial t}\right),$$

and $\kappa = m_o c/\hbar$, show that the components ψ_1, ψ_2 satisfy

$$i\hbar\partial\Psi/\partial t = H\Psi$$

where

$$\psi = \begin{bmatrix}\psi_1 \\ \psi_2\end{bmatrix},$$

$$H = (\tau_3 + i\tau_2)\frac{\vec{p}^2}{2\kappa} + \kappa\tau_3,$$

and

$$\tau_1 = \begin{bmatrix}0 & 1 \\ 1 & 0\end{bmatrix}, \quad \tau_2 = \begin{bmatrix}0 & -i \\ i & 0\end{bmatrix}, \quad \tau_3 = \begin{bmatrix}1 & 0 \\ 0 & -1\end{bmatrix}.$$

3 Let $H = c\vec{\alpha}\cdot\vec{p} + \beta m_o c^2 + V(r)$ be the Dirac Hamiltonian for an electron moving in a central potential $V(r)$. The orbital angular momentum L is defined by $\vec{L} = \vec{r} \times \vec{p}$.
(i) Show that $[\vec{L},\vec{H}] \neq 0$, and conclude that orbital angular momentum is not a constant of the motion.
(ii) Define $\vec{\sigma}'$ by $\vec{\sigma}' = \begin{bmatrix}\vec{\sigma} & 0 \\ 0 & \vec{\sigma}\end{bmatrix}$, and show that $[\vec{L} + \frac{1}{2}\hbar\,\vec{\sigma}', H] = 0$
Conclude that the total angular momentum $\vec{J}$ (which must be a constant of the motion) is given by $\vec{J} = \vec{L} + \frac{1}{2}\hbar\,\vec{\sigma}'$.

4 Prove equation (33).

5 Prove that the symmetry tensors

$$YF_{k_1 k_2 k_3}, \quad k_j = 1,2,3;\ j = 1,2,3,$$

where Y is the Young operator associated with the tableaux

1
2
3

1	3
2	

1	2	3

have respectively 1, 8, and 10 independent components.

6 (a) Considering an infinitesimal Lorentz transformation

$L^{\alpha}_{\mu} = \delta^{\alpha}_{\mu} + \varepsilon^{\alpha}_{\mu}$ show that

$$\varepsilon^{\beta}_{\nu}\,\eta^{\alpha\nu} + \varepsilon^{\alpha}_{\mu}\,\eta^{\mu\beta} = 0.$$

(b) Using part (a) show that

$$A^{\mu\nu}\,L^{\alpha}_{\mu}\,L^{\beta}_{\nu} = A^{\alpha\beta} \Rightarrow A^{\alpha\beta} = a(x)\,\eta^{\alpha\beta}.$$

(c) Show that the only linear homogeneous second-order differential equation for a scalar field ϕ, which is form invariant under Poincaré transformations, is (setting $C = 1$)

$$\nabla^2\phi - \frac{\partial^2\phi}{\partial t^2} + \lambda\phi = 0$$

for some constant λ. This is the Klein-Gordon equation for particles with mass $m = \hbar\sqrt{-\lambda}$. (Hint: Start by writing the most general linear homogeneous second-order differential equation for ϕ. That is $A^{\mu\nu}\partial^2\phi/\partial x^{\mu}\partial x^{\nu} + B^{\mu}\partial\phi/\partial x^{\mu} + c\phi = 0$.)

7 Discuss the transformation properties of the following 16 quantities under Lorentz transformations:

(i) $\bar{\psi}(x)\psi(x)$,

(ii) $\bar{\psi}(x)\gamma_{\mu}\psi(x)$,

(iii) $\bar{\psi}(x)\gamma_5\psi(x)$, where $\gamma_5 = \gamma_1\gamma_2\gamma_3\gamma_4$,

(iv) $\bar{\psi}(x)\gamma_5\gamma_{\mu}\psi(x)$,

(v) $\bar{\psi}(x)\,\dfrac{\gamma_{\mu}\gamma_{\nu} - \gamma_{\nu}\gamma_{\mu}}{2}\,\psi(x)$,

where $\psi(x)$ is a Dirac bispinor and $\bar{\psi}(x) = \psi^{+}\gamma_4$.

10
Constructs in Riemannian geometry

10.1. INTRODUCTION

We return now to the discussion of Riemannian geometry which was begun in Chapter 7. In Chapter 7 our interest was to broaden the conceptual view of geometry and to show how it accommodates such divergent notions as curvilinear coordinates, dynamics of a multiparticle system in the space of generalized coordinates, and the space-time framework of special relativity. In point of fact, special relativity required only a minimal broadening of the Euclidean viewpoint, since the extension of Euclidean 3-space to the 4-dimensional space-time of special relativity could still be accommodated in a flat, though non-Euclidean geometry. The distinction between covariant and contravariant tensors or tensor fields, so essential to Riemannian geometry, enabled us, via the Lorentz metric, to use real space-time coordinates and to distinguish clearly through the metric tensor between space and time coordinates.

Our interest now represents a more serious and fundamental departure from the Euclidean world-view. In general relativity, to which we shall turn briefly in Chapter 11, the mere presence of matter or energy warps the space-time reference frame, and in such a profound manner that the dynamical complications of the system under study are transferred to the geometry of the space-time framework. As we will see in the next chapter, this beautifully profound view of Einstein represents a theory of gravitation as much as it does a theory of relativity.

The present chapter is, in part, a prelude to the brief introduction to general relativity presented in Chapter 11; but in part it stands on its own as a thought-provoking and disciplined subject which may lend itself to a wider application in the world of non-linear phenomena than has been generally appreciated. The central idea of Einstein of embedding the parametric equations of space, time, and matter in a suitable geometric continuum necessarily has application to a wide range of parametric relationships between the physical variables of non-linear systems.

The mathematics of Riemannian geometry involves straightforward analysis of the type encountered in courses on advanced calculus. What is new and different is the conceptual framework. The central idea of a tensor quantity is preserved, namely, if a geometric continuum has more than one parametrization, one must have ways of translating from one description to the other. The most convenient formulation of the geometry is one in which the equations fundamental to its description have the same form in every permissible parametrization. Because the space is curved, one can expect simple geometric relationships only locally, that is in the immediate neighbourhood of the point in question. Relationships between geometric quantities at distant points can generally be obtained only by complicated integrations, where suitable integrability conditions must in any case be satisfied.

Section 2 is concerned with the problem of covariant differentiation in Riemannian geometry. This is not a unique process, but can be made unique by an appropriate choice of a geometric construct called a <u>connection</u>. Section 3 introduces the notion of geodesic coordinates as the simplest choices for the elucidation of the geometry. Geodesic coordinates defined at a point correspond to a vanishing connection at that point. In Section 4, Christoffel symbols of the second kind are shown to be the unique choice of connection for which the torsion tensor (antisymmetric part of the connection) vanishes and the covariant derivative of the metric tensor is zero.

In Section 5 the concept of parallel transport is introduced whereby a tensor at a point can be transported along any curve in such a manner as to keep the covariant derivative (generalized gradient) perpendicular to the tangent at all points along the curve. Parallel transport enables one to make useful comparisons between the values of a tensor at different points in the geometric continuum. In Section 6 it is shown that geodesics are autoparallel curves (i.e. curves which parallel transport their own tangents) using Christoffel symbols of the second kind as a connection. Section 7 gives as an example the 2-dimensional curved surface of the sphere. Sections 8 and 9 explore the properties of the Riemann-Christoffel curvature tensors preparatory to the introduction to general relativity presented in Chapter 11. Finally, in Section 10, it is shown that even in Riemannian geometry one can usefully define local basis vectors just as in the theory of curvilinear coordinates.

10.2. THE AFFINE CONNECTION AND COVARIANT DERIVATIVES

The covariant derivative of a scalar $\Phi(u^1,\ldots,u^n)$ field is denoted by $\Phi_{;i}$ and has components

(1) $$\Phi_{;i} = \partial\Phi/\partial u^i, \qquad i = 1,\ldots,n.$$

Thus the covariant derivative of Φ is its gradient and evidently transforms as a covariant vector.

Let a^i be an arbitrary contravariant vector field. From equation (58) of Chapter 7, the components a^i, $i = 1,\ldots,n$, transform in the manner

(2) $$\bar{a}^m = a^j\left(\partial\bar{u}^m/\partial u^j\right)$$

under change of parametrization from unbarred to barred coordinates. Taking the partial derivative of equation (2) with respect to $\bar{u}^p$ we obtain

$$\frac{\partial\bar{a}^m}{\partial\bar{u}^p} = \frac{\partial a^j}{\partial\bar{u}^p}\frac{\partial\bar{u}^m}{\partial u^j} + a^j\,\frac{\partial^2\bar{u}^m}{\partial\bar{u}^p\partial u^j}\,.$$

Using the chain rule the above equation becomes

(3) $$\frac{\partial \bar{a}^m}{\partial \bar{u}^p} = \frac{\partial a^j}{\partial u^k} \frac{\partial u^k}{\partial \bar{u}^p} \frac{\partial \bar{u}^m}{\partial u^j} + a^j \frac{\partial^2 \bar{u}^m}{\partial u^j \partial u^k} \frac{\partial u^k}{\partial \bar{u}^p} .$$

Because of the last double sum on the right-hand side of the above equation $\partial \bar{a}^m/\partial \bar{u}^p$ does not transform as a mixed second-rank tensor. Let G^i_{sj} be a set of n^3 functions such that the quantities $a^i_{;k}$ defined by

(4) $$a^i_{;k} = \frac{\partial a^i}{\partial u^k} + G^i_{jk} a^j, \quad i,k = 1,\ldots,n,$$

transform as the components of a mixed second-rank tensor; that is

(5) $$\bar{a}^m_{;p} = a^i_{;k} \frac{\partial \bar{u}^m}{\partial u^i} \frac{\partial u^k}{\partial \bar{u}^p} .$$

We call $a^i_{;k}$ the <u>covariant derivative</u> of a^i with respect to G^i_{jk}.

We now seek an explicit expression to show how the quantities G^i_{jk} must transform in order that the covariant derivative (4) transforms properly. Equation (4) defines the components of $a^i_{;k}$ in any (allowable) coordinate system. Hence one can write in the barred system

(6) $$\bar{a}^m_{;p} = \frac{\partial \bar{a}^m}{\partial \bar{u}^p} + \bar{G}^m_{jp} \bar{a}^j .$$

Putting equations (4) and (6) into equation (5) we get

$$\frac{\partial \bar{a}^m}{\partial \bar{u}^p} + \bar{G}^m_{jp} \bar{a}^j = \left(\frac{\partial a^i}{\partial u^k} + G^i_{jk} a^j \right) \frac{\partial \bar{u}^m}{\partial u^i} \frac{\partial u^k}{\partial \bar{u}^p} .$$

Using equations (3) and cancelling an identical term on each side, we obtain

$$\bar{G}^m_{jp} \bar{a}^j + a^j \frac{\partial^2 \bar{u}^m}{\partial u^j \partial u^k} \frac{\partial u^k}{\partial \bar{u}^p} = G^i_{jk} a^j \frac{\partial \bar{u}^m}{\partial u^i} \frac{\partial u^k}{\partial \bar{u}^p} .$$

Since $\bar{a}^j = a^r \partial \bar{u}^j / \partial u^r$ by the covariant transformation law for vectors,

$$\bar{G}^m_{jp} \frac{\partial \bar{u}^j}{\partial u^r} a^r = \left(G^i_{jk} \frac{\partial \bar{u}^m}{\partial u^i} \frac{\partial u^k}{\partial \bar{u}^p} - \frac{\partial^2 \bar{u}^m}{\partial u^j \partial u^k} \frac{\partial u^k}{\partial \bar{u}^p} \right) a^j .$$

Since r and j are dummy indices (i.e. summed over) we may interchange r and j on the left-hand side of the above equation to get

$$\bar{G}^m_{rp} \frac{\partial \bar{u}^r}{\partial u^j} = G^i_{jk} \frac{\partial \bar{u}^m}{\partial u^i} \frac{\partial u^k}{\partial \bar{u}^p} - \frac{\partial^2 \bar{u}^m}{\partial u^j \partial u^k} \frac{\partial u^k}{\partial \bar{u}^p}$$

where we have used the fact that the a_j, $j = 1,\ldots,n$, were arbitrary. Multiplying both sides by

$$\left(\frac{\partial u^n}{\partial \bar{u}^m}\right) \left(\frac{\partial \bar{u}^p}{\partial u^s}\right),$$

summing on all repeated indices, and using the expressions

$$\frac{\partial u^n}{\partial \bar{u}^m} \frac{\partial \bar{u}^m}{\partial u^i} = \delta^n_i \quad \text{and} \quad \frac{\partial u^k}{\partial \bar{u}^p} \frac{\partial \bar{u}^p}{\partial u^s} = \delta^k_s ,$$

we get the explicit transformation law for the quantities G^h_{jq} which involves only derivatives of barred and unbarred coordinates:

$$(7) \qquad G^h_{jq} = \left(\bar{G}^m_{rp} \frac{\partial \bar{u}^r}{\partial u^j} \frac{\partial \bar{u}^p}{\partial u^q} + \frac{\partial^2 \bar{u}^m}{\partial u^j \partial u^q} \right) \frac{\partial u^h}{\partial \bar{u}^m} .$$

Thus the necessary and sufficient condition that the covariant derivative of an arbitrary contravariant vector with respect to G^i_{jk} transform as a mixed second-rank tensor is that the n^3 functions G^i_{jk}, $i,j,k = 1,\ldots,n$, transform under change of parametrization in the fashion given by equation (7). Any ordered set of n^3 functions G^m_{rp} which transform under change of parametrization from barred to unbarred coordinates as in equation (7) is called a <u>connection</u>.

Proceeding in a similar fashion we can define the <u>covariant derivative</u> of a covariant vector a_i with respect to the connection G^j_{ik} by

(8) $$a_{i;k} = \frac{\partial a_i}{\partial u^k} - G^j_{ik} a_j,$$

where, by virtue of the transformation properties of G^j_{ik} expressed in (7), the quantities $a_{i;k}$ transform as the components of a covariant second-rank tensor.

These ideas can be generalized to tensors of higher rank. The covariant derivative of an arbitrary tensor $A^{q_1 \ldots q_s}_{\ell_1 \ldots \ell_r}$ with respect to the connection G^i_{jk} is denoted $A^{q_1 \ldots q_s}_{\ell_1 \ldots \ell_r;i}$ and defined by

(9) $$A^{q_1 \ldots q_s}_{\ell_1 \ldots \ell_r;i} = \frac{\partial A^{q_1 \ldots q_s}_{\ell_1 \ldots \ell_r}}{\partial u^i} - \sum_{m=1}^{r} G^k_{\ell_m i} A^{q_1 \ldots q_s}_{\ell_1 \ldots \ell_{m-1} k \ell_{m+1} \ldots \ell_r} + \sum_{n=1}^{s} G^{q_n}_{pi} A^{q_1 \ldots q_{n-1} p q_{n+1} \ldots q_s}_{\ell_1 \ldots \ell_r} .$$

Covariant differentiation obeys a product rule much like ordinary differentiation. If S and T are two tensors then

(10) $$\nabla(S \otimes T) = (\nabla S) \otimes T + S \otimes (\nabla T)$$

where ∇A <u>denotes the covariant derivative</u> of a tensor A with respect to some connection G^i_{jk}. If A has components $A^{j_1 \ldots j_m}_{s_1 \ldots s_p}$, $j_1, \ldots, j_m, s_1, \ldots, s_p = 1, 2, \ldots, n$, and $\vec{v}$ is a contravariant vector with components v^k, $k = 1, 2, \ldots, n$, the covariant derivative of A in the direction $\vec{v}$ is denoted $\nabla_{\vec{v}} A$ and is the tensor with components $A^{j_1 \ldots j_n}_{s_1 \ldots s_p;k} v^k$, $j_1, \ldots, j_m, s_1, \ldots, s_p = 1, 2, \ldots, n$. Directional covariant derivatives have the following properties:

(11a) $$\nabla_{a\vec{v}+b\vec{u}} S = (a\nabla_{\vec{v}} S \oplus b\nabla_{\vec{u}} S),$$

(11b) $\nabla_{\vec{u}}(aS \oplus bT) = a\nabla_{\vec{u}}S \oplus b\nabla_{\vec{u}}T,$

(11c) $\nabla_{\vec{u}}(S \otimes T) = [(\nabla_{\vec{u}}S) \otimes T] \oplus [S \otimes \nabla_{\vec{u}}T],$

where S,T are tensors, $\vec{v},\vec{u}$ are contravariant vectors, and a,b are real numbers.

A connection G^k_{ij} is not in general a tensor; however, if we write G^k_{ij} as

(12) $G^k_{ij} = S^k_{ij} + A^k_{ij},$

where

(13a) $S^k_{ij} = \frac{1}{2}(G^k_{ij} + G^k_{ji}),$

(13b) $A^k_{ij} = \frac{1}{2}(G^k_{ij} - G^k_{ji}),$

the n^3 functions A^k_{ij}, k,i,j = 1,2,...n, transform as the components of a third-rank tensor. The quantities A^k_{ij} are skew-symmetric with respect to the subscripts and the quantities S^k_{ij} are symmetric with respect to the subscripts. We call A^k_{ij} the <u>torsion tensor</u> of the given connection and if it is zero the connection is called <u>symmetric</u>.

10.3. GEODESIC COORDINATES

It is always possible to find a local coordinate system in which the components of a symmetric connection G^i_{jk} vanish at a particular point. Let one system of coordinates for a neighbourhood of a point P_o be denoted by u^i. Let the coordinates of the point P_o itself be denoted $_ou^i$. Introduce new coordinates $\bar{u}^i$ defined by

(14) $\bar{u}^i = u^i - {}_ou^i + \frac{1}{2}\, {}_oG^i_{mn}(u^m - {}_ou^m)(u^n - {}_ou^n).$

A subscript 0 denotes that a quantity is evaluated at the point P_o. Differentiating the above equation with respect to u^j gives

$$(15)\qquad \frac{\partial \bar{u}^i}{\partial u^j} = \delta^i_j + {}_oG^i_{jn}(u^n - {}_ou^n)$$

where the symmetry $G^i_{mn} = G^i_{nm}$ of the connection has been used. Evaluating equation (15) at the point P_o gives ${}_o(\partial\bar{u}^i/\partial u^j) = \delta^i_j$ from which it follows also that ${}_o(\partial u^i/\partial\bar{u}^j) = \delta^i_j$. Thus the Jacobian of the transformation from unbarred to barred coordinates is different from zero at P_o and the $\bar{u}^i$ as defined by (14) are indeed 'good' coordinates in some neighbourhood of the point P_o.

Multiplying equation (15) by $\partial u^j/\partial\bar{u}^k$ and summing on j yields

$$(16)\qquad \delta^i_k = \frac{\partial u^i}{\partial \bar{u}^k} + {}_oG^i_{jn}(u^n - {}_ou^n)\,\frac{\partial u^j}{\partial \bar{u}^k}$$

which, upon differentiation with respect to $\bar{u}^h$, becomes

$$0 = \frac{\partial^2 u^i}{\partial \bar{u}^h \partial \bar{u}^k} + {}_oG^i_{jn}\,\frac{\partial u^n}{\partial \bar{u}^h}\,\frac{\partial u^j}{\partial \bar{u}^k} + {}_oG^i_{jn}(u^n - {}_ou^n)\,\frac{\partial^2 u^j}{\partial \bar{u}^k \partial \bar{u}^h}.$$

Hence at P_o

$$(17)\qquad {}_o\left(\frac{\partial^2 u^i}{\partial \bar{u}^h \partial \bar{u}^k}\right) = -{}_oG^i_{jn}\,\delta^n_h\,\delta^j_k = -{}_oG^i_{kh}.$$

From equation (7), with interchange of barred and unbarred quantities, we have

$$(7a)\qquad \bar{G}^h_{jq} = \left(G^m_{rp}\,\frac{\partial u^r}{\partial \bar{u}^j}\,\frac{\partial u^p}{\partial \bar{u}^q} + \frac{\partial^2 u^m}{\partial \bar{u}^j \partial \bar{u}^q}\right)\frac{\partial \bar{u}^h}{\partial u^m}.$$

Evaluating equation (7a) at P_o and substituting from equation (17) we obtain the interesting result

$$(18)\qquad {}_o\bar{G}^h_{jq} = ({}_oG^m_{rp}\,\delta^r_j\,\delta^p_q - {}_oG^m_{qj})\,\delta^h_m = 0.$$

Thus in the particular barred coordinate system defined in equation (14), the components of the symmetric connection G^i_{jk}

vanish at the point P_o. A coordinate system in which the components of a connection vanish at a point is called a <u>geodesic coordinate system</u> and the point at which the components of the connection vanish is called a <u>pole</u>. If a connection G^i_{jk} is not symmetric then there does not exist a geodesic coordinate system for that connection (see problem 1).

Geodesic coordinates are a useful tool for proving tensor identities with a minimum amount of calculation. Consider, for example, equation (10) in the case where covariant derivatives are taken with respect to a symmetric connection. At any point P_o we may choose a geodesic coordinate system such that all the components of the connection are zero at that point. Then at P_o covariant differentiation reduces to ordinary partial differentiation. Thus, by the product rule for partial differentiation, in the geodesic coordinates

(19) $${}_o\nabla(S \otimes T) - {}_o[(\nabla S) \otimes T) \oplus (S \otimes (\nabla T)] = 0.$$

However, if a tensor is zero in one coordinate system it is zero in all coordinate systems (this follows from the multilinear nature of the transformation law for tensor components). Thus in any coordinate system the above equation is true. However, the point P_o was arbitrarily chosen so at all points in the space we have the result in equation (10):

(10) $$\nabla(S \otimes T) = [(\nabla S) \otimes T] \oplus [S \otimes (\nabla T)].$$

10.4. THE CHRISTOFFEL SYMBOLS

We call the n^3 functions $[ij,k]$, $i,j,k = 1,\dots,n$, defined by

(20) $$[ij,k] = \frac{1}{2}\left(\frac{\partial g_{ik}}{\partial u^j} + \frac{\partial g_{jk}}{\partial u^i} - \frac{\partial g_{ij}}{\partial u^k}\right)$$

<u>Christoffel symbols of the first kind</u> and the n^3 functions Γ^ℓ_{ij}, $\ell,i,j = 1,\dots,n$, defined by

(21) $$\Gamma^\ell_{ij} = g^{\ell k}[ij,k]$$

Christoffel symbols of the second kind. From equations (20) and (21) and the symmetry of the metric tensor g_{jk} it follows that

(22a) $[ij,k] = [ji,k]$,

(22b) $\Gamma^{\ell}_{ij} = \Gamma^{\ell}_{ji}$,

and

(22c) $\partial g_{ik}/\partial u^j = [ij,k] + [kj,i]$.

The Christoffel symbols of the first kind are not tensor components since

$$\overline{[ij,k]} = \frac{1}{2}\left(\frac{\partial \bar{g}_{ik}}{\partial \bar{u}^j} + \frac{\partial \bar{g}_{jk}}{\partial \bar{u}^i} - \frac{\partial \bar{g}_{ij}}{\partial \bar{u}^k}\right)$$

$$= \frac{1}{2}\left[\frac{\partial}{\partial \bar{u}^j}\left(\frac{\partial u^\ell}{\partial \bar{u}^i}\frac{\partial u^m}{\partial \bar{u}^k} g_{\ell m}\right)\right.$$

$$\left. + \frac{\partial}{\partial \bar{u}^i}\left(\frac{\partial u^\ell}{\partial \bar{u}^j}\frac{\partial u^m}{\partial \bar{u}^k} g_{\ell m}\right) - \frac{\partial}{\partial \bar{u}^k}\left(\frac{\partial u^\ell}{\partial \bar{u}^i}\frac{\partial u^m}{\partial \bar{u}^j} g_{\ell m}\right)\right]$$

which becomes, using the product rule for differentiation,

$$\overline{[ij,k]} =$$

$$\frac{1}{2}\left[\frac{\partial u^\ell}{\partial \bar{u}^i}\frac{\partial u^m}{\partial \bar{u}^k}\frac{\partial u^n}{\partial \bar{u}^j}\frac{\partial g_{\ell m}}{\partial u^n} + \frac{\partial u^\ell}{\partial \bar{u}^j}\frac{\partial u^m}{\partial \bar{u}^k}\frac{\partial u^n}{\partial \bar{u}^i}\frac{\partial g_{\ell m}}{\partial u^n} - \frac{\partial u^\ell}{\partial \bar{u}^i}\frac{\partial u^m}{\partial \bar{u}^j}\frac{\partial u^n}{\partial \bar{u}^k}\frac{\partial g_{\ell m}}{\partial u^n}\right]$$

$$+ \frac{1}{2}\left[\frac{\partial}{\partial \bar{u}^j}\left(\frac{\partial u^\ell}{\partial \bar{u}^i}\frac{\partial u^m}{\partial \bar{u}^k}\right) + \frac{\partial}{\partial \bar{u}^i}\left(\frac{\partial u^\ell}{\partial \bar{u}^j}\frac{\partial u^m}{\partial \bar{u}^k}\right) - \frac{\partial}{\partial \bar{u}^k}\left(\frac{\partial u^\ell}{\partial \bar{u}^i}\frac{\partial u^m}{\partial \bar{u}^j}\right)\right] g_{\ell m}.$$

Renaming indices and expanding the second term on the right-hand side of the above equation, we get

(23) $$\overline{[ij,k]} = [\ell m,n]\frac{\partial u^\ell}{\partial \bar{u}^i}\frac{\partial u^m}{\partial \bar{u}^j}\frac{\partial u^n}{\partial \bar{u}^k} + g_{\ell m}\frac{\partial u^\ell}{\partial \bar{u}^k}\frac{\partial^2 u^m}{\partial \bar{u}^j \partial \bar{u}^i},$$

which is not the transformation law for tensor components.

Multiplying equation (23) by

$$\bar{g}^{ks}\partial u^t/\partial \bar{u}^s$$

and summing over k, one has

$$\overline{[ik,j]}\ \bar{g}^{ks}\frac{\partial u^t}{\partial \bar{u}^s} = [\ell m,n]\ \frac{\partial u^\ell}{\partial \bar{u}^i}\frac{\partial u^m}{\partial \bar{u}^j}\left(\frac{\partial u^n}{\partial \bar{u}^k}\frac{\partial u^t}{\partial \bar{u}^s}\bar{g}^{ks}\right)$$

$$+\ g_{\ell m}\left(\bar{g}^{ks}\ \frac{\partial u^\ell}{\partial \bar{u}^k}\frac{\partial u^t}{\partial \bar{u}^s}\right)\frac{\partial^2 u^m}{\partial \bar{u}^i\partial \bar{u}^j}$$

which can be re-expressed, using equation (21), as

$$\frac{\partial u^t}{\partial \bar{u}^s}\ \bar{\Gamma}^s_{ij} = \Gamma^t_{\ell m}\ \frac{\partial u^\ell}{\partial \bar{u}^i}\frac{\partial u^m}{\partial \bar{u}^j} + \frac{\partial^2 u^t}{\partial \bar{u}^i\partial \bar{u}^j}\ ,$$

where we have also used the transformation properties of the tensor components g^{nt} and $\bar{g}^{\ell t}$ (see Chapter 7). Multiplying by $\partial \bar{u}^h/\partial u^t$ and summing over t, we obtain finally

$$(24)\qquad \bar{\Gamma}^h_{ij} = \left(\Gamma^t_{\ell m}\ \frac{\partial u^\ell}{\partial \bar{u}^i}\frac{\partial u^m}{\partial \bar{u}^j} + \frac{\partial^2 u^t}{\partial \bar{u}^i\partial \bar{u}^j}\right)\frac{\partial \bar{u}^h}{\partial u^t}\ .$$

Comparison with equation (7), noting equation (22b), shows that the Christoffel symbols of the second kind are a <u>symmetric connection</u>. The covariant derivative of the metric tensor g_{ij} with respect to the Christoffel symbols of the second kind is zero since

$$g_{ij;\ell} = \frac{\partial g_{ij}}{\partial u^\ell} - g_{kj}\Gamma^k_{i\ell} - g_{ik}\Gamma^k_{j\ell},$$

which from equations (21) and (22c)

$$= [i\ell,j] + [j\ell,i] - g_{kj}g^{ks}[i\ell,s] - g_{ik}g^{ks}[j\ell,s]$$

$$= [i\ell,j] + [j\ell,i] - \delta^s_j[i\ell,s] - \delta^s_i[j\ell,s]$$

$$= 0.$$

Let $G^s_{i\ell}$ be a connection for which the torsion tensor is zero and the covariant derivative of the metric tensor is zero. Then

(25) $$0 = g_{ij;\ell} = \frac{\partial g_{ij}}{\partial u^\ell} - g_{kj}G^k_{i\ell} - g_{ik}G^k_{j\ell}$$

and

(26a) $$\frac{\partial g_{ij}}{\partial u^\ell} = g_{kj}G^k_{i\ell} + g_{ik}G^k_{j\ell}.$$

Renaming indices in equation (26a) we also have

(26b) $$\frac{\partial g_{j\ell}}{\partial u^i} = g_{k\ell}G^k_{ji} + g_{jk}G^k_{\ell i}$$

and

(26c) $$\frac{\partial g_{i\ell}}{\partial u^j} = g_{k\ell}G^k_{ij} + g_{ik}G^k_{\ell j}.$$

Adding equations (26a) and (26b) and subtracting equation (26c) gives

(27) $$\left(\frac{\partial g_{ij}}{\partial u^\ell} + \frac{\partial g_{j\ell}}{\partial u^i} - \frac{\partial g_{i\ell}}{\partial u^j}\right) = g_{kj}(G^k_{i\ell} + G^k_{\ell i}) + g_{k\ell}(G^k_{ji} - G^k_{ij}) + g_{ik}(G^k_{j\ell} - G^k_{\ell j})$$

where we have used the symmetry of the metric tensor g_{jk}. However, we also have that $G^k_{i\ell}$ is a symmetric connection. Hence the above equation becomes

$$g_{kj}G^k_{i\ell} = \frac{1}{2}\left(\frac{\partial g_{ij}}{\partial u^\ell} + \frac{\partial g_{j\ell}}{\partial u^i} - \frac{\partial g_{i\ell}}{\partial u^j}\right) = [i\ell,j].$$

Multiplying by g^{sj} and summing over j, we get

(28) $$G^s_{i\ell} = g^{sj}[i\ell,j] = \Gamma^s_{i\ell}.$$

Hence the Christoffel symbols of the second kind are the unique connection for which the torsion tensor vanishes and the covariant derivative of the metric tensor is zero.

The covariant derivative with respect to the Christoffel symbol of the second kind of the Kronecker delta is zero:

$$\delta^i_{j;\ell} = \delta^s_j \Gamma^j_{sj} - \delta^j_s \Gamma^s_{i\ell} = \Gamma^j_{i\ell} - \Gamma^j_{i\ell} = 0.$$

Taking the covariant derivative with respect to the Christoffel symbols of the second kind of the equation

$$g^{ij} g_{j\ell} = g^i_\ell$$

gives $g^{ij}{}_{;s}\, g_{j\ell} = 0$,

which implies that

$$g^{ij}{}_{;s} = 0. \tag{29}$$

A coordinate system in which the Christoffel symbols of the second kind vanish at a point P_o is called locally inertial (with respect to P_o). We shall see in the following chapter that in the general theory of relativity an observer using a locally inertial coordinate system (with respect to a point P_o) feels no gravitational field at the point P_o.

10.5. PARALLEL TRANSPORT

Let ${}_oT$ be a tensor at some point P_o (with coordinates ${}_ou^i$) and let C be a curve $u^i(s)$ parametrized by its arc length s which originates at the point P_o (i.e. $u^i(0) = {}_ou$). From ${}_oT$ we may define a tensor field $T(u^1(s),\ldots,u^n(s)) = T(s)$ on the curve C as the solution to the covariant differential equations

$$\nabla_{\vec{u}} T = 0 \tag{30a}$$

with the initial conditions

(30b) $T({}_ou^1,\ldots,{}_ou^n) = T(0) = {}_oT$

where $\vec{u}$ is the tangent vector to the curve (i.e. the vector with components du^i/ds, $i = 1,2,\ldots,n$) and the covariant differentiation is done with respect to the connection G^i_{jk}.

If T is an rth-rank tensor, equation (30a) is n^r first-order, linear, coupled, ordinary differential equations for T(s) and equation (30b) is n^r initial conditions. It follows that there exists a unique tensor field T(s) that is a solution to the 'initial' value problem given by equations (30a) and (30b). The tensor field T(s) is said to be the result of parallel transport of the given tensor ${}_oT$ along C, and the tensors in the field are said to be parallel with respect to C in the sense of the connection G^i_{jk}.

Parallel transport provides a means for comparing tensors at different points. If ${}_oT$ is a tensor at the point P_o and ${}_1S$ is a tensor at a point P_1 different from P_o, we can compare the two tensors by parallel transporting ${}_oT$ to the point P_1 along some curve joining the points P_o and P_1. Of course the result will in general depend on what curve is chosen, but it is nonetheless a useful mode of comparison. For example, suppose $A(u^1,\ldots u^n)$ is a tensor field and we wish to get an indication of the rate of change of the tensor $A(u^1,\ldots,u^n$ as we go along a curve C defined by $u^i = u^i(s)$. Note that the tensor field $A(u^1,\ldots,u^n)$ is not necessarily parallel with respect to the curve C. From our experience with calculus it is evident that a useful measure of the rate of change of $A(u^1,\ldots,u^n)$ along C at a point P_o (with coordinates $u^i(s_o)$) is given by the following limit:

$$\lim_{s_1\to s_o} \frac{[A(u^1(s_1),\ldots,u^n(s_1))]_{\substack{\text{parallel}\\ \text{transported to}\\ P_o \text{ along } C}} - A(u^1(s_o),\ldots,u^n(s_o))}{s_1 - s_o} \tag{31}$$

As a specific example, we consider a second-rank mixed tensor field with components $A^i_j(u^1,\ldots,u^n)$, $i,j = 1,2,\ldots,n$. The tensor field that results from parallel transport of $A^i_j(u^1(s_1),\ldots,u^n(s_1))$ satisfies, according to the definition (30a), the set of equations

$$A^i_{j;k}\,\frac{du^k}{ds} = 0$$

along the curve C. From the definition (9), one then has

$$\frac{\partial A^i_j}{\partial u^k}\,\frac{du^k}{ds} = (G^p_{jk}\,A^i_p - G^i_{pk}\,A^p_j)\,\frac{du^k}{ds}\,,$$

from which it follows that

$$(32)\qquad dA^i_j = [G^p_{jk}\,A^i_p - G^i_{pk}\,A^p_j]\,(du^k).$$

Here dA^i_j is the change in the components of $A^i_j(u^1(s_1),\ldots,u^n(s_1))$ as it is parallel transported from $u^i(s_1)$ to $u^i(s_o)$ in the limit where $|s_1 - s_o|$ is small, that is

$$(33)\qquad dA^i_j = [A^i_j(u^1(s_1),\ldots,u^n(s_1))]\ \text{parallel transported to } P_o \text{ along C} - A^i_j(u^1(s_1),\ldots,u^n(s_1)).$$

Further, du^k is the change in the coordinates as one goes from $u^k(s_1)$ to $u^k(s_o)$ in the limit where $|s_1 - s_o|$ is small, namely

$$(34)\qquad du^i = u^k(s_o) - u^k(s_1).$$

It follows from equations (32), (33), and (34) that

$$(35)\qquad \lim_{s_1\to s_o} \frac{[A^i_j(u^1(s_1),\ldots,u^n(s_1)]_{\text{parallel transported to } P_o \text{ along c}} - A^i_j(u^1(s_o),\ldots,u^n(s_o))}{s_1 - s_o}$$

$$= \lim_{s_1 \to s_o} \frac{A^i_j(u^1(s_1),\dots,u^n(s_1)) + (G^i_{pk}A^p_j - G^p_{jk}A^i_p)(u^k(s_1) - u^k(s_o))}{s_1 - s_o} - \frac{A^i_j(u^1(s_o),\dots,u^n(s_o))}{s_1 - s_o}$$

$$= \lim_{s_1 \to s_o} \frac{A^i_j(u^1(s_1),\dots,u^n(s_1)) - A^i_j(u^1(s_o),\dots,u^n(s_o))}{s_1 - s_o}$$

$$+ {}_o(G^k_{pk}\ A^p_j - G^p_{jk}\ A^i_p) \lim_{s_1 \to s_o} \left(\frac{u^k(s_1) - u^k(s_0)}{s_1 - s_o} \right)$$

$$= {}_o\left(\frac{dA^i_j}{ds} + (G^i_{pk}\ A^p_j - G^p_{jk}\ A^i_p) \frac{du^k}{ds} \right) = {}_o\left(A^i_{j;k} \frac{du^k}{ds} \right).$$

Since (35) holds for every pair of tensor indices i and j, we can write

(36) $\quad {}_o(\nabla_{\vec{u}} A) =$

$$\lim_{s_1 \to s_o} \frac{[A(u^1(s_1),\dots,u^n(s_1))]_{\substack{\text{parallel}\\ \text{transported}\\ \text{to } P_o \text{ along } c}} - A(u^1(s_o),\dots,u^n(s_o))}{s_1 - s_o}.$$

Equation (36) is valid for an arbitrary tensor field A (with differentiability conditions placed on its components) even though it was proved only for the case of a mixed second-rank tensor field. It provides a geometric interpretation of covariant directional derivatives.

If $\vec{\nu}$ and $\vec{\omega}$ are two vector fields (with components ν^j and ω^j, $j = 1,2,\ldots,n$, respectively) that are parallel with respect to a curve C defined by $u^i = u^i(s)$ in the sense of the Christoffel symbols, and if the scalar product $\vec{\nu}\cdot\vec{\omega}$ is a constant along C, then

$$(37) \qquad 0 = \frac{d}{ds}(\nu_m\omega^m) = \frac{\partial}{\partial u^j}(\nu_m\omega^m)\frac{du^j}{ds} = \frac{\partial}{\partial u^j}(g_{\ell m}\nu^\ell\omega^m)\frac{du^j}{ds}$$

$$= g_{\ell m;j}\nu^\ell\omega^m\frac{du^j}{ds} + g_{\ell m}(\nu^\ell_{;j}\omega^m + \omega^m_{;j}\nu^\ell)\frac{du^j}{ds},$$

where we have noted the implications of equation (10). However, since $\vec{\nu}$ and $\vec{\omega}$ are parallel with respect to C,

$$\nu^\ell_{;j}\frac{du^j}{ds} = \omega^\ell_{;j}\frac{du^j}{ds} = 0$$

and the above equation becomes

$$(38) \qquad (g_{\ell m;j})\frac{du^j}{ds}\nu^\ell\omega^m = 0.$$

If equation (38) is to hold for arbitrary vectors and curves, we must have $g_{\ell m;j} = 0$. Thus the Christoffel symbols of the second kind are the unique symmetric connection for which the inner product between any two vector fields parallel to an arbitrary curve C remains constant along C. Since the magnitude of a vector and the angle between two vectors are expressible in terms of inner products, these quantities are also constant along C if the Christoffel symbols of the second kind are used as the connection.

10.6. AUTOPARALLEL CURVES AND GEODESICS

An autoparallel curve is one which parallel transports its own tangent vector. Thus if a curve $u^i(s)$ is autoparallel it satisfies the differential equations

$$(39a) \qquad \nabla_{\vec{u}}\vec{u} = 0$$

where $\vec{u}$ is the tangent vector field du^k/ds. In component form equation (39a) reads

$$\text{(39b)} \quad \frac{d^2u^i}{ds^2} + G^i_{jk}\frac{du^j}{ds}\frac{du^k}{ds} = 0.$$

The set of equations (39b) consists of n ordinary differential equations for an autoparallel curve. They are second-order differential equations, and 2n conditions are required to specify a particular autoparallel curve. The initial value problem where a point on the curve and a tanget vector at that point are specified has a unique solution; however, the boundary value problem where two points on the autoparallel curve are specified may not have a unique solution.

Consider the configuration space of a conservative system of N particles each of mass m_i, $i = 1,2,\ldots,N$, and position $\vec{r}_i$, $i = 1,2,\ldots,N$. Let $u^1,\ldots,u^n$ be a set of generalized coordinates. If we do not endow the space with the metric given by equation (54) of Chapter 7 but instead use

$$\text{(40)} \quad g_{jk} = T \sum_{i=1}^{N} M_i \frac{\partial\vec{r}_i}{\partial u^j} \cdot \frac{\partial\vec{r}_i}{\partial u^k}$$

where T is the kinetic energy, the system will trace out a curve in configuration space such that the length is an extremum.

In a n-dimensional Riemannian space a curve whose length is an extremum is called a <u>geodesic</u>. Let C be the curve $u^i(t)$ (parametrized by an arbitrary parameter t) which joins the points P_o and P_1, such that $u^i(t_o)$ are the coordinates of P_o and $u^i(t_1)$ are the coordinates of P_1. Then the distance between P_o and P_1 measured along the curve C is

$$\text{(41)} \quad s = \int_{t_o}^{t_1} \sqrt{e g_{ij}(u^k)\frac{du^i}{dt}\frac{du^j}{dt}}\, dt.$$

Suppose C is a geodesic. Let $\underline{C}$ be an arbitrary curve near C defined by $\underline{u}^i(t) = u^i(t) + \delta u^i(t)$ where $\delta u^i(t_o) = \delta u^i(t_1) = 0$ so that $\underline{C}$ also joins the points P_o and P_1. The distance between the points P_o and P_1 measured along $\underline{C}$ is given by

$$(42) \qquad \underline{s} = \int_{t_o}^{t_1} \sqrt{e g_{ij}(\underline{u}^k) \frac{d\underline{u}^i}{dt} \frac{d\underline{u}^j}{dt}}\, dt.$$

From the definition of the barred coordinates

$$g_{ij}(\underline{u}^k) \frac{d\underline{u}^i}{dt} \frac{d\underline{u}^j}{dt}$$

$$\simeq \left(g_{ij}(u^k) + \frac{\partial g_{ij}}{\partial u^s} \delta u^s\right)\left(\frac{du^i}{dt} + \frac{d(\delta u^i)}{dt}\right)\left(\frac{du^j}{dt} + \frac{d(\delta u^j)}{dt}\right)$$

$$\simeq g_{ij} \frac{du^i}{dt} \frac{du^j}{dt} + 2g_{ij} \frac{d(\delta u^i)}{dt} \frac{du^j}{dt} + \left(\frac{\partial g_{ij}}{\partial u^s} \delta u^s\right) \frac{du^i}{dt} \frac{du^j}{dt},$$

where $g_{ij} = g_{ij}(u^k)$ and we have expressed the result to first order in small quantities. Using this result in equation (42), we get

$$\underline{s} = \int_{t_o}^{t_1} \sqrt{e g_{ij} \frac{du^i}{dt} \frac{du^j}{dt}}$$

$$\times \left[1 + \frac{2g_{ij} \frac{d(\delta u^i)}{dt} \frac{du^j}{dt} + \frac{\partial g_{ij}}{\partial u^s} (\delta u^s) \frac{du^i}{dt} \frac{du^j}{dt}}{g_{ij} \frac{du^i}{dt} \frac{du^j}{dt}}\right]^{\frac{1}{2}} dt.$$

Finally, using the approximation $(1 + x)^{\frac{1}{2}} \simeq 1 + x/2$ for $x << 1$, we obtain

$$\underline{s} \approx \int_{t_o}^{t_1} \sqrt{e g_{ij} \frac{du^i}{dt} \frac{du^j}{dt}} \times \left[1 + \frac{g_{ij} \dfrac{d(\delta u^i)}{dt} \dfrac{du^j}{dt} + \dfrac{1}{2} \dfrac{\partial g_{ij}}{\partial u^s} \delta u^s \dfrac{du^i}{dt} \dfrac{du^j}{dt}}{g_{ij} \dfrac{du^i}{dt} \dfrac{du^j}{dt}} \right] dt.$$

The difference in distance between the paths C and C̲ is given by

(43)
$$\delta s = \underline{s} - s = \int_{t_o}^{t_1} \frac{g_{ij} \dfrac{d(\delta u^i)}{dt} \dfrac{du^j}{dt} + \dfrac{1}{2} \dfrac{\partial g_{ij}}{\partial u^s} (\delta u^s) \dfrac{du^i}{dt} \dfrac{du^j}{dt}}{\sqrt{e g_{ij} \dfrac{du^i}{dt} \dfrac{du^j}{dt}}} dt.$$

Choosing the parameter of integration t to be s, the arc length along the curve C, and using equation (48) of Chapter 7, we have

$$\delta s = \underline{s} - s = \int_{s_o}^{s_1} \left(g_{ij} \frac{d(\delta u^i)}{ds} \frac{du^j}{ds} + \frac{1}{2} \frac{\partial g_{ij}}{\partial u^k} \delta u^k \frac{du^i}{ds} \frac{du^j}{ds} \right) ds.$$

Integration by parts of the first term in the above expression then yields

$$\delta s = \underline{s} - s = \int_{s_o}^{s_1} \left[-\delta u^i \frac{d}{ds}\left(g_{ij} \frac{du^j}{ds} \right) + \frac{1}{2} \frac{\partial g_{ij}}{\partial u^k} \delta u^k \frac{du^i}{ds} \frac{du^j}{ds} \right] ds + g_{ij} \frac{du^j}{ds} \delta u^i \Bigg|_{s_o}^{s_1}$$

But $\delta u^i(s_1) = \delta u^i(s_o) = 0$. Therefore

$$\delta s = \int_{s_o}^{s_1} \left[-\delta u^i \frac{\partial g_{ij}}{\partial u^k} \frac{du^k}{ds} \frac{du^j}{ds} - \delta u^i g_{ij} \frac{d^2u^j}{ds^2} + \frac{1}{2} \frac{\partial g_{ij}}{\partial u^k} \delta u^k \frac{du^i}{ds} \frac{du^j}{ds} \right] ds.$$

Renaming summation indices, this equation can be rewritten as

$$(44) \qquad \delta s = \int_{s_o}^{s_1} -\delta u^i \left[\frac{\partial g_{ij}}{\partial u^k} \frac{du^k}{ds} \frac{du^j}{ds} - \frac{1}{2} \frac{\partial g_{kj}}{\partial u^i} \frac{du^k}{ds} \frac{du^j}{ds} + g_{ij} \frac{d^2u^j}{ds^2} \right] ds.$$

Since the length of C is an extremum, $\delta s = 0$ for arbitrary δu^i (i.e. for arbitrary nearby curves C̲). Hence the integrand itself must vanish:

$$\left[\frac{\partial g_{ij}}{\partial u^k} \frac{du^k}{ds} \frac{du^j}{ds} - \frac{1}{2} \frac{\partial g_{kj}}{\partial u^i} \frac{du^k}{ds} \frac{du^j}{ds} + g_{ij} \frac{d^2u^j}{ds^2} \right] = 0$$

or, equivalently,

$$(45) \qquad \frac{1}{2} \frac{\partial g_{ij}}{\partial u^k} \frac{du^k}{ds} \frac{du^j}{ds} + \frac{1}{2} \frac{\partial g_{ik}}{\partial u^j} \frac{du^j}{ds} \frac{du^k}{ds} - \frac{1}{2} \frac{\partial g_{kj}}{\partial u^i} \frac{du^k}{ds} \frac{du^j}{ds} + g_{ij} \frac{d^2u^j}{ds^2} = 0.$$

Using the definition of the Christoffel symbols of the first kind, this can be expressed as

$$(46a) \qquad g_{ij} \frac{d^2u^j}{ds^2} + [kj,i] \frac{du^k}{ds} \frac{du^j}{ds} = 0$$

or, multiplying by $g^{\ell i}$ and summing over i, as

$$(46b) \qquad \frac{d^2u^\ell}{ds^2} + \Gamma^\ell_{kj} \frac{du^k}{ds} \frac{du^j}{ds} = 0.$$

Hence <u>geodesics are autoparallel curves</u> with the <u>Christoffel symbols of the second kind</u> as the connection.

In Euclidean space with coordinates x^i and metric tensor

$$g_{ij} = 0 \text{ for } i \neq j, \quad g_{ij} = 1 \text{ for } i = j,$$

equation (46b) reduces to $d^2x^\ell/ds^2 = 0$, which implies that $x^\ell = A^\ell s + D^\ell$. Thus the geodesics are straight lines. The constants A^ℓ and D^ℓ are determined by the initial or boundary conditions of the geodesic.

A curve of zero length is called minimal. A minimal curve need not be a geodesic. Consider, for example, a 3-dimensional Riemannian space with metric

$$g_{ij} = 0 \text{ for } i \neq j, \quad g_{11} = g_{22} = -g_{33} = 1.$$

Then $[kj,i] = 0$ and the differential equations for a geodesic are again given by

$$\frac{d^2u^\ell}{ds^2} = 0, \quad \ell = 1,2,3.$$

The curve

$$u^1(s) = \int_{s_o}^{s} f(t)\,dt, \quad u^2(s) = \int_{s_o}^{s} g(t)\,dt,$$

(47)

$$u^3(s) = \int_{s_o}^{s} \sqrt{f^2(t) + g^2(t)}\,dt$$

is minimal for arbitrary $f(t)$ and $g(t)$ but a geodesic only if

$$\frac{df(t)}{dt} = \frac{dg(t)}{dt} = 0.$$

10.7. GEODESICS ON THE SURFACE OF A SPHERE (a worked example)

The 2-dimensional surface of a unit sphere is defined by

$$x_1^2 + x_2^2 + x_3^2 = 1$$

where x_1, x_2, x_3 are Cartesian coordinates of a 3-dimensional Euclidean space in which the sphere is embedded. Points on the surface of the sphere may be put into a correspondence with ordered pairs (u^1, u^2) where u^1 is the colatitude and u^2 the longitude (see Figure 10.1). Explicitly,

$$\begin{aligned} x_1 &= \phi_1(u^1,u^2) = \sin u^1 \cos u^2, \\ x_2 &= \phi_2(u^1,u^2) = \sin u^1 \sin u^2, \qquad (48) \\ x_3 &= \phi_3(u^1,u^2) = \cos u^1. \end{aligned}$$

The metric tensor g_{jk} is defined by equation (7) of Chapter 7. In the present case, we obtain

$$(49) \qquad g_{11} = 1, \quad g_{22} = \sin^2 u^1, \quad g_{ij} = 0,\ i \neq j.$$

For any 2-dimensional Riemannian space with $g_{21} = g_{12} = 0$ equations (45) for a geodesic reduce to the following two equations:

$$g_{11}\left(\frac{d^2u^1}{ds^2}\right) - \frac{1}{2}\frac{\partial g_{22}}{\partial u^1}\left(\frac{du^2}{ds}\right)^2 + \frac{\partial g_{11}}{\partial u^2}\left(\frac{du^1}{ds}\right)\left(\frac{du^2}{ds}\right) + \frac{1}{2}\frac{\partial g_{11}}{\partial u^1}\left(\frac{du^1}{ds}\right)^2 = 0,$$

$$g_{22}\left(\frac{d^2u^2}{ds^2}\right) - \frac{1}{2}\frac{\partial g_{11}}{\partial u^2}\left(\frac{du^1}{ds}\right)^2 + \frac{\partial g_{22}}{\partial u^1}\left(\frac{du^1}{ds}\right)\left(\frac{du^2}{ds}\right) + \frac{1}{2}\frac{\partial g_{22}}{\partial u^2}\left(\frac{du^2}{ds}\right)^2 = 0.$$

For the 2-dimensional surface of a sphere these equations become

$$(50a) \qquad \frac{d^2u^1}{ds^2} - \sin 2\theta\left(\frac{du^2}{ds}\right)^2 = 0,$$

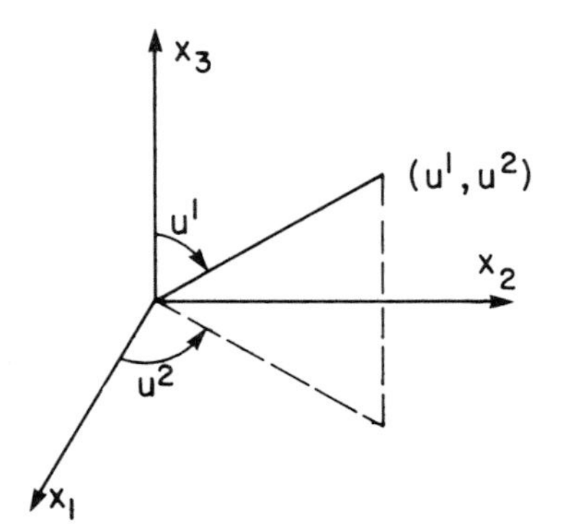

Figure 10.1. Colatitude and longitude on the surface of a sphere.

(50b) $\frac{d}{ds}\left[\left(\sin u^1\right)^2 \frac{du^2}{ds}\right] = 0.$

Integrating the second of these equations once gives

$$\left(\sin u^1\right)^2 \left(\frac{du^2}{ds}\right) = \frac{1}{h}$$

where h is an arbitrary constant. Therefore

(51) $ds = h(\sin u^1)^2\, du^2.$

From the general expression for a length differential (equation (46), Chapter 7)

(52) $ds^2 = (du^1)^2 + (\sin u^1)^2(du^2)^2.$

Combining expressions (51) and (52) we obtain

(53) $du^1 = \sin u^1 \sqrt{(h \sin u^1)^2 - 1}\,.\, du^2.$

Integrating equation (53) one obtains a parametric equation for the geodesic involving u^1 and u^2:

$$u^2 = {}_0u^2 + \cos^{-1}(k \cot u^1),$$

where $k = 1/\sqrt{h^2 - 1}$ and ${}_0u^2$ is an arbitrary constant. Rearranging this equation and replacing u^1 and u^2 by the more familiar angles θ, ϕ for the colatitude and longitude, we obtain

$$\text{(54)} \qquad k \cos\theta = \cos\phi_o \cos\phi + \sin\phi_o \sin\phi$$

for the equation of a geodesic on the surface of a sphere.

A great circle is a curve that results from the intersection of a plane passing through the centre of the sphere with the surface of the sphere. Such a plane is defined by its normal $\hat{n}$. Points (x_1,x_2,x_3) in the plane then satisfy

$$\text{(55)} \qquad n_1x_1 + n_2x_2 + n_3x_3 = 0, \qquad n_1^2 + n_2^2 + n_3^2 = 1.$$

The intersection of the plane with the surface of the sphere

$$\text{(56)} \qquad x_1^2 + x_1^2 + x_3^2 = 1$$

yields the great circle given parametrically by equations (55) and (56). Replacing the Cartesian coordinates in (55) and (56) by their spherical polar equivalents yields the parametric equation

$$\text{(57a)} \qquad n_1\sin\theta \cos\phi + n_2\sin\theta \sin\phi + n_3\cos\theta = 0$$

or, alternatively,

$$\text{(57b)} \qquad \frac{n_1}{\sqrt{n_1^2 + n_2^2}} \cos\phi + \frac{n_2}{\sqrt{n_1^2 + n_2^2}} \sin\phi + \sqrt{\frac{1}{n_1^2 + n_2^2} - 1}\, \cot\theta = 0,$$

which is just equation (54) for a geodesic with

$$\text{(58)} \qquad \phi_o = \tan^{-1}\left(\frac{n_2}{n_1}\right), \qquad k = \sqrt{\frac{1}{n_1^2 + n_2^2} - 1}\ .$$

10.8. THE RIEMANN-CHRISTOFFEL CURVATURE TENSOR

The second-order covariant derivative with respect to the Christoffel symbols of the second kind of an arbitrary covariant vector v_k is denoted as $v_{k;i;j}$ and given by

$$v_{k;i;j} = \frac{\partial}{\partial u^j}\left(\frac{\partial v_k}{\partial u^i} - v_\ell \Gamma^\ell_{ki}\right) - \left(\frac{\partial v_s}{\partial u^i} - v_\ell \Gamma^\ell_{si}\right)\Gamma^s_{kj} - \left(\frac{\partial v_k}{\partial u^s} - v_\ell \Gamma^\ell_{ks}\right)\Gamma^s_{ij}$$

or

$$(59) \quad v_{k;i;j} = \frac{\partial^2 v_k}{\partial u^j u^i} - \frac{\partial v_\ell}{\partial u^j}\Gamma^\ell_{ki} - v_\ell \frac{\partial}{\partial u^j}\Gamma^\ell_{ki} - \frac{\partial v_s}{\partial u^i}\Gamma^s_{kj} + v_\ell \Gamma^\ell_{si}\Gamma^s_{kj} - \frac{\partial v_k}{\partial u^s}\Gamma^s_{ij} + v_\ell \Gamma^\ell_{ks}\Gamma^s_{ij}.$$

If we subtract from this expression a similar expression with indices i and j interchanged after renaming some dummy indices we get the expression

$$(60) \quad v_{k;i;j} - v_{k;j;i} = \left[\frac{\partial}{\partial u^i}\Gamma^\ell_{kj} - \frac{\partial}{\partial u^j}\Gamma^\ell_{ki} + \Gamma^\ell_{is}\Gamma^s_{kj} - \Gamma^\ell_{js}\Gamma^s_{ki}\right]v_\ell.$$

Thus, unlike ordinary partial derivatives, second-order covariant derivatives do not necessarily commute. We now define the expression

$$(61) \quad R^\ell_{kij} = \frac{\partial}{\partial u^i}\Gamma^\ell_{kj} - \frac{\partial}{\partial u^j}\Gamma^\ell_{ki} + \Gamma^\ell_{is}\Gamma^s_{kj} - \Gamma^\ell_{js}\Gamma^s_{ki}$$

to simplify the appearance of equation (60):

$$(62) \quad v_{k;i;j} - v_{k;j;i} = R^\ell_{kij}\, v_\ell.$$

From the quotient theorem it follows that R^ℓ_{kij} is a tensor. It is called the Riemann-Christoffel curvature tensor. Since R^ℓ_{kij} is dependent on the metric tensor it must be influenced by the 'curved nature' of the space. In fact a Riemannian space is flat if and only if the Riemann-Christoffel curvature tensor is identically zero. Some insight into the relationship between R^ℓ_{kij} and the 'curvature' of a Riemannian space can be gained by considering two nearby geodesics $u^i(s)$ and $u^i(s) + \delta u^i(s)$. Let $\delta\vec{n}$ be the vector with components $\delta u^i(s)$. Then

$\nabla_{\vec{u}}\nabla_{\vec{u}}\delta\vec{n}$ is a measure of the relative 'acceleration' with respect to u^i of the two nearby geodesics towards each other. To emphasize this, we denote the components of $\nabla_{\vec{u}}\nabla_{\vec{u}}\delta\vec{n}$ by $D^2(\delta u^i)/Ds^2$, $i = 1,2,\ldots,n$. Since $u^i(s)$ and $u^i(s) + \delta u^i(s)$ are geodesics, we have

(63) $$0 = \frac{d^2 u_i}{ds^2} + \Gamma^i_{\nu\lambda}(u^j)\,\frac{du^\nu}{ds}\,\frac{du^\lambda}{ds},$$

(64) $$0 = \frac{d^2}{ds^2}(u^i + \delta u^i) + \Gamma^i_{\nu\lambda}(u^j + \delta u^j)\,\frac{d}{ds}(u^\nu + \delta u^\nu)\,\frac{d}{ds}(u^\lambda + \delta u^\lambda).$$

Taking the difference between the above equations to first order in δu^i gives

(65) $$0 = \frac{d^2\delta u^i}{ds^2} + \frac{\partial\Gamma^i_{\nu\lambda}}{\partial u^p}\,\delta u^p\,\frac{du^\nu}{ds}\,\frac{du^\lambda}{ds} + 2\Gamma^i_{\nu\lambda}\frac{du^\nu}{ds}\,\frac{d\delta u^\lambda}{ds}.$$

Now

$$\frac{D^2}{Ds^2}\,\delta u^i = \left[\delta u^i{}_{;k}\,\frac{du^k}{ds}\right]_{;q}\left(\frac{du^q}{ds}\right)$$

$$= \frac{d^2\delta u^i}{ds^2} + \frac{\partial\Gamma^i_{\lambda k}}{\partial u^p}\left(\frac{du^k}{ds}\right)\left(\frac{du^p}{ds}\right)\delta u^\lambda + 2\Gamma^i_{pq}\left(\frac{du^q}{ds}\right)\left(\frac{d\delta u^p}{ds}\right)$$

$$+ \Gamma^i_{\lambda k}\,\delta u^\lambda\,\frac{d^2u^k}{ds^2} + \Gamma^p_{jk}\Gamma^i_{pq}\left(\frac{du^k}{ds}\right)\left(\frac{du^q}{ds}\right)\delta u^j.$$

Substituting into the above equation the value of $d^2\delta u^2/ds^2$ given by equation (65) and the value of d^2u^i/ds^2 given by equation (63) yields

(66) $$\frac{D^2}{Ds^2}\,\delta u^i = R^i_{\nu\mu p}\,\delta u^\mu\,\frac{du^\nu}{ds}\,\frac{du^p}{ds}.$$

Thus the Riemann-Christoffel curvature tensor is the multilinear operator that maps the tangent vectors of two nearby geodesics and their separation vector into a vector that represents the acceleration of the two geodesics towards each other. The acceleration of nearby geodesics towards each other is a measure of the 'curvature' of a space. Consider, for example, the 2-dimensional surface of a sphere of radius R. In this space

$$ds^2 = R^2(du^1) + R^2\sin^2 u^1(du^2)^2$$

and the geodesics are great circles. The larger the radius of the sphere the slower two nearby geodesics accelerate towards each other. But the 'curvature' of a sphere is determined directly by its radius. Thus the relative acceleration of nearby geodesics is indeed a measure of the curved nature of the space. In Euclidean space geodesics are straight lines and hence do not 'accelerate' towards each other.

10.9. PROPERTIES OF THE RIEMANN-CHRISTOFFEL CURVATURE TENSOR

The components of the Riemann-Christoffel curvature tensor R^{ℓ}_{ijk} are called Riemann symbols of the second kind and the components of the associate tensor R_{hijk} defined by

$$(67) \qquad R_{hijk} = g_{\ell h}R^{\ell}_{ijk}$$

are called Riemann symbols of the first kind. From equations (67) and (61) it follows that

$$R_{hijk} = g_{\ell h}\left[\frac{\partial}{\partial u^j}\Gamma^{\ell}_{ik} - \frac{\partial}{\partial u^k}\Gamma^{\ell}_{ij} + \Gamma^{\ell}_{js}\Gamma^{s}_{ik} - \Gamma^{\ell}_{ks}\Gamma^{s}_{ij}\right].$$

Using equation (21) this becomes

$$R_{hijk} = g_{\ell h}\left[\frac{\partial}{\partial u^j}\Gamma^{\ell}_{ik} - \frac{\partial}{\partial u^k}\Gamma^{\ell}_{ij}\right] + [js,h]\Gamma^{s}_{ik} - [ks,h]\Gamma^{s}_{ij}.$$

But

$$g_{\ell h}\frac{\partial}{\partial u^j}\Gamma^{\ell}_{ik} = \frac{\partial}{\partial u^j}(g_{\ell h}\Gamma^{\ell}_{ik}) - \Gamma^{\ell}_{ik}\frac{\partial g_{\ell h}}{\partial u^j}.$$

Applying equations (21) and (22) to the above equation we get

$$g_{\ell h} \frac{\partial}{\partial u^j} \Gamma^{\ell}_{ik} = \frac{\partial}{\partial u^j} [ik,h] - \Gamma^{\ell}_{ik} [\ell j,h] - \Gamma^{\ell}_{ik}[hj,\ell].$$

Thus

$$R_{hijk} = \frac{\partial}{\partial u^j}[ik,h] - \Gamma^{\ell}_{ik}([\ell j,h] + [hj,\ell]) - \frac{\partial}{\partial u^k}[ij,h]$$

$$+ \Gamma^{\ell}_{ij}([\ell k,h] + [hk,\ell]) + [js,h]\ \Gamma^{s}_{ik} - [ks,h]\ \Gamma^{s}_{ij}.$$

On changing dummy index s to ℓ and using (22a) this equation simplifies to

$$R_{hijk} = \frac{\partial}{\partial u^j} [ik,h] - \frac{\partial}{\partial u^k} [ij,h] + \Gamma^{\ell}_{ij}[hk,\ell] - \Gamma^{\ell}_{ik}[hj,\ell].$$

In view of equations (20) and (21) we can also write

$$(68)\qquad R_{hijk} = \frac{1}{2}\left[\frac{\partial^2 g_{hk}}{\partial u^i \partial u^j} + \frac{\partial^2 g_{ij}}{\partial u^k \partial u^h} - \frac{\partial^2 g_{hj}}{\partial u^i \partial u^k} - \frac{\partial^2 g_{ik}}{\partial u^j \partial u^h}\right]$$

$$+ g^{\ell m}([ij,m][hk,\ell] - [ik,m][hj,\ell]).$$

A number of symmetry properties of R_{hijk} follow from the form of equation (68)

$$(69)\qquad R_{hijk} = -R_{ihjk} = -R_{hikj} = R_{jkhi}$$

and

$$(70)\qquad R_{hijk} + R_{hjki} + R_{hkij} = 0.$$

From equation (69) it follows that no more than two indices of a Riemann symbol of the first kind may be alike without the symbol vanishing.

In geodesic coordinates with a pole at the point P_0 covariant differentiation with respect to the Christoffel symbols of the second kind reduces to ordinary partial differentiation (at the point P_o). Thus

$$_o\left[R^{\ell}_{kij;p} - \frac{\partial}{\partial u^p} R^{\ell}_{kij}\right] = 0.$$

From the definition of R^{ℓ}_{kij} it follows that

$$_o\left[R^{\ell}_{kij;p} - \frac{\partial^2 \Gamma^{\ell}_{kj}}{\partial u^i \partial u^p} + \frac{\partial^2 \Gamma^{\ell}_{ki}}{\partial u^j \partial u^p}\right] = 0.$$

Interchanging i, j, and p cyclically and adding gives

$$_o\left[R^{\ell}_{kij;p} + R^{\ell}_{kjp;i} + R^{\ell}_{kpi;j}\right] = 0.$$

But if a tensor is zero in one coordinate system it is zero in all coordinate systems. Hence the above equation is valid in all coordinate systems. Since the point P_o was arbitrary

(71) $$R^{\ell}_{kij;p} + R^{\ell}_{kjp;i} + R^{\ell}_{kpi;j} = 0.$$

Multiplying the above equation by $g_{\ell h}$ and summing over ℓ yields the <u>Bianchi identity</u>

(72) $$R_{hkij;p} + R_{hkjp;i} + R_{hkpi;j} = 0.$$

Contraction of the Riemann-Christoffel curvature tensor can only occur in one way if it is to yield a non-zero tensor since R_{ijkp} is antisymmetric on interchange of i with j and k with p. The tensor R_{jn} defined by

(73) $$R_{jn} = R^{\ell}_{jn\ell} = g^{\ell s} R_{sjn\ell}$$

is called the Ricci tensor and the scalar R defined by

(74) $$R = g^{jn} R_{jn}$$

is called the curvature invariant (see problem 11). Taking the covariant derivative with respect to the Christoffel symbols of the second kind of the curvature invariant and using the Bianchi identity, one obtains

$$R_{;r} = g^{jn}R_{jn;r} = g^{jn}g^{\ell s}R_{sjn\ell;r}$$

$$= g^{jn}g^{\ell s}(-R_{sj\ell r;n} - R_{sjrn;\ell})$$

which, from equations (69), after relabelling dumming indices, can be written

(75) $$R_{;r} = 2g^{jn}R_{jr;n}.$$

The Einstein tensor denoted G^i_j is defined by

(76) $$G^i_j = g^{i\ell}R_{j\ell} - \tfrac{1}{2}R\delta^i_j.$$

Taking the covariant derivative of G^i_j and contracting on i yields

$$G^i_{j;i} = g^{i\ell}R_{j\ell;i} - \tfrac{1}{2}R_{;j}$$

$$= g^{i\ell}R_{j\ell;i} - \tfrac{1}{2}(2g^{\ell i}R_{\ell j;i}) = g^{i\ell}(R_{j\ell;i} - R_{\ell j;i}).$$

But from equations (60) $R_{j\ell;i} = R_{\ell j;i}$ and thus

(77) $$G^i_{j;i} = 0.$$

Hence, the Einstein tensor is a constant with respect to covariant differentiation with the Christoffel symbols of the second kind providing the connection.

10.10. BASIS VECTORS

In Chapter 2 the concept of a position vector in 2- and 3-dimensional Euclidean space was introduced. It was written as the sum of components times basis vectors. The basis vectors may have been pictured as rigid rods of unit length (and vanishing width) lying tangent to the coordinate axis. Later a definition of a vector in terms of its components only was given. In Chapter 7 the definition of a vector was again given in terms of its components; however, the concept of basis vectors is still a useful one even though the notion of a position vector must be abandoned in Riemannian geometry.

In the case of a 2-dimensional surface of a sphere embedded in a 3-dimensional Euclidean space, at a point P_o the basis vectors may still be envisaged as rigid rods of unit length (and vanishing width) which are tangent to the coordinate curves at P_o. The 2-dimensional surface of a sphere, however, is a manifold by itself and there is no need to embed it in a higher-dimensional Euclidean space. If we do not picture the surface of the sphere as existing embedded in a 3-dimensional Euclidean space the notion of basis vectors as rigid rods breaks down. Fortunately the definition of basis vectors can be generalized so that it makes sense even in this case.

The basis vectors for the contravariant vectors in a n-dimensional Riemannian space are denoted $\hat{e}_1,\ldots,\hat{e}_n$ and defined as the directional derivatives of the coordinate curves. That is

$$(78) \qquad \hat{e}_i = \frac{\partial}{\partial u^i}, \qquad i = 1,2,\ldots,n.$$

Suppose $\vec{v}$ is an arbitrary contravariant vector with components v^i, $i = 1,2,\ldots,n$. Then

$$(79) \qquad \vec{v} = v^i\hat{e}_i = v^i\partial/\partial u^i.$$

Under change of parametrization from u^i to the new coordinates $\bar{u}^i$

(80) $$\frac{\partial}{\partial u^i} = \frac{\partial \bar{u}^j}{\partial u^i} \frac{\partial}{\partial \bar{u}^j}$$

so that the components of a contravariant vector transform in a different fashion from the basis vectors. If the definition of the basis vectors $\hat{e}_i$, $i = 1,2,\ldots,n$, is to be consistent then $\vec{v}$ must be invariantly defined independently of the parametrization of the space:

$$\vec{v} = v^i \frac{\partial}{\partial u^i} = \bar{v}^k \left(\frac{\partial u^i}{\partial \bar{u}^k}\right)\left(\frac{\partial \bar{u}^j}{\partial u^i}\right) \frac{\partial}{\partial \bar{u}^j} = \bar{v}^j \frac{\partial}{\partial \bar{u}^j} = \vec{\bar{v}} .$$

The action of a contravariant vector $\vec{v}$ on a scalar function $f = f(u^1,\ldots,u^n)$ is given by

(81) $$\vec{v}(f) = v^j \partial f/\partial u^j .$$

Let df denote the covariant vector defined by

(82) $$df(\vec{v}) = \vec{v}(f)$$

where $\vec{v}$ is an arbitrary contravariant vector. Comparing equations (81) and (82), we have that df is the gradient of f (or the covariant derivative of f).†

Let $\vec{\omega}^*$ be an arbitrary covariant vector with components ω_j, $j = 1,2,\ldots,n$. If we write

$$\vec{\omega}^* = \omega_j \hat{e}^{*j}$$

then

$$\omega^i v_i = \vec{\omega}^*(\vec{v}) = \vec{\omega}^*(v^j \hat{e}_j) = v^j \vec{\omega}^*(\hat{e}_j) = v^j \omega^i \hat{e}^{*i}(\hat{e}_j) .$$

Therefore

(83) $$\hat{e}^{*i}(\hat{e}_j) = \delta^i_j .$$

---

†We use the notation df instead of $\vec{\nabla} f$ so that equation (84) has the familiar form $df = (\partial f/\partial u^j) du^j$ instead of $\nabla f = (\partial f/\partial u^j) \nabla u^j$.

But

$$du^i(\hat{e}_j) = \hat{e}_j(u^i) = \partial u^i/\partial u^j = \delta^i_j.$$

Thus the basis vectors $\hat{e}^{*j}$, $j = 1,2,\ldots,n$, for the covariant vectors are the du^j, $j = 1,2,\ldots,n$. In particular

$$\text{(84)} \qquad df = \frac{\partial f}{\partial u^j}\, du^j = \frac{\partial f}{\partial u^j}\, \hat{e}^{*j}.$$

The scalar product between two contravariant vectors $\vec{v} = v^i\hat{e}_i$ and $\vec{p} = p^i\hat{e}_i$ is given by

$$\vec{v}\cdot\vec{p} = g_{ij}v^ip^j = v^ip^j\hat{e}_i\cdot\hat{e}_j.$$

Thus

$$\text{(85a)} \qquad g_{ij} = \hat{e}_i\cdot\hat{e}_j$$

and a similar analysis using covariant vectors gives

$$\text{(85b)} \qquad \hat{e}^{*i}\cdot\hat{e}^{*j} = g^{ij}.$$

The covariant directional derivative of a contravariant vector field $\vec{v} = v^j\hat{e}_j$† in the direction of $\hat{e}_\ell$ is given by

$$\text{(86)} \qquad \nabla_{\hat{e}_\ell}\vec{v} = \nabla_{\hat{e}_\ell}(v^j\hat{e}_j) = \frac{\partial v^j}{\partial u^\ell}\,\hat{e}_j + v^j\nabla_{\hat{e}_\ell}\,\hat{e}_j$$

where we have used equation (11c). Comparing the above equation to equation (4) implies that

$$\text{(87)} \qquad \nabla_{\hat{e}_\ell}\hat{e}_j = G^k_{j\ell}\hat{e}_k$$

where $G^k_{j\ell}$ is the connection.

Given two contravariant vector fields

†Here $v^j = v^j(u^1,\ldots,u^n)$ is a function of position in the Riemannian space.

$$\vec{x} = x^i(u^1,\ldots,u^n)\,\frac{\partial}{\partial u^i} \quad\text{and}\quad \vec{y} = y^i(u^1,\ldots,u^n)\,\frac{\partial}{\partial u^i}$$

the Lie derivative of $\vec{y}$ with respect to $\vec{x}$, denoted $\mathcal{D}_{\vec{x}}\vec{y}$, is the contravariant vector field defined by the commutator:†

(88) $$\begin{aligned}\mathcal{D}_{\vec{x}}\vec{y} &= [\vec{x}, \vec{y}] \\ &= \left[x^i(u^1,\ldots,u^n)\,\frac{\partial}{\partial u^i}\,,\; y^j(u^1,\ldots,u^n)\,\frac{\partial}{\partial u^j}\right] \\ &= x^i\,\frac{\partial}{\partial u^i}\left(y^j\,\frac{\partial}{\partial u^j}\right) - y^j\,\frac{\partial}{\partial u^j}\left(x^i\,\frac{\partial}{\partial u^i}\right) \\ &= x^i y^j\,\frac{\partial^2}{\partial u^i\partial u^j} + x^i\,\frac{\partial y^j}{\partial u^i}\,\frac{\partial}{\partial u^j} - y^j x^i\,\frac{\partial^2}{\partial u^j\partial u^i} - y^j\,\frac{\partial x^i}{\partial u^j}\,\frac{\partial}{\partial u^i} \\ &= \left[\left(\frac{\partial y^j}{\partial u^i}\right)x^i - \left(\frac{\partial x^j}{\partial u^i}\right)y^i\right]\frac{\partial}{\partial u^j}\,.\end{aligned}$$

It is easy to verify that the commutator satisfies the Jacobi identity

(89) $$[\vec{x}, [\vec{y}, \vec{z}]] + [\vec{y}, [\vec{z}, \vec{x}]] + [\vec{z}, [\vec{x}, \vec{y}]] = 0$$

for any three contravariant vector fields $\vec{x}$, $\vec{y}$, $\vec{z}$. If one uses a symmetric connection then

(90) $$\mathcal{D}_{\vec{x}}\vec{y} = \nabla_{\vec{x}}\vec{y} - \nabla_{\vec{y}}\vec{x}\,.$$

By convention the Lie derivative of a scalar field $\phi(u^1,\ldots,u^n)$ with respect to the contravariant vector field $\vec{x}$ is defined by

(91) $$\mathcal{D}_{\vec{x}}\phi = \vec{x}(\phi) = x^j\partial\phi/\partial u^j.$$

This definition permits us to define the Lie derivative of any covariant vector field $\vec{w}^* = w_i(u^1,\ldots,u^n)\,du^i$ as the covariant vector $\mathcal{D}_{\vec{x}}\vec{w}^*$ satisfying

†The commutator of two operators was defined in Chapter 6.

(92) $$(\mathcal{D}_{\vec{x}}\vec{w}^*)(\vec{y}) = \vec{x}(\vec{w}^*(\vec{y})) - \vec{w}^*([\vec{x}, \vec{y}])$$

for any contravariant vector field $\vec{y}$. It follows from equation (92) that

(93) $$\mathcal{D}_{\vec{x}}\vec{w}^* = \left[x^j \frac{\partial w_i}{\partial u^j} - w_j \frac{\partial x^j}{\partial u^i}\right] du^i.$$

Combining equations (91) and (92) gives

(94) $$\mathcal{D}_{\vec{x}}(\vec{w}^*(\vec{y})) = (\mathcal{D}_{\vec{x}}\vec{w}^*)(\vec{y}) + \vec{w}^*(\mathcal{D}_{\vec{x}}\vec{y}).$$

This suggests that the Lie derivative of an arbitrary tensor field T (with components $T_{i_1\ldots i_s}{}^{j_1\ldots j_r}$) with respect to a contravariant vector field $\vec{x}$ be defined as the tensor field $\mathcal{D}_{\vec{x}}T$ satisfying

(95) $$(\mathcal{D}_{\vec{x}}T)(\vec{y}_1,\ldots,\vec{y}_s, \vec{w}_1^*,\ldots,\vec{w}_r^*) = \mathcal{D}_{\vec{x}}(T(\vec{y}_1,\ldots,\vec{y}_s, \vec{w}_1^*,\ldots,\vec{w}_r^*))$$
$$- T(\mathcal{D}_{\vec{x}}\vec{y}_1,\ldots,\vec{y}_s, \vec{w}_1,\ldots,\vec{w}_r^*) - \ldots$$
$$- T(\vec{y}_1,\ldots,\vec{y}_s, \vec{w}_1^*,\ldots, \mathcal{D}_{\vec{x}}\vec{w}_r^*),$$

for arbitrary contravariant vector fields $\vec{y}_1,\ldots,\vec{y}_s$ and covariant vector fields $\vec{w}_1^*,\ldots,\vec{w}_r^*$.

The curvature tensor field R is defined by (see problem 6)

(96) $$R(\vec{w}^*,\vec{x},\vec{y},\vec{z}) = \vec{w}^*(K(\vec{x},\vec{y})(\vec{z}))$$

where

(97) $$K(\vec{x}, \vec{y}) = \nabla_{\vec{x}}\nabla_{\vec{y}} - \nabla_{\vec{y}}\nabla_{\vec{x}} - \nabla_{[\vec{x},\vec{y}]}$$

for arbitrary contravariant vector fields $\vec{x},\vec{y},\vec{z}$ and an arbitrary covariant vector field $\vec{w}^*$.

REFERENCES

1 H.S.M. Coxeter, Introduction to Geometry. John Wiley and Sons, New York, 1969
2 L.P. Eisenhart, Riemannian Geometry. Princeton University Press, Princeton, 1960
3 H. Goldstein, Classical Mechanics. Addison-Wesley, Cambridge, Mass., 1950
4 Erwin Kreyszig, Differential Geometry and Riemannian Geometry. University of Toronto Press, Toronto, 1968
5 D. Lawden, Tensor Calculus and Relativity. Methuen and Co. Ltd., London, 1967
6 D. Lovelock and H. Rund, Tensors, Differential Forms, and Variational Principles. John Wiley and Sons, New York, 1975
7 C.W. Misner, K.S. Thorne, and J.A. Wheeler, Gravitation. W.H. Freeman and Company, San Francisco, 1973
8 E. Schroedinger, Space-Time Structure. Cambridge at the University Press, 1950
9 B. Spain, Tensor Calculus. Oliver and Boyd, New York, 1965
10 Michael Spivak, Calculus on Manifolds. W.A. Benjamin Inc., New York, 1965
11 I.S. Sokolnikoff, Tensor Analysis and Applications to Geometry and Mechanics of Continua. John Wiley and Sons, New York, 1964
12 A. Trautman, F.A.E. Pirani, and H. Bondi, Brandeis Summer Institute in Theoretical Physics, Volume One: Lectures on General Relativity. Prentice-Hall Inc., Englewood Cliffs, N.J., 1964
13 C.E. Weatherburn, Riemannian Geometry and the Tensor Calculus. Cambridge at the University Press, 1950

PROBLEMS

1 Explain why it is not possible to find geodesic coordinates for a connection that is not symmetric.

2 Verify equation (89).

3 (a) Show that in the 4-dimensional Riemannian space with metric tensor

$$g_{ij} = 0 \text{ for } i \neq j, \quad g_{11} = g_{22} = g_{33} = -g_{44} = 1,$$

any curve of the form

$$u^1 = \int R\cos\theta\cos\Psi ds, \quad u^3 = \int R\sin\theta ds,$$

$$u^2 = \int R\cos\theta\sin\Psi ds, \quad u^4 = \int R ds$$

(where R, θ, and Ψ are functions of the arc length s) is minimal.

(b) When is such a curve a geodesic?

4 (a) Prove that the vectors $\hat{e}_j = \partial/\partial u^j$, $j = 1,2,3$, are linearly independent.

(b) Prove that the vectors $\overset{*}{\hat{e}}{}^j = du^j$, $j = 1,2,3$, are linearly independent.

(c) Prove that δ^i_j is a tensor field.

(d) Show that $[\hat{e}_j, \hat{e}_k] = 0$, $j,k = 1,2,\ldots,n$.

5 Explain why the following proofs are incorrect:

(a) Let T^{jk} and S^{jk} be tensor fields. In geodesic coordinates with a pole at the point P_o the equation

$$_o\left[\frac{\partial}{\partial u^\ell}(T^{jk}S_{ji}) - (T^{jk}{}_{;\ell}S_{ji}) - (T^{jk}S_{ji;\ell})\right] = 0$$

holds since partial differentiation is the same as covariant differentiation with respect to a symmetric connection. But if a tensor is zero in one coordinate system it is zero in all coordinate systems at that point. The point P_o was arbitrary; hence

$$\frac{\partial}{\partial u^\ell}(T^{jk}S_{ji}) = T^{jk}{}_{;\ell}S_{ji} + T^{jk}S_{ji;\ell}$$

where the covariant differentiation is done with respect to a symmetric connection.

(b) Let a^i be a contravariant vector field. Then under a change of coordinates we have

$$\frac{\partial \bar{a}^m}{\partial \bar{u}^p} = \frac{\partial a^i}{\partial \bar{u}^p}\frac{\partial \bar{u}^m}{\partial u^i} + a^i\frac{\partial^2 \bar{u}^m}{\partial \bar{u}^p \partial u^i}\,.$$

But second-order partial differentials commute. Therefore

$$\frac{\partial \bar{a}^m}{\partial \bar{u}^p} = \frac{\partial a^i}{\partial \bar{u}^p} \frac{\partial \bar{u}^m}{\partial u^i} + a^i \frac{\partial^2 \bar{u}^m}{\partial u^i \partial \bar{u}^p}$$

$$= \frac{\partial a^i}{\partial u^s} \frac{\partial u^s}{\partial \bar{u}^p} \frac{\partial \bar{u}^m}{\partial u^i} + a^i \frac{\partial}{\partial u^i} \left(\frac{\partial \bar{u}^m}{\partial \bar{u}^p} \right)$$

$$= \frac{\partial a^i}{\partial u^s} \frac{\partial u^s}{\partial \bar{u}^p} \frac{\partial \bar{u}^m}{\partial u^i} + a^i \frac{\partial}{\partial u^i} \delta^m_p$$

$$= \frac{\partial a^i}{\partial u^s} \frac{\partial u^s}{\partial \bar{u}^p} \frac{\partial \bar{u}^m}{\partial u^i} .$$

Hence $\partial a^i/\partial u^s$ is a second-rank mixed tensor field.

6 Show that the tensor R defined by equations (96) and (97) has components R^ℓ_{kij} given by equation (61).

7 (a) Prove that $g^{ij} \dfrac{\partial g_{kj}}{\partial u^\ell} + g_{kj} \dfrac{\partial g^{ij}}{\partial u^\ell} = 0$.

(b) Show that $\partial g^{im}/\partial u^\ell = -(g^{ij}\Gamma^m_{\ell j} + g^{im}\Gamma^i_{\ell j})$.

(c) Let g denote the determinant of the n x n matrix with components g_{ij}, $i,j = 1,2,\ldots,n$. Show that

$$\frac{\partial g}{\partial u^\ell} = g g^{ij} \frac{\partial g_{ij}}{\partial u^\ell}$$

(d) Show that $\dfrac{\partial g}{\partial u^\ell} = -g g_{ij} \dfrac{\partial g^{ij}}{\partial u^\ell}$

(e) Hence prove that $\partial(\log\sqrt{g})/\partial u^\ell = \Gamma^i_{i\ell}$.

8 (a) Prove that $[\vec{x},[\vec{y}, \vec{z}]] + [\vec{y},[\vec{z}, \vec{x}]] + [\vec{z}, [\vec{x}, \vec{y}]] = 0$ for any three contravariant vector fields $\vec{x}$, $\vec{y}$, $\vec{z}$.
(b) Derive equation (93) from (92).

9 A Riemannian space where $R_{ij} = \phi g_{ij}$ at all points and ϕ is a scalar is called an Einstein space.
(a) Show that in an Einstein space $R_{ij} = Rg_{ij}/n$ where n is the dimension of the space and R is the curvature invariant.
(b) Show that every 2-dimensional Riemannian space is an Einstein space.

10 Find the equation for geodesics on the 2-dimensional surface defined by

$$x = u^1 \cos u^2, \quad y = u^1 \sin u^2, \quad z = u^1 \cos\alpha,$$

where x,y,z are the Cartesian coordinates of a point in 3-dimensional Euclidean space and α is a constant.

11 For the 2-dimensional surface of a sphere of radius r evaluate (a) the curvature tensors $R^i_{jk\ell}$ and (b) the curvature invariant R.

12 Show that parallel transport of an arbitrary contravariant vector ${}_o\vec{v}$ at a point P_o around a small closed curve C in general yields a different vector ${}_o\vec{v}^1$ unless the Riemann-Christoffel curvature tensor $R^i_{jk\ell}$ is zero at P_o.

11
General relativity - an example of a non-linear theory

11.1. INTRODUCTION

The purpose of this chapter is to expose general relativity as an example of a highly non-linear theory which seems essential to our understanding of the universe and to our explanation of the special role of gravitation in the scheme of things. We are not out to examine the theory in detail or to review its historical development. In parallel with our brief expositions of quantum mechanics and special relativity, we present general relativity on a postulational basis without claim either to the uniqueness or to the completeness of the postulates, but rather because, within strict economies of time, this approach exposes to the reader the basic assumptions of the theory and brings him to the point where he can begin his own calculations.

In Einstein's special theory, the relativity of space and time blurs the boundaries separating these concepts. What is a space interval to one observer is partly a space interval, but also partly a time interval, to a second observer in uniform relative motion with respect to the first. Similarly for time intervals. Hence the idea that time is a 'fourth dimension,' albeit a dimension with a difference. Space-time is viewed as a 'flat' world, a 4-dimensional continuum characterized by a Lorentz (indefinite) metric. In general relativity, Einstein challenged conventional views still further by hypothesizing that the space-time continuum is warped by the presence of matter; alternatively, that the fundamental manifestation of matter is, in fact, curvature in the space-time continuum. In effect, general relativity is a covariant theory of the gravitational interaction, with covariance referring to a broad class of space-time coordinate transformations including transformations to accelerated reference systems. Just as special relativity blurs the conceptual boundaries between space and time, so does general relativity blur the boundaries between the form and existence of a space-time continuum on the one hand and the presence of matter in it on the other.

The central idea of general relativity stems from the principle of equivalence, which states that gravitational and inertial forces are fundamentally indistinguishable; alternatively it is a statement about the equivalence of inertial and gravitational mass. All matter possesses the property of resistance to acceleration in inertial frames of references; inertial mass is the proportionality constant in Newtonian mechanics between applied forces and resultant accelerations. Likewise all matter gravitates, and the proportionality constant between the gravitational force and the gravitational potential is called the gravitational mass. It struck Einstein as remarkable that these two masses are in fact identical (Eötvös experiment; more recently Dicke and collaborators).† Einstein pursued this argument to a logical conclusion, for example in his famous elevator experiment according to which an accelerated observer in an elevator cannot distinguish an increased upward acceleration from an increased strength of the gravitational field itself.

Einstein embodied the principle of equivalence in a geometric formulation of the laws of mechanics. Previous attempts at geometricization of the dynamical laws using an absolute time scale had been abortive. What made the new attempt successful was in part due to the dynamics inherent in a space-time geometry; moreover, the principle of equivalence made it possible to transform the gravitational field away by a suitable choice of coordinates corresponding to a locally inertial reference frame.

---

†R.V. Eötvös, D. Pekar, and E. Fekete, Annalen der Physik 68:11 (1922); P.G. Roll, R. Krotkov, and R.H. Dicke, Ann. Phys. 26:442 (1964).

11.2. POSTULATES OF GENERAL RELATIVITY

I. The world is a 4-dimensional space-time continuum whose basic geometry is Riemannian. That is, there exists at each point in space-time a metric tensor $g_{\mu\nu}$ and sets of coordinates x^{ν} such that the square of the space-time interval between adjacent world events

$$ds^2 = g_{\mu\nu}dx^{\mu}dx^{\nu}$$

is invariant under coordinate transformations. (See Chapter 7 for a discussion of how coordinates are parametrizations of space-time in a neighbourhood of the world events in question.)

II. At each point in space-time, by a suitable choice of coordinates the gravitational field can be reduced to zero. In other words, for each point in space-time there exists locally an inertial system of coordinates with a Lorentz metric.

Postulate II embodies the principle of equivalence in a natural manner. Let P_o be a world event (i.e. a specific space-time point) and let O denote a coordinate system which is locally inertial with respect to P_o (see Chapter 10). In accord with our considerations in Chapter 8 on special relativity, we can write the metric for coordinate system O at point P_o as

$$(1) \qquad {}_o g_{\mu\nu} = \eta_{\mu\nu}, \qquad \mu,\nu = 0,1,2,3,$$

where $\eta_{\mu\nu}$ is the diagonal tensor illustrated in equation (6) of Chapter 8. In coordinate system O, gravitational forces do not appear at the point P_o. Let x^{μ} denote the coordinates of points in the neighbourhood of P_o as seen by O, and $\bar{x}^{\mu}$ the coordinates of such points in any other admissible coordinate system. Since space-time is Riemannian, the metric tensor in $\bar{O}$ is related to $\eta_{\mu\nu}$ through the transformation equations

$$(2) \qquad {}_o \bar{g}_{\mu\nu} = \left(\frac{\partial x^{\alpha}}{\partial \bar{x}^{\mu}}\right)_o \left(\frac{\partial x^{\beta}}{\partial \bar{x}^{\nu}}\right)_o \eta_{\alpha\beta}$$

where again the subscript zero denotes evaluation at the point (world event) P_o.

However, equation (2) does not itself determine the metric tensor at the point P_o given some particular distribution of matter in the space-time continuum. One needs an additional postulate which relates the curvature to the distribution of matter density at each point in space-time. The choice of relationship is by no means unique and some criteria must be adduced for making the most judicious choice. We shall not pursue this interesting but difficult question in its entirety, but simply state a set of three criteria based on which Einstein's choice was made.

Our criteria are as follows:

(i) The principle of covariance must be satisfied. This criterion guarantees that the form of the equations of motion is independent of the choice of coordinate system, so that no preferred coordinate systems exist.

(ii) The field equations are required to be second order in the space-time derivatives of the metric tensor. This is the simplest choice which guarantees the weak-field (or classical) limit. The reasoning is as follows: in the weak-field limit (as we shall see in equation (24)) the time-time component g^{44} of the metric tensor is linearly dependent on ν, the classical gravitational potential. Since ν in the classical case satisfies a Poisson equation with respect to the mass density distribution ρ,

$$\nabla^2\nu = +4\pi\rho H, \quad H = \text{gravitational constant},$$

we anticipate that g^{44} and, by covariance all $g^{\mu\nu}$, similarly should satisfy second-order equations.

(iii) The field equations should be quasi-linear (i.e. linear in the second-order derivatives of the metric tensor $g_{\mu\nu}$). This criterion guarantees that the solutions are uniquely determined up to a coordinate transformation by the values of $g_{\mu\nu}$ and its first derivatives (initial and boundary values). In turn, this means that the solutions are uniquely determined up to a coordinate transformation by the initial and boundary values of the matter distribution and its gradient and velocity fields.

These three criteria lead us to a third general postulate of general relativity:

III. The Einstein field equations expressing the relationship between the metric tensor and the energy-momentum stress tensor describing the matter distribution and its dynamics in space-time are given by

$$(3) \qquad G_{\mu\nu} = -\frac{8\pi H}{c^4} T_{\mu\nu},$$

where $T_{\mu\nu}$ is the matter stress tensor and $G_{\mu\nu}$ the covariant form of the Einstein tensor whose dependence on the metric was expressed in equation (76) of Chapter 10. H is the gravitational constant and c the velocity of light.

The matter energy-momentum stress tensor may be decomposed into kinetic and potential parts:

$$(4) \qquad T_{\mu\nu} = \Theta_{\mu\nu} + P_{\mu\nu}$$

where $\Theta_{\mu\nu}$ is the kinetic part and $P_{\mu\nu}$ the potential part. For a continuous distribution of matter with a rest mass density ρ and 4-velocity $u^\nu = dx^\nu/ds$ the kinetic part of the stress tensor is taken to be

$$(5) \qquad \Theta^{\mu\nu} = \rho u^\mu u^\nu.$$

The potential part $P_{\mu\nu}$ of the stress energy-momentum tensor gives the potential energy-momentum of the matter distribution excluding, of course, gravitational interactions which are viewed in the framework of general relativity as a manifestation of the curvature of space-time itself. In the case of electromagnetic interactions we can regard $P_{\mu\nu}$ as the analogue of the electromagnetic stress tensor $S_{\mu\nu}$ in equation (73) of Chapter 8.

The Einstein field equations expressed in equation (3) are a set of coupled second-order, quasi-linear (i.e. linear in the second derivatives of $g_{\mu\nu}$) differential equations for the metric. Once the metric is determined, the equations of motion can be deduced. For a test particle whose interference with the metric locally can be ignored, the equations of motion are geodesics as we shall shortly demonstrate. This circumstance is a dramatic success of the Einstein theory from a conceptual viewpoint, since the motion of test particles moving under gravitational forces arises from the variational principle of requiring that the motion minimize the space-time displacement. The motion is completely reflected in the space-time geometry imposed by the universal distribution of gravitating matter acting upon the test particle.

11.3. THE MOTION OF A TEST PARTICLE

We now consider a universe† characterized by a uniform, weakly interacting distribution of matter (i.e. a gas) of density ρ. Under these circumstances, the potential energy-momentum stress tensor can be set equal to zero and the field equations written

$$(6) \qquad G^{\mu\nu} = -\frac{8\pi H}{c^4}\,\Theta^{\mu\nu}.$$

In Chapter 10 (equation (77)) it was shown that the covariant derivative of the Einstein field tensor with respect to index ν vanishes. Taking the covariant derivative of both sides of equation (6) then yields

$$(7) \qquad G^{\mu\nu}{}_{;\nu} = 0 = -\frac{8\pi H}{c^4}\,\Theta^{\mu\nu}{}_{;\nu}$$

and hence, from (5), using the differentiation chain rule,

$$(8) \qquad u^{\mu}{}_{;\nu}\,u^{\nu} + u^{\mu}u^{\nu}{}_{;\nu} = 0.$$

Multiplying by u_{μ} and summing over μ in equation (8) then yields

†Universe here refers to a sufficiently large region of space-time such that outside influences may be regarded as small throughout most of its interior.

(9) $u^{\mu}{}_{;\mu} = 0$

by virtue of the fact that $u^{\nu}u_{\nu} = -1$ implies that $u_{\mu}u^{\mu}{}_{;\nu} = 0$. Hence, equation (8) can be re-expressed as

(10) $u^{\mu}{}_{;\nu}\, u^{\nu} = 0 = \nabla_{\vec{u}}\, \vec{u}$.

Equation (10) expresses the covariant derivative of u^{μ} in the 'direction' $\vec{u}$ (see equations (10) and (11) of Chapter 10) with respect to the connection $\Gamma^{\mu}_{\nu\sigma}$ and, hence, from equation (46b) of Chapter 10, represents the equation for a <u>geodesic</u>. Physically this means that the 'gas' particles move along geodesics of the space-time manifold to the extent that their mutual interaction energy can be ignored (i.e. $P_{\mu\nu} = 0$).

We now generalize to an arbitrary distribution of matter with a test particle moving through it. From the principle of equivalence, as discussed above, we can find a coordinate system which is locally inertial at any given point P_o of the space-time continuum. For such a coordinate system the metric tensor is locally Lorentzian

(11) ${}_{o}g_{\mu\nu} = \eta_{\mu\nu}$

and the Christoffel symbols vanish (i.e. we are using geodesic coordinates with a pole at point P_o). Hence a test particle at the point P_o in the space-time manifold will move according to the equations

(12) ${}_{o}(d^2x^{\mu}/ds^2) = 0$,

where s measures the distance (see equation (39b) of Chapter 10) and the x^{μ} are the space-time coordinates of the test particle in the locally inertial reference frame.

Equation (12) is not a tensor equation in its present form; however, noting that the Christoffel symbols vanish at P_o, it may easily be cast in tensor form (i.e. in a form which is manifestly covariant with respect to arbitrary coordinate transformations):

(13) $$\left(\frac{du^\mu}{ds} + \Gamma^\mu_{\rho\sigma} u^\rho u^\sigma\right)_o = \left(u^\mu{}_{;\nu} u^\nu\right)_o = (\nabla_{\vec{u}} \vec{u})_o = 0,$$

where $u^\mu = dx^\mu/ds$ is the usual 4-velocity. Equation (13) is manifestly covariant and hence true in any coordinate system; using the same argument as for equation (10) and noting that the point P_o was arbitrary, it shows that the test particle in an arbitrary gravitational field moves along geodesic curves in the space-time continuum.

11.4. WEAK GRAVITATIONAL FIELDS

Generally speaking, it is not possible to choose a global coordinate system to describe the motion of a test particle via equation (13) since geodesics are curves and inertial coordinates chosen at a point P_o are only locally valid. However, if the gravitational field is sufficiently weak, we can introduce Cartesian-like coordinates

(14) $$x^\nu \equiv (x, y, z, ict)$$

such that the metric is quasi-inertial (i.e. approximately inertial):

(15) $$g_{\mu\nu} = \delta_{\mu\nu} + h_{\mu\nu}.$$

The sixteen quantities $h_{\mu\nu}$, $\mu,\nu = 1,2,3,4$, represent a small perturbation on the Minkowski metric of flat space. Using this metric, the Christoffel symbols of the first kind are given by

(16a) $$[\mu\nu,\lambda] = \frac{1}{2}\left(\frac{\partial h_{\nu\lambda}}{\partial x^\mu} + \frac{\partial h_{\mu\lambda}}{\partial x^\nu} - \frac{\partial h_{\mu\nu}}{\partial x^\lambda}\right)$$

and the Christoffel symbols of the second kind are given approximately by

(16b) $$\Gamma^\lambda_{\mu\nu} = g^{\lambda\alpha}[\mu\nu,\alpha] \approx \delta^{\lambda\alpha}[\mu\nu,\alpha] = [\mu\nu,\lambda].$$

The distance s can be interpreted as the proper time τ (times c) of the test particle, so that one can write

$$\frac{dx^i}{d\tau} = \gamma \frac{dx^i}{dt} = \gamma v^i, \quad i = 1,2,3,$$

$$\frac{dx^4}{d\tau} = ic\gamma, \quad \sqrt{1 - v^2/c^2} = \gamma^{-1},$$

(17)
$$\frac{d^2x^i}{d\tau^2} = \frac{1}{c^2} \gamma^4 v \frac{dv}{dt} v^i + \gamma^2 \frac{d^2x^i}{dt^2}, \quad i = 1,2,3,$$

and

$$\frac{d^2x^4}{d\tau^2} = \frac{i\gamma^3 v}{c} \frac{dv}{dt}$$

as the equations of motion.

Consider the simple case of a test particle at rest at the instant of consideration. Hence

(18)
$$\frac{1}{c} \frac{dx^i}{d\tau} = \frac{dx^i}{ds} = 0, \quad i = 1,2,3,$$

$$\frac{dx^4}{ds} = \frac{1}{c} \frac{dx^4}{d\tau} = i,$$

and the equations of motion (13) become

$$\frac{d^2x^\nu}{ds^2} - \Gamma^\nu_{44} = 0, \tag{19}$$

which to first order in small quantities can be simplified to read

$$\frac{d^2x^\nu}{ds^2} = \frac{\partial h_{4\nu}}{\partial x^4} - \frac{1}{2} \frac{\partial h_{44}}{\partial x^\nu} . \tag{20}$$

For $v = 0$, $ds^2 = c^2 d\tau^2 = c^2 dt^2$; moreover for a static metric

$$\frac{\partial h_{4i}}{\partial x^4} = 0, \tag{21}$$

so that one obtains

(22) $$\frac{d^2x^i}{dt^2} = -\frac{c^2}{2}\frac{\partial h_{44}}{\partial x^i}, \qquad i = 1,2,3.$$

In classical Newtonian physics, the acceleration of a test particle is related to the potential ν of the gravitational field via

(23) $$\frac{d^2x^i}{dt^2} = -\frac{\partial \nu}{\partial x^i}.$$

Hence, if general relativity is to reduce to classical mechanics in the limit of weak gravitational fields, we require

(24) $$h_{44} = \frac{2}{c^2}\nu$$

so that

(25) $$g_{44} = 1 + \frac{2}{c^2}\nu.$$

In this limit we can see at once how the Newtonian gravitational force appears as a result of the curvature of a weak gravitational field.

11.5. THE SCHWARZSCHILD SOLUTION

As a simple example, we calculate in this section the metric due to a spherically symmetric body at rest. We assume that no other matter is close enough to deform the metric, so that a test particle senses only a static, spherically symmetric metric field.

Let x^i, $i = 1,2,3$, be the Cartesian-like coordinates for a 3-dimensional space with origin at the centre of spherical symmetry. The form of the metric follows from two observations: first, that the only differential expressions in the x^i which are invariant under rotations are $(dx^1)^2 + (dx^2)^2 + (dx^3)^2$ and $x^1dx^1 + x^2dx^2 + x^3dx^3$; second, since the metric is static, ds^2 is invariant under the transformation $t \rightarrow -t$. It can then be easily shown that

(26) $$ds^2 = a(r)(dr)^2 + r^2[(d\theta)^2 + \sin^2\theta(d\phi)^2] - b(r)c^2dt^2$$

where r, θ, ϕ are spherical polar coordinates. It follows that the components of the metric tensor are then given by

$$g_{11} = a(r), \quad g_{22} = r^2, \quad g_{33} = r^2\sin^2\theta,$$

$$g_{44} = -c^2b(r), \quad g_{\mu\nu} = 0 \quad \text{for } \mu \neq \nu.$$

Thus, one also has

$$g^{11} = 1/a(r), \quad g^{22} = 1/r^2, \quad g^{33} = 1/(r^2\sin^2\theta),$$

$$g^{44} = -1/c^2b(r), \quad g^{\mu\nu} = 0 \quad \text{for } \mu \neq \nu.$$

We can immediately calculate the non-zero Christoffel symbols of the second kind as follows:

$$\Gamma^1_{11} = a'/2a, \quad \Gamma^2_{12} = \Gamma^2_{21} = 1/r,$$

$$\Gamma^1_{22} = -r/a, \quad \Gamma^3_{13} = \Gamma^3_{31} = 1/r,$$

$$\Gamma^1_{33} = -r\sin^2\theta/a, \quad \Gamma^2_{33} = -\sin\theta\cos\theta,$$

$$\Gamma^3_{23} = \Gamma^3_{32} = \cot\theta,$$

$$\Gamma^4_{14} = \Gamma^4_{41} = b'/2b, \quad \Gamma^1_{44} = c^2b'/2a.$$

Here a' and b' denote da/dr and db/dr.

The non-zero components of the Ricci tensor can also be calculated at once:

$$R_{11} = \frac{b''}{2b} - \frac{(b')^2}{4b^2} - \frac{a'b'}{4ab} - \frac{a'}{ra},$$

$$R_{22} = \frac{rb'}{2ab} - \frac{ra'}{2a^2} + \frac{1}{a} - 1,$$

$$R_{33} = R_{22}\sin^2\theta,$$

$$R_{44} = c^2\left(-\frac{b''}{2a} + \frac{(b')^2}{4ab} + \frac{a'b'}{4a^2} - \frac{b'}{ra}\right).$$

Outside the spherically symmetric body there is no matter contributing to the gravitational field; hence $T_{\mu\nu} = 0$ in this region and Einstein's field equations become

$$G_{\mu\nu} = R_{\mu\nu} - \frac{1}{2} g_{\mu\nu} R = 0. \tag{27}$$

Multiplying by $g^{\mu\nu}$ and summing on μ and ν gives $R = 0$, so that the field equations simplify to

$$R_{\mu\nu} = 0. \tag{28}$$

From the values of the $R_{\mu\nu}$ given above, it follows that equations (28) constitute a set of three ordinary differential equations for the functions $a(r)$ and $b(r)$ in the Schwarzschild metric. In fact, only two of the three equations are independent, and they yield the following solution:

$$b = 1/a = (1 - 2m/r), \tag{29}$$

where m is a constant of integration.

It follows that the metric outside the spherically symmetric body is given by

$$ds^2 = \frac{(dr)^2}{1 - 2m/r} + r^2[(d\theta)^2 + \sin^2\theta(d\phi)^2] - c^2(1 - 2m/r)(dt)^2. \tag{30}$$

The value of the constant of integration m is given by combining the equation†

$$g_{44} = -\left(1 + \frac{2\upsilon}{c^2}\right)c^2 \tag{25}$$

†The appearance of the additional c^2 and minus sign is due to the fact that in this case we have $x^4 = t$ whereas previously we had $x^4 = ict$.

with the classical expression for the potential ν due to a mass M at a distance r from its centre and outside the mass distribution:

(31) $\nu = -HM/r.$

Hence

(32) $g_{44} = -(1 - 2HM/rc^2)c^2$

and the constant of integration turns out to be

(33) $m = HM/c^2.$

11.6. CONSEQUENCES OF GENERAL RELATIVITY

We comment briefly here on the consequences of general relativity which relate to the main themes of the present work. General relativity is actually a theory of gravitation as well as a theory which casts the equations of motion into a coordinate-independent (covariant) form. From the point of view of the equivalence principle, relativity and gravitation are inseparable concepts so that a satisfactory theory must embody both. The price paid for this unification is a highly non-linear set of equations for describing the large scale of the universe; nonetheless, in most astrophysical or astronomical problems, the density of matter in the large is so small, and the gravitational attraction so weak, that linear approximations contain the essential physics necessary to understand them.

Whether general relativity is an ultimately correct theory is a question yet to be answered. What is already certain is that it has useful conceptual aspects, that it is capable of some quantitative predictions, such as the bending of light grazing the sun,† the precession of the perihelion of mercury, and the slowing down of clocks in a strong gravitational field.

---

†It is sometimes claimed that this result can be explained within the framework of special relativity by endowing the photon

Aside from its quantitative successes, general relativity predicts some remarkable phenomena which might play a crucial role in the structure and dynamical history of the universe. Prominent among these is the prediction of black holes for spherical objects which are sufficiently massive that the so-called Schwartzchild radius (where the radial portion of the line element in equation (30) becomes singular, viz. $r = 2m$) falls outside the matter radius of the object. Under such conditions it can be shown that matter captured by the star is irreversibly captured once it has penetrated the Schwarzschild radius. For an external observer, penetration takes an indefinitely long time. Before the discovery of neutron stars it appeared that this was only a mathematical curiosity of the theory; there is now, however, speculation that some large neutron stars may, in fact, be 'black holes.' There is also some speculation as to the association of elementary particles with microscopic gravitational singularities. Speculations of the latter type have little substance as yet since no evidence has accumulated as to the nature of gravitational phenomena at microscopic separations of interacting particles.

Attempts to unify all physical laws under the umbrella of a single theory have either not been successful or are in their infancy. Unified field theories embracing both gravitation and electromagnetism have been moderately successful,† but the relationship of gravitation to the microscopic manifestation of elementary particles and their interactions remains well beyond the reach of present conceptual schemes.

---

with an inertia $m = h\nu/c^2$. Aside from being wrong by a factor 2, this argument is based not only on special relativity but employs the classical theory of gravitation as well. By contrast, in general relativity gravitational and inertial effects are treated in a self-consistent manner and are an inherent part of the theory.

†For a recent example, see J.W. Moffat and D.H. Boal, Phys. Rev. D11:1375 (1975).

REFERENCES

1 A. Einstein, The Meaning of Relativity. Princeton University Press, Princeton, 1953
2 D.F. Lawden, An Introduction to Tensor Calculus and Relativity. Methuen & Co. Ltd., London, 1967
3 H.A. Lorentz, A. Einstein, H. Minkowski, and H. Weyl, The Principle of Relativity. Dover Publications Inc., New York, 1952
4 C.W. Misner, K.S. Thorne, and J.A. Wheeler, Gravitation. W.H. Freeman and Co., San Francisco, 1973
5 W. Pauli, Theory of Relativity. Pergamon Press, New York, 1958
6 S. Weinberg, Gravitation and Cosmology: Principles and Applications of the General Theory of Relativity. John Wiley and Sons, New York, 1972

PROBLEMS

1 (i) Using units where c = 1 and is dimensionless show that the equations for a geodesic may be put in the form

$$\frac{d}{d\tau}(g_{\mu\nu}U^{\mu}) - \frac{1}{2}g_{\alpha\beta,\mu}U^{\alpha}U^{\beta} = 0, \quad \text{where } g_{\alpha\beta,\mu} = \frac{\partial g_{\alpha\beta}}{\partial x^{\mu}},$$

with $U^{\alpha} = dx^{\alpha}/d\tau$, $U^{\alpha}U_{\alpha} = -1$. Thus if $g_{\alpha\beta}$ is independent of some coordinate x^{α} then $d(g_{\alpha\nu}U^{\nu})/d\tau = 0$ along a geodesic and $g_{\alpha\nu}U^{\nu}$ is a constant of the motion for a test particle.
(ii) Show that for a test particle moving in the Schwarzschild geometry

$$E \equiv \left(1 - \frac{2m}{r}\right)\frac{dt}{d\tau}, \quad J \equiv r^2(\sin^2\theta)\,\frac{d\phi}{d\tau}$$

are constants of the motion. Justify the interpretation of E as the energy per unit mass and J as the angular momentum per unit mass.
(iii) For a test particle moving in the Schwarzschild geometry show that $g_{\alpha\beta}U^{\alpha}U^{\beta} = -1$ may be written as

$$\left(\frac{du}{d\phi}\right)^2 = \frac{E^2 - (1 - 2u)(1 + L\,u^2)}{L^2}$$

where $u = m/r$, $L = J/m$.

(iv) Differentiate the above equation with respect to ϕ to get

$$\frac{d^2u}{d\phi^2} + u = \frac{1}{L^2} + 3u^2,$$

showing that the relativistic correction to the Newtonian equation

$$\frac{d^2u}{d\phi^2} + u = \frac{1}{L^2}$$

is the $3u^2$ term.

2 A test particle of velocity $\vec{\beta}$ (again we use units where c = 1 and is dimensionless) flies past a mass M at an impact parameter b so great that the deflection is small. Show that the deflection is

$$\Delta\phi = \frac{2HM}{b\beta^2}(1 + \beta^2).$$

3 Show that the Einstein field equations for a uniform weakly interacting gas of density ρ reduce to Poisson's equation

$$\nabla^2\nu = +4\pi\rho H$$

for weak, static gravitational fields.

12
Horizons

12.1. INTRODUCTION

Throughout this book we have stressed the natural flow of ideas from experience to physical conceptualization to mathematical description. Naturally, this flow pattern has feedback loops: a mathematical equation has predictive value which stimulates new experimentation leading to deeper insights and eventually to more sophisticated mathematical modelling. Our emphasis has been on the interplay between concept formation and mathematical structure. For the most part illustrations have been drawn from areas of physical enquiry which, though still active, possess a comfortable degree of permanence in the sense that future theories will incorporate present ones without serious conflict or contradiction. In the present chapter we examine some frontier areas of modern discovery and try to anticipate some directions of future development.

Scientific activity involves discovery, modelling, explanation,and prediction. In Section 2 we discuss some aspects of explanation and understanding which are relevant to the conceptual process which relates mathematical modelling to scientific discovery. Section 3 is concerned with the nature of scientific discovery and with the search for new knowledge on the frontier of science. Finally, in Section 4, we return to the theme of the book and examine some new directions in the interplay between physical concepts and mathematical structures.

12.2. EXPLANATION AND UNDERSTANDING

The nature of explanation is elusive; whole books have been

written on the subject. We must content ourselves here with a few remarks of particular relevance to the task in hand. In large part, understanding is familiarity. To say, for example, that one 'understands' the wave function in quantum theory is not quite the same as to say one understands why seven sixes are forty-two. If we use the wave function often enough in successful calculations, we are inclined to play down its essential mystery and to regard it as a commonplace tool of the trade. We shall take explanation, however, to be something more than mere familiarity; in the scientific context it is usually couched in mathematical relationships between interpreted quantities and has predictive value.

There are evidently levels of explanation and degrees of understanding. Nuclear physicists are inclined to say that they do not understand nuclear forces; yet nuclear power stations, which operate efficiently and on a massive scale, are designed on the basis of our understanding of nuclear force laws. To a hockey player momentum conservation is an intuitive fact of life; to a quantum field theorist it is a consequence of translational invariance. Even among scientists strong disagreements occur as to what constitutes a satisfactory explanation of a given phenomenon, even when there is agreement on a set of underlying axiomatic relationships. To the atomic physicist, Coulomb's law may appear to explain all of chemistry and most of biology, but the developmental biologist finds Coulomb's law useless in trying to explain why regeneration occurs in a severed amphibian limb but not in a mammalian limb. To take an example closer to home, quantum theorists using the relativistic Dirac equation will disagree as to whether electron 'spin' is associated with actual dynamical rotation, or whether it arises as a kinematic effect out of a doubling of three-dimensional space at the microscopic level.

Understanding, in scientific parlance, is usually predicated on the assumption that there exists an objective reality which can be perceived, analysed, and described. Indeed, it is impressive how the scientific world, given time, arrives at a rather common and universal explanation of any particular phenomenon. Supporting the notion of objective reality is the incredible fact that events, ranging over 40 orders of magnitude in space and time variables, from the microcosmic world of elementary particles to the cosmic world of myriad galaxies, find an explanation in the same set of basic physical laws. These considerations are even more impressive when we realize that objective reality is a highly personal thing, born within our individual minds on the basis of sensory information which is transformed and translated on its way to the brain and processed there by ways and means we do not understand. Yet despite differing prejudices and often fierce argumentation, we eventually come to accept an essentially common reality.

Ultimately, of course, one has to deal with the nature of consciousness itself. Most practising scientists are either unaware of the problem or choose to ignore it. Distinguished scientists such as Wigner[†] and Wheeler,[††] however, have argued that it cannot be ignored. According to this view the human consciousness is an integral part of the measuring process in quantum theory, not in the accepted sense that the uncertainty principle expresses an interaction between observer and observed, but in a deeper and more profound sense. In a related context Niels Bohr has speculated[†††] whether a complementarity principle operates which prevents us from completely knowing ourselves, perhaps from understanding our own awareness of understanding.

[†] E.P. Wigner, 'Epistemological Perspective on Quantum Theory,' in D. Hooker, Contemporary Research in the Foundations and Philosophy of Quantum Theory (Reidel Publishing Co., Dordrecht, Holland, 1973).

[††] J.A. Wheeler, 'The Quantum and the Universe,' Sanible Symposium (1977), to be published in the International Journal of Quantum Chemistry.

[†††] N. Bohr, 'Physical Science and the Problem of Life,' as reprinted in Atomic Physics and Human Knowledge (John Wiley & Sons, New York, 1958), pp. 94-101. See also J. Mehra, 'Quantum Mechanics and the Explanation of Life,' American Scientists 61: 722 (1973).

As an analytical science, physics stands at the fore. Over the past two centuries physicists and chemists have had remarkable success in interpreting nature, using an essentially reductionist philosophy in which one tries to understand a global phenomenon by dissecting it and studying its parts. This success is symbolized and codified in mathematical structures such as differential equations, which describe events locally, but which by integration yield global properties; it is epitomized in the atomic theory in which one builds atoms from elementary particles, molecules from atoms, matter from molecules, and the universe from matter.

Nonetheless, it is clear that a totally reductionist philosophy is not sufficient for total explanation and understanding, even within the physical sciences, and certainly not in biology. In physical science for example, the exclusion principle is ruthlessly holistic since it asserts that electrons form a world collective with a totally antisymmetric wave function; even the slightest disturbance of one electron affects, in principle, all other electrons in the universe. The exclusion principle cannot be couched in terms of the properties of a single electron or in terms of the interaction properties of a pair or a triple - it is in this sense an irreducible concept applying to the totality of electrons in the universe.

The limits to reductionism are also illustrated by modern elementary particle experiments conducted with high energy particle beams. To investigate the structure of an elementary particle, one needs a probe which is smaller in size. Small probes mean short de Broglie wavelengths, thus requiring that projectiles move at high energy - the greater the energy, the smaller the probe. However, the description of a simple collision process involving a projectile and a target particle at high energies is rendered ineffective by the process of creation of new particles, so that the original design of a two-body experiment is converted into a many-particle experiment. Expansion in Feynman diagrams symbolizes the attempt to stay within a reductionist analysis, but the difficulties which abound in modern quantum field calculations, such as divergence problems, suggest we may be using a procedure which corresponds to a modern version of the epicycle calculations of planetary orbits carried out in the pre-Copernican (Ptolemaic) era.

It is particularly clear in the biological context that reductionism is a limited philosophy. Just as in physical science, there are levels of explanation in biology: at the atomic level, at the macromolecular level, at the cellular level, at the tissue level, at the organ level, at the body level, at the herd level, at the eco-system level. Succeeding levels are characterized by specific emergent properties which cannot be predicted[†] from the properties of systems at the lower levels. A dramatic example of an emergent property is the human consciousness, which is associated with the global nervous system but cannot be predicted from the properties of individual neural cells (neurons) which compose the system, or from a knowledge of their interactions.

The advantage of reductionism, when it works, is that it is clearly analytical and lends itself to mathematical description and quantification. Holistic properties are often more difficult to describe and to quantify. Thom[††] has argued that we have an obsession with quantification - it is our security blanket - yet many of the important things in life, for example, a Beethoven symphony or a beautiful tree, are qualitative rather than quantitative.

12.3. SEARCH AND DISCOVERY

It is not possible in a few short paragraphs to touch upon every frontier of discovery in modern physical science. Rather we confine our remarks to a few general features and some specific examples. In fact, we are living in revolutionary times in

---

[†]Both by reason of complexity and by what one might call 'irreducible emergence.' The exclusion principle in physics is an example of an irreducible emergence.

[††]R. Thom, Structural Stability and Morphogenesis (W.A. Benjamin Inc., New York, 1975).

which modern technology makes possible an ever-increasing range of sophisticated experimentation. The basis of this revolution is the widespread use of instruments based on the principles of quantum optics and quantum electronics: low-temperature technology provides low-noise detectors and a variety of superconducting devices; laser optics provides unique high-intensity, high-Q electromagnetic oscillators as sources and detectors; finally, modern solid state technology provides quick-response, stable, electronic amplification and fast and efficient data-handling systems. The result has been dramatic developments in both experiment and theory in physics, chemistry, and biology, as separate disciplines but also in combination.

It is difficult to find an area of physical enquiry, either classical or quantum, which has not received the magic touch of this new technology. For example, more has been published on classical fluid mechanics in the last decade than in the total previous history of the subject. The barriers between various fields of physics which restricted physicists to narrow specialties have tended to break down and interdisciplinary research involving physics, chemistry, biology, and mathematics has tended to increase. Despite natural pressures for specialization there has been some movement back towards a unitary natural philosophy, wherein science is regarded as a many-sided activity, its exposition requiring various complementary areas for search and discovery.

We consider several examples of frontier research which are likely to broaden the horizons of knowledge in the next few decades: two examples are from astronomy - black holes and the big bang; one example is taken from elementary particle physics, one from biology, and the final example from electromagnetism.

In the last two or three decades, astronomers have discovered whole new worlds through microwave and x-ray astronomy. Such exotic stellar objects as quasars and neutron stars have been added to the compendium of curious astronomical objects. It is now conjectured that even more curious objects exist, namely, black holes which are thought to occur when stars several times the mass of the sun collapse at the end of their nuclear burning cycles. Black holes reveal themselves to the outside world only through their gravitation (mass and angular momentum) but they possess the property of sucking in to their unobservable interiors whatever matter falls within the pull of their gravitational fields. Black holes are still in the conjectural stage, but most physicists believe they exist, and a promising candidate has been found as an x-ray source in the constellation Cygnus.

Our second example illustrates a rather curious aspect of astronomy which sets it apart from the other physical sciences: measurements relate to events in the distant past with a relevant time scale stretching back over billions of years, as perceived in our reference frame. The sky is in effect a history book. Present measurements are generally consistent with the view that cosmic history had a common point in space and time some twenty to thirty billion years ago - the 'big bang.' According to this view, in the resulting expansion the universe cooled off to its present temperature of about 3°K. The structure of the world today as we see it is a direct result of this expansion process. A particularly active area in modern astronomy is the search for evidence in the cosmic debris which would bear upon the nature of the crucial events that took place in the first few minutes following the catastrophic birth of the universe.

At the microcosmic end of the scale one has the world of elementary particles. Most physicists now believe that a fundamental group of particles called quarks exists, whether or not they can be experimentally isolated. Discovery in the laboratory has been reported by Fairbank and his co-workers,† though convincing proof will eventually require confirming measurements in several other laboratories. Aside from the interesting questions whether quarks exist, and whether they can be isolated, is

---

†G.S. LaRue, W.M. Fairbank, and A.F. Hebard, Phys. Rev. Letters, 38: 1011 (1977).

the question whether quarks are limited to 5 (or probably 6) 'flavours' - up, down, strange, charm, truth, and beauty - and 3 'colours' - red, blue, and yellow - or whether the variety is more extensive and possibly even endless. The recent discovery of the tau lepton at around 1800 MeV raises the probable number of leptons also to 6 (electron, muon, tauon, plus their associated neutrinos) and suggests a strong association between quark flavours and lepton varieties. The profound mystery in elementary particle physics which has persisted for 30 years, namely, what the distinction between electron and muon is, is now heightened by the appearance of the tauon since, presumably, it also behaves like an electron except for its larger mass.

No real dynamical theory yet exists to explain the behaviour of quarks. The current belief is that quarks interact by exchanging massless particles (called gluons). 'Colour' plays a similar role to charge for the electromagnetic field where the exchange particle is the photon. Unlike photons, however, which do not carry electric charge, the gluons do carry colour. The resulting theory is called quantum chromodynamics (QCD) in rough analogy to quantum electrodynamics (QED). Physicists are divided into those who believe that all the important particles have now been discovered and those who believe the list is still incomplete and possibly endless.

Despite the sweeping distinctions that separate physics and biology there is strong reason to believe that there also exists an underlying unity. It is intellectually intoxicating to realize that biological edifices of fantastic beauty and incredible complexity, like star fish, whooping cranes, and Douglas firs, have, at an elemental level of description, a common basis in Coulomb's law. The vast range of forms, sizes, and colours that occur in the biological world is, in some reasonably well-defined sense, a manifestation of the range of possible solutions of the Maxwell-Lorentz (coupled field-matter) equations. This does not mean that one should seek to understand biology by diligent application of Coulomb's law; the level of complexity is such that one must start at another level of description. Once we know, however, that structures like DNA exist, it is sensible to contemplate the question whether these operate in harmony with the laws of electromagnetism, and to design experiments to test this hypothesis. In short, it is essential to distinguish between the statements that physics explains biology (untrue) and biological phenomena operate in harmony with physical laws (probably true).

The interface between biology and physics provides an almost endless frontier for search and discovery. An exotic example would be the ancient mind-body problem, particularly as this involves the quantum theory of measurement mentioned in Section 2. A more immediate example is the challenge presented by a wide range of biological phenomena to the study of the statistical mechanics of non-linear, non-equilibrium systems. The whole question, for example, of how dissipative biological systems self-organize, grow, and regulate themselves is largely unanswered at the present time. This question will be pursued again in Section 4.

We consider finally another kind of frontier in which one does not seek new fundamental equations but rather complex and intricate solutions of perfectly well-known equations. The equations describing the electromagnetic field were written down by Maxwell more than one century ago; they are still valid today. Yet only recently have we achieved in the laboratory the coherent solutions necessary for laser action. In other words writing down the basic equations constitutes only the beginning of the story. Another example is the problem of finding whether or not solutions of Maxwell's equations exist which can confine the ultra-hot plasmas necessary for the nuclear fusion process and which are sufficiently stable for a practical energy machine. Even if such configurations do exist, they represent only a kindergarten level of complexity compared to the electromagnetic configurations which hold macromolecules like DNA together and which guide their complex mechanical action as functioning machines in biological processes.

12.4. CONCEPTS AND STRUCTURES

The twentieth-century success paradigm in physics rests primarily on five basic, interlocking theoretical structures: the classical mechanics of moving point particles, the Maxwell-Lorentz equations for the electromagnetic field coupled to matter, the ensemble theory for stochastic systems, the quantum theory of microscopic processes, and Einstein's theory of relativity and gravitation. These five structures display a great variety of conceptual patterns and use a wide range of mathematical designs. Much of what we observe about the universe is encompassed by them, either singly or jointly. Nonetheless, without major modifications and additions they do not account for many of the recent observations made in such frontier fields as astronomy, elementary particle physics, and biology. We raise here the question whether the kinds of conceptual patterns and mathematical designs employed in these theories are sufficient for present and future developments, or whether new patterns and designs will be required as we move beyond the present horizons of scientific discovery. In this section we examine a number of conceptual questions and puzzles and discuss briefly some of the mathematical constructs which have been proposed to deal with them.

The reductionist tendency to seek an explanation for large structures in terms of the interactions of their subunits is paralleled by our desire to identify certain primitive concepts as a basis for complex ideas. David Bohm† has suggested order as the basic concept underlying our general ideas on form, hierarchy, organization, complexity, etc. Order is concerned with elemental distinctions or comparisons, such as between left and right or objects successively disposed. Organization implies a differential system with different parts performing different tasks, but with some overall unity or purpose - for

---

†D. Bohm, 'Some Remarks on the Notion of Order,' in Towards a Theoretical Biology, vol. 2 (Aldine Pub. Co., Chicago, 1969).

example, a plant or animal cell or a city. Clearly, an organization is highly ordered and has a significant degree of complexity and hierarchy. Generally speaking, in physics one is more concerned with order, in biology with organization. One speaks of a crystal as ordered and a gas as disordered. Such biological properties as self-reproduction, self-regulation, and consciousness, however, require a high level of organization for their explication.

Associated with the concepts of order and discreteness, but at a more structural level, is the concept of symmetry. Symmetry is a pervasive and extraordinarily useful concept in both physics and biology. In standard analysis† symmetry corresponds to form invariance under a group of point transformations. For example, the equilateral triangle is invariant under the dihedral group of reflections about its diagonals and rotations of $2\pi n/3$, $n = 0, \pm 1, \pm 2, \ldots$, about an axis perpendicular to the face and passing through its centre. As we have discussed at some length, symmetry in quantum systems is intimately tied in with conservation laws. For example, for isolated systems invariance with respect to translations (translational symmetry) implies conservation of linear momentum, while invariance under the rotation group (rotational symmetry) implies conservation of angular momentum. It is important in this context to realize that specific solutions of a differential equation usually possess a lower symmetry than that possessed by the equation itself. For example, the Schroedinger equation for the hydrogen atom is invariant with respect to the full rotation-inversion group referred to its centre-of-mass. An S state also possesses this symmetry, but higher angular momentum states possess successively lower symmetry. Reduced symmetry is often associated with greater complexity, where the degree of symmetry is measured by the order of the symmetry group and complexity refers to the

---

†See, for example, A.V. Shubnikov and V.A. Kopstick, Symmetry in Science and Art (Plenum Press, New York, 1974).

number of parameters necessary to describe the detailed form of a structure.

Cases where the solutions have a symmetry reduced from that of the basic equations occur elsewhere. An interesting example is the attempt by Türing[†] and more recently Kaufman et al.[††] to use the solutions of a 2-component reaction-diffusion system to represent morphogenetic development in biological species. The reaction-diffusion equations are invariant under the full rotation-inversion group, but their solutions for given boundary conditions will, in general, possess a lower order of symmetry, but with an associated complexity appropriate to the differentiated structure of the biological system.

Closely related to the concept of symmetry is the notion of pattern. Pattern recognition problems crop up in such diverse circumstances as aerial identification photography, event selection in automated measurements on bubble chamber photographs in high-energy physics, visual processing systems in animals, and animal memory access and recall systems. Finally, we remark that the notion of pattern is related to morphology, the theory of forms, which is clearly of interest in the growth and development of biological systems. Arising out of morphology and its associated structures in biology is the exercise of function. The question whether function precedes structure or whether structure precedes function is a bit of a chicken-and-egg problem, but there can be no doubt of the extreme importance of the relationship between structure and function in every aspect of biological organization.

An interesting dichotomy of concept, which arises out of our experience and finds convenient expression in mathematics, is the setting out of the discrete from the continuous. Nuts and bolts are best measured digitally, whereas water and natural gas lend themselves to continuous measure. This elemen-

---

[†] A.M. Türing, 'The Chemical Basis of Morphogenesis,' Phil. Trans. Roy. Soc. B 237: 37 (1952).

[††] S.A. Kaufman, R.M. Shymko, and K. Trabert, 'Control of Sequential Compartment Formation in Drosophila,' Science 199: 259 (1978)

tal dichotomy of object characterization is reflected in the mathematical notions of integers and the real numbers. Modern point set theory provides a convenient umbrella for discrete and continuous conceptual schemes; nonetheless in both physics and biology there is a stark reality to the distinction. The real line is divisible into finer and finer, but equivalent segments; a frog just does not divide in this way. Sometimes systems lend themselves to both discrete and continuous descriptions. For example, setting aside the possible presence of field divergences at the points where particles are located, the electric field in an ordered crystal viewed along a line of lattice sites may be regarded as continuous. Its periodicity, however, lends it a discreteness of pattern or form: the crystal possesses translational invariance with respect to the discrete subgroup corresponding to shifts which are multiples of the lattice constant. Clearly, electric fields which possess this element of discretion tell us something interesting about the system.

An important aspect of scientific investigation, closely tied in with conceptualization, is the need for suitable mathematical structures. For example, a central theme in this book has been to emphasize the degree of success attained with the use of linear structures - linear operators on a linear vector space, with appropriate physical interpretation and identification. But we have already seen that, successful as linear theory has been, it is not adequate to handle the problems of hydrodynamics and gravitation; it is certainly not adequate to handle the highly non-linear problems which arise pervasively in chemistry and biology. No systematic theory of non-linear differential equations has been developed despite heroic efforts to do so. Even in simple cases like a two-species predator-prey model in population dynamics, one has to rely on computer simulations to examine the stability of certain kinds of solutions. Intensive effort will presumably be devoted to finding methods of solving coupled, non-linear, partial differential equations; it is possible, however, that one will have to abandon such attempts to seek realistic models, and find alternative ways of extracting essential results in the absence of detailed solutions. One such possible alternative which has received some attention is the construction of a biological statistical mechanics.†

Another approach to non-linear mechanics is to give up at the outset the quest for detailed solutions, and to concentrate instead on overall form and structure. In this connection the work of René Thom is particularly noteworthy.†† Frequently in the natural and physical sciences one is primarily interested in certain critical points or regions where the behaviour of the system changes abruptly when some control variable is changed, for example when water changes to ice as the temperature is reduced or when an animal's behaviour changes from fear to anger as it is provoked beyond some threshold of tolerance. In a specific set of limited cases Thom's analysis shows how to identify and characterize such critical points. Great arguments have raged recently on the question whether Thom's work is empty or profound. Certainly it provides a useful classification scheme, just as group theory does in another context. A particularly beautiful application of Thom's work which emphasizes its relationship to stability theory has been cited by Berry and Mackley:††† hydrodynamic flow patterns around six rotating, parallel cylinders show transitions between stable flow configurations which correspond to the unfolding of the elliptical umbilic. This example makes it clear how important is the work of René Thom in generating a comprehensive approach to the analysis of the stability properties of non-linear mechanical, electrical, and chemical systems.

---

†For a detailed discussion of this general problem and some computer studies of non-linear systems see C.J. Lumsden, 'On the Dynamics of Biological Ensembles: Canonical Theory and Computer Simulation,' Ph.D. Thesis, University of Toronto, 1977.

††R. Thom, loc. cit.

†††M.V. Berry and M.R. Mackley, Phil. Trans. Roy. Soc. London, 287: 1 (1977).

The Prigogine school[†] has been particularly active in raising the question whether all biological structure, in its myriad variations, in both genotypic and phenotypic developments, arises out of instabilities inherent in dissipative systems. From this point of view one regards biological organizations as essentially dissipative systems, maintained by a resource environment of energy and metabolites, which grow, differentiate, reproduce, and perform all the higher biological functions. According to this view, evolution itself is but a manifestation of the behaviour of complex, dissipative non-linear systems whereby new structures in space and time form as a consequence of the development of instabilities in old structures. Whether such sweeping claims are justifiable or not, they present a challenge to conventional thinking and stimulate a whole new train of thought about the origins and development of hierarchical biological structures. Models such as the Brusselator[††] tend to lend theoretical support to the proposition that complexity is closely associated with instabilities in dissipative, non-linear systems. The now famous Belousov-Zhabotinsky reactions,[†††] which generate spontaneously (using red-blue acid-base indicators) intricate spatial configurations and spontaneous time oscillations as a result of several interacting chemical species, is an impressive demonstration of the power of this kind of argument.

The history of mathematics[*] records an obsession among philosophers and mathematicians with the nature of geometry; the predominant view in pre-Riemannian times was that the universe was basically Euclidean and that other geometries were derived from Euclidean geometry by projective techniques; indeed, all metric geometries can be derived in

---

[†]G. Nicolis and I. Prigogine, Self-organization in Non-equilibrium Systems: from Dissipative Structures to Order through Fluctuations (Wiley and Sons, Toronto, 1977).

[††]G. Nicolis and I, Prigogine, loc. cit.

[†††]See, e.g., R.M. Noyes and R.J. Field, Annual Review of Physical Chemistry 25: 95 (1974).

[*]M. Kline, Mathematical Thought from Ancient to Modern Times (Oxford University Press, Oxford, 1972).

this way. In the post-Riemann period, particularly following the development of non-Riemannian geometries, the absolute role of Euclidean geometry declined and many mathematicians adopted the view that number theory and analysis were more fundamental than geometry in the clarity and simplicity of their axioms. This view does not detract from the value of geometry per se, and one could argue that the success of Einstein's geometrization of the dynamics of gravitating bodies again raises the possibility that geometry is fundamental in the scheme of things. It is, of course, clear that the various fields of mathematics like set theory, analysis, algebra, group theory, geometry topology are all interlinked. Still the question remains: in what directions should mathematics develop to deal most effectively with modern problems in physics and biology? The twentieth-century tendency seems to be for mathematicians to work with abstract structures somewhat divorced from explicit realization in physical problems.

Perhaps the greatest challenges to mathematicians today come from the biologists. Aside from the challenge we have already discussed concerning the classification and behaviour of non-linear and dissipative systems, there is the general challenge of dealing qualitatively and quantitatively with morphological problems - form, shape, symmetry, pattern. How are such problems best attacked, through algebra or geometry? or by new schemes as yet undevised? How does one deal explicitly with biological development, where an organism grows and by differentiation attains a complicated structural form and function? Aside from understanding why these changes occur, even the problem of describing the changes is massive. Symmetry is clearly involved, and a group theoretical approach to symmetry classification is highly developed,† but this approach is an on or off affair - either an object has invariance under a group or it has not; degrees of symmetry are not tolerated. Such an approach is evidently of limited use when it comes to describing

†A.V. Shubnikov and V.A. Kopstick, loc. cit.

the morphology of the heart or the morphology of a neural network in the cerebral cortex.

To take a modest example, recent x-ray crystallographic studies[†] of the rather small transfer RNA molecule show a detailed structure that has few symmetries. Nonetheless, the molecular structure is highly specific in form and in function; in broad terms it is L-shaped with a clover-leaf substructure. How does one describe these structures and relate their functional behaviour to such principles as the principle of minimum free energy?

If the information content in such problems is too high for practical purposes, how can it be effectively reduced? In classical dynamics it is possible to identify certain primary constants of the motion without resort to a detailed solution of the equations of motion. Moreover, for systems of large numbers of more-or-less identical particles, one can invoke statistical methods for identifying average behaviour at equilibrium. For more complex systems like those involved in chemistry and biology, where the need is greater, no such general procedures exist, though interesting attempts have been made to establish them.[††]

Man has for many generations been keenly interested in space-filling shapes or designs in two and three dimensions.[†††] Regular tesselations occur in many biological organisms, and are a pervasive component of many art forms, for example, wallpaper design or Escher's drawings.[*] The question has been raised many times whether the microscopic structure of space and time might in fact display such cellular structure. A related question is whether the possible states of matter somehow

---

[†]A. Alexander and S.H. Kim, Scientific American, 238, No. 1 (1978), page 52.

[††]For a detailed discussion of these and related points, see C.J. Lumsden, loc. cit.

[†††]M. Kline, Mathematics in Western Culture (Oxford University Press, Oxford, 1953).

[*]A.V. Shubnikov and V.A. Kopstick, loc. cit.

reflect possible number systems; in short, is the universe geometrical and/or arithmetical, thus giving basic justification for the relevance of mathematical reasoning? As an example, we refer briefly to a suggestion first made by Pais and later explored by Günaydin and Gürsey.†

The essential observation is that only four number fields have the property that the norm of a product is given by the product of the norms, namely, the real numbers, the complex numbers, the quaternions, and the octonions (called Cayley numbers after their discoverer). Only the real numbers are required in classical mechanics although the complex numbers can be used as a convenience in some applications. In non-relativistic point-particle quantum mechanics, the wave function is complex-valued. Solutions of the relativistic Dirac equation which include electron spin are most naturally regarded as spinors. In the spinor analysis, dynamical variables appear as 2 x 2 matrices in a quaternion basis, as we have seen in Chapter 3.

The question naturally arises of why real numbers, complex numbers, and quaternions occur in these schemes and not octonions, for example. The quaternion basis expressed as 2 x 2 Hermitian matrices is manifested by a two-ness in nature; for example, it was precisely the doublet structure of the spectra of hydrogen-like atoms that led Goudsmit and Uhlenbeck to propose the existence of electron spin. An octonion basis would require 3 x 3 Hermitian matrices, which should show up as a three-ness in nature. In fact, a pronounced three-ness exists in terms of unitary spin or its concomitant in the quark model, and also in the three-ness of quark colouring. However, we are left with open-ended considerations which may eventually be resolved, namely, real and complex numbers are commutative and associative with respect to multiplication; quaternions are associative but non-commutative; octonions are neither associative nor commutative. In quantum mechanics, one can live

---

†A. Pais, Phys. Rev. Letters 7: 291 (1976); M. Günaydin and F. Gürsey, Phys. Rev. D9: 3387 (1974).

with non-commutivity; it is part of the theory. But non-associativity presents problems since matrices do associate; in fact, it was just this property of combining spectra that led Heisenberg to his matrix formulation. For the interested reader, this whole question is related to Jordan algebras and to an interesting attempt to reformulate quantum mechanics by Jordan, Von Neumann, and Wigner.†

The algebraic approach to quantum systems is used at many levels. At the level of modelling it has been used, for example, in nuclear collective motion.†† If the low-energy spectrum of a many-particle system arises out of collective motions (usually vibrations and rotations) one can construct a model Hamiltonian whose variables are the collective coordinates and whose low-energy spectrum simulates the experimentally measured spectrum. The motion of the actual system can then be modelled in terms of the dynamical symmetries of the model Hamiltonian; these symmetries are generated by the elements of the so-called spectrum-generating algebras. Whether such methods might apply to dynamical problems in biology is an open question.

A recent attempt to unify quantum theory and gravitation provides yet another example of an algebraic approach.††† The Poincaré group is extended by adding four new generators which behave as spinors and which vary as the square root of the translation operator. The scheme is referred to as supersymmetry since an irreducible representation of the extended group includes both fermions and bosons and hence unifies the connection between spin and statistics. Researchers involved with supersymmetry are hopeful that the scheme, when appropriately

---

†P. Jordan, J. Von Neumann, and E.P. Wigner, Ann. Math. 35: 39 (1934)

††See e.g. D.J. Rowe and G. Rosensteel, 'The Nuclear Collective Model and the Symplectic Group,' Proceedings, VI International Colloquium on Group Theoretical Methods in Physics (Tubingen, 1977).

†††See Physics Today 30, No. 10 (October 1977), p. 17.

modified, will achieve a unity between quantum theory and gravitation. It is by no means the only attempt to do so and we mention it by way of illustration only.

Finally we mention a recent development which seems to give a unique status to three field theories of current interest, namely Maxwell's electromagnetic field, Einstein's gravitational field, and the Yang-Mills field for quark-quark interaction. The essential idea† is that if one restricts the possible Hamiltonians which take a field from a given initial to a given final configuration by imposing a certain 'requirement of embedability,' then the unique vector field is Maxwell's electrodynamics, the unique second-rank tensor field is Einstein's gravitation, and the unique isotopic spin vector field is the Yang-Mills field for binding quarks. This is an example of the physicist's dream of mirroring the unique world of objective reality in terms of unique mathematical descriptions based on a set of fundamental axioms. The dream can only be partially realized since, as Godel†† has shown, no finite set of axioms is ever complete in the sense that for any given set, questions can be raised which cannot be answered without additional axioms. In this sense then our struggle to describe the universe in mathematical terms, however beautiful and interesting, will always be incomplete.

---

†S. Hojman, K. Kuchar, and C. Tretelboim, Nature of Physical Sciences 245: 97 (1973); Ann. of Physics 76: 88 (1976); J.A. Wheeler, 'The Quantum and the University,' loc. cit.

††E. Nagel and J.R. Newman, 'Godel's Proof,' in M. Kline, Mathematics in the Modern World (W.H. Freeman and Co., San Francisco, 1968).

APPENDICES

APPENDIX A
Linear and multilinear algebra

A set V is called a vector space over a field F if there exist two operations, one called vector addition, which maps V x V into V and is denoted by +, and the other scalar multiplication, which maps F x V into V and is denoted by $\cdot$, such that:

(i) V is closed under both scalar multiplication and vector addition,

(ii) $\vec{v}_1 + \vec{v}_2 = \vec{v}_2 + \vec{v}_1$ for all $\vec{v}_1, \vec{v}_2 \in V$;

(iii) $\vec{v}_1 + (\vec{v}_2 + \vec{v}_3) = (\vec{v}_1 + \vec{v}_2) + \vec{v}_3$ for all $\vec{v}_1, \vec{v}_2, \vec{v}_3 \in V$;

(iv) there exists an $\vec{o} \in V$ such that $\vec{o} + \vec{v} = \vec{v}$ for all $\vec{v} \in V$;

(v) for all $\vec{v} \in V$, there exists a unique $\vec{v}^{-1} \in V$ such that $\vec{v} + \vec{v}^{-1} = \vec{o}$;

(vi) $a\cdot(\vec{v}_1 + \vec{v}_2) = a\cdot\vec{v}_1 + a\cdot\vec{v}_2$ for all $\vec{v}_1, \vec{v}_2 \in V$ and all $a \in F$;

(vii) $(a + b)\cdot\vec{v} = a\cdot\vec{v} + b\cdot\vec{v}$ for all $a,b \in F$ and all $\vec{v} \in V$;

(viii) $a\cdot(b\cdot\vec{v}) = (a\cdot b)\cdot\vec{v} = b\cdot(a\cdot\vec{v})$ for all $a,b \in F$ and all $\vec{v} \in V$;

(ix) $1\cdot\vec{v} = \vec{v}$ for all $\vec{v} \in V$, where 1 is the multiplication identity of the field F.

Elements of a vector space are called vectors.

The set of all ordered n-tuples of real numbers forms a vector space over the field of real numbers if vector addition is defined by

$$(v_1,\ldots,v_n) + (u_1,\ldots,u_n) = (v_1 + u_1,\ldots,v_n + u_n)$$

and scalar multiplication is defined by

$$c\cdot(v_1,\ldots,v_n) = (cv_1,\ldots,cv_n).$$

The set of all infinite sequences of real numbers also forms a vector space over the reals with vector addition defined by

$$(v_1,\ldots,v_j,\ldots) + (u_1,\ldots,u_j,\ldots) = (v_1 + u_1,\ldots, v_j + u_j,\ldots)$$

and scalar multiplication defined by

$$c\cdot(v_1,\ldots,v_j,\ldots) = (cv_1,\ldots,cv_j,\ldots).$$

A field F forms a vector space over itself with the definition of scalar multiplication and vector addition the same as multiplication and addition in the field.

Finally note that the set of all functions of a single real variable which are continuous forms a vector space over the field of real numbers with addition of functions and multiplication of a function by a real number defined in the usual manner.

From the definition of a vector space it follows that $\vec{v}^{-1} = -1\cdot\vec{v}$ and $o\cdot\vec{v} = \vec{o}$ for every $\vec{v} \varepsilon V$ where o is the zero of the field over which V is a vector space.

If W is a subset of a vector space V and the elements of W form a vector space with the same definition of vector addition and scalar multiplication as was used to define V, then W is called a vector subspace of V. Let X and Y be two vector subspaces of V. Then their intersection $X \cap Y = \{\vec{z} \varepsilon V| \vec{z} \varepsilon X \text{ and } \vec{z} \varepsilon Y\}$ and their sum $X + Y = \{\vec{x} + \vec{y}| \vec{x} \varepsilon X \text{ and } \vec{y} \varepsilon Y\}$ are also vector subspaces of V. If $X \cap Y = \{\vec{o}\}$ we call X + Y the direct sum of X and Y and denote it by $X \oplus Y$.

An important subspace of the set of real sequences $(v_1,\ldots,v_j,\ldots)$ is one for which $\sum_{j=1}^{\infty} (v_j)^2$ converges. This vector space is called ℓ_2 space.

The span of a set of vectors $\{\vec{v}_i\}$, i ε d, where d is some index set, is denoted $S_p\{\vec{v}_i\}$ and defined as the set of all vectors $\vec{v}$ which can be written as a finite linear combination of the $\vec{v}_i$; that is the set of all $\vec{v} = \sum_i a_i\vec{v}_i$ where $a_i \in F$ and all but a finite number of the a_i are zero. $S_p\{\vec{v}_i\}$ when $\vec{v}_i \in V$ is a vector subspace of V.

A set $\{\vec{v}_i\}$, $i \in d$, where $\vec{v}_i \in V$, is called linearly dependent if there exist elements a_i of the field F (over which V is a vector space) not all zero such that $\sum_i a_i\vec{v}_i = 0$ where all but a finite number of the a_i are zero. A set which is not linearly dependent is called linearly independent. The set $\{\vec{v}_i\}$, $i \in d$, is called a basis for V if

(i) $\{\vec{v}_i\}$, $i \in d$, is linearly independent,

(ii) $V = S_p\{\vec{v}_i\}$.

It can be shown that if a vector space V has a basis with a finite number of elements n then all bases of V have n elements; the integer n is called the dimension of V.

Let V and U be two vector spaces over the field F. An operator L which maps V into U is called linear if

$$L(a_1\vec{v}_1 + a_2\vec{v}_2) = a_1L(\vec{v}_1) + a_2L(\vec{v}_2) \quad \text{for all } \vec{v}_1,\vec{v}_2 \in V \text{ and all } a_1,a_2 \in F. \tag{1}$$

The set of all linear operators which map V into U is a vector space over the field F denoted by L(V;U) with addition of linear operators defined by

$$(L_1 + L_2)(\vec{v}) = L_1(\vec{v}) + L_2(\vec{v}) \quad \text{for all } L_1,L_2 \in L(V;U) \text{ and all } \vec{v} \in V,$$

and multiplication by a scalar defined by

$$(a\cdot L)(\vec{v}) = a\cdot(L(\vec{v})) \quad \text{for all } L \in L(V;U), \text{ all } a \in F, \text{ and all } \vec{v} \in V.$$

Of particular importance is the case where U is the field F itself. We call L(V;F) the dual space of V and denote it by the special symbol V*. If V is finite-dimensional, then V* has the same dimension as V.

For example, if V is the two-dimensional vector space of ordered pairs (x_1,x_2), where $x_1,x_2 \in R$, then any linear map L of V into the field of real numbers R may be represented by a column matrix

$$\begin{bmatrix} L_1 \\ L_2 \end{bmatrix}, \quad L_1,L_2 \in R$$

whose action on any vector $\vec{v} = (x_1,x_2)$ is given by matrix multiplication. That is

$$L(\vec{v}) = \begin{bmatrix} x_1 & x_2 \end{bmatrix} \begin{bmatrix} L_1 \\ L_2 \end{bmatrix} = L_1x_1 + L_2x_2.$$

The dual space $L(V;R)$ is also two-dimensional with basis $\left\{\begin{bmatrix} 1 \\ 0 \end{bmatrix}, \begin{bmatrix} 0 \\ 1 \end{bmatrix}\right\}$, for example.

Let V, U, W be three vector spaces over the same field F. An operator L which maps V x U into W is called bilinear if

(2a) $L(\vec{v}, a_1\vec{u}_1 + a_2\vec{u}_2) = a_1L(\vec{v},\vec{u}_1) + a_2L(\vec{v},\vec{u}_2)$,

for all $a_1,a_2 \in F$, all $\vec{v} \in V$,

and all $\vec{u}_1,\vec{u}_2 \in U$,

and

(2b) $L(a_1\vec{v}_1 + a_2\vec{v}_2, \vec{u}) = a_1L(\vec{v}_1,\vec{u}) + a_2L(\vec{v}_2,\vec{u})$

for all $a_1,a_2 \in F$, all $\vec{u} \in U$,

and all $\vec{v}_1,\vec{v}_2 \in V$.

The set of all bilinear operators which map V x U into W forms a vector space over F (with the usual definition for scalar multiplication and addition of operators) which we denote by $L(V,U;W)$.

Suppose that V and U are vector spaces of dimension n and m respectively over the same field F. The outer product space or tensor product space of V and U is defined as the dual of the vector space $L(V,U;F)$ and denoted by $V \otimes U$. Since $L(V,U;F)$ has dimension mn it follows that $V \otimes U$ also has dimension mn.

Let $\vec{u}$ and $\vec{v}$ be two fixed elements of U and V respectively. The element $\vec{\beta}$ of $V \otimes U$ defined by

$$\vec{\beta}(\vec{\alpha}) = \vec{\alpha}(\vec{v},\vec{u}) \quad \text{for all } \vec{\alpha} \in L(V,U;F) \tag{3}$$

is denoted by $\vec{v} \otimes \vec{u}$. Hence

$$[\vec{v} \otimes \vec{u}](\vec{\alpha}) = \vec{\alpha}(\vec{v},\vec{u}) \quad \text{for all } \vec{\alpha} \in L(V,U;F), \text{ all } \vec{v} \in V, \text{ and all } \vec{u} \in U. \tag{4}$$

Note that not every element of $V \otimes U$ can be written as $\vec{v} \otimes \vec{u}$ for some $\vec{v} \in V$ and $\vec{u} \in U$. From the bilinearity of $\vec{\alpha}$ we have

$$(a_1\vec{v}_1 + a_2\vec{v}_2) \otimes \vec{u} = a_1(\vec{v}_1 \otimes \vec{u}) + a_2(\vec{v}_2 \otimes \vec{u}), \tag{5a}$$

$$\vec{v} \otimes (a_1\vec{u}_1 + a_2\vec{u}_2) = a_1(\vec{v} \otimes \vec{u}_1) + a_2(\vec{v} \otimes \vec{u}_2) \tag{5b}$$

for all $\vec{v},\vec{v}_1,\vec{v}_2 \in V$, all $\vec{u},\vec{u}_1,\vec{u}_2 \in U$, and all $a_1,a_2 \in F$.

Let $\{\vec{v}_j\}$, $j = 1,\ldots,n$, and $\{\vec{u}_j\}$, $j = 1,2,\ldots,m$, respectively be bases for V and U. Then arbitrary elements $\vec{v}$ and $\vec{u}$ of V and U respectively may be written as

$$\vec{v} = \sum_{j=1}^{n} a_j\vec{v}_j \quad \text{and} \quad \vec{u} = \sum_{k=1}^{m} b_k\vec{u}_k.$$

The a_j, $j = 1,\ldots,n$ are called the components of $\vec{v}$ with respect to the basis $\{\vec{v}_i\}$ and the b_k, $k = 1,\ldots,m$ are called the components of $\vec{u}$ with respect to the basis $\{\vec{u}_j\}$. From equations (5) it follows that

$$\vec{v} \otimes \vec{u} = \sum_{j=1}^{n} \sum_{k=1}^{m} a_j b_k(\vec{v}_j \otimes \vec{u}_k). \tag{6}$$

In fact the set of vectors $\{\vec{v}_j \otimes \vec{u}_k\}$, $j = 1,\ldots,n$, $k = 1,\ldots,m$, is a basis for the vector space $V \otimes U$ so that the nm quantities $a_j b_k$, $j = 1,\ldots,n$, $k = 1,\ldots,m$, are the components of $\vec{v} \otimes \vec{u}$ with respect to the basis $\{\vec{v}_j \otimes \vec{u}_k\}$.

Let $V^1,\ldots,V^p$ and W be vector spaces over the same field F. An operator L which maps $V^1 \times \ldots \times V^p$ into W is called p-linear if

$$(7) \quad L(\vec{v}^1, \ldots, \vec{v}^{j-1}, a_1\vec{v}^j_1 + a_2\vec{v}^j_2, \ldots, \vec{v}^p)$$

$$= a_1 L(\vec{v}^1,\ldots,\vec{v}^j_1,\ldots,\vec{v}^p) + a_2 L(\vec{v}^1,\ldots\vec{v}^j_2,\ldots,\vec{v}^p)$$

for all $\vec{v}^k \in V^k$, $k = 1,\ldots,p$; all $\vec{v}^j_1, \vec{v}^j_2 \in V^j$, $j = 1,\ldots,p$;

and all $a_1, a_2 \in F$.

The set of all p-linear maps of $V^1 \times \ldots \times V^p$ into W forms a vector space over the field F (with the usual definition for scalar multiplication and addition of operators) denoted by $L(V^1,\ldots,V^p;W)$.

Suppose that $V^1,\ldots,V^p$ are vector spaces, of dimension $n_1,\ldots,n_p$ respectively, over the same field F. The outer product space or tensor product space of $V^1,\ldots,V^p$ is defined as the dual of $L(V^1,\ldots,V^p;F)$ and denoted by $V^1 \otimes \ldots \otimes V^p$.

The results for the tensor product of two vector spaces of finite dimension may be applied to the case of the tensor product of p finite-dimensional vector spaces if the obvious modifications are made.

The notion of an outer product of vector spaces may be extended to infinite-dimensional vector spaces,† but the definition becomes much more abstract. Most of the results given for finite-dimensional vector spaces go over to the infinite-dimensional case if the obvious modifications are made.

†See, for example, W.H. Greub, Multilinear Algebra (Springer-Verlag, New York, 1967).

PROBLEMS

1 (i) Show that the set of all ordered n-tuples of complex numbers $(z_1,\ldots,z_n)$ forms a vector space over the field of complex numbers with addition and scalar multiplication defined in the usual manner. This vector space is usually denoted by C^n.
(ii) Show that the set of all infinite sequences of real numbers $(a_1,a_2,\ldots,a_n,\ldots)$ such that

$$\sum_{i=1}^{\infty} a_i^2 \text{ converges}$$

forms a vector space over the reals with vector addition of two sequences defined by

$$(a_1,\ldots,a_n,\ldots) + (b_1,\ldots,b_n,\ldots)$$
$$= (a_1 + b_1,\ldots, a_n + b_n, \ldots)$$

and a scalar multiplication defined by

$$a\cdot(b_1,\ldots,b_n,\ldots) = (ab_1,\ldots ab_n,\ldots).$$

2 (i) Prove that $(a_1\vec{v}_1 + a_2\vec{v}_2) \otimes \vec{u} = a_1(\vec{v}_1 \otimes \vec{u}) + a_2(\vec{v}_2 \otimes \vec{u})$, for all $\vec{v}_1,\vec{v}_2 \in V$, all $\vec{u} \in U$, and all $a_1,a_2 \in F$, where V and U are finite-dimensional vector spaces over the same field F.
(ii) Prove that $\{\vec{v}_j \otimes \vec{u}_k\}$, $j=1,\ldots,n$, $k=1,\ldots,m$, is a basis for $V \otimes U$ if $\{\vec{v}_j\}$, $j=1,\ldots,n$, is a basis for V and $\{\vec{u}_k\}$, $k=1,\ldots,m$, is a basis for U.

APPENDIX B
Linear transformations on a vector space - orthogonal and unitary matrices

We define the inner product between two ordered n-tuples of complex numbers $\vec{u} = (u_1, \ldots, u_n)$ and $\vec{v} = (v_1, \ldots, v_n)$, denoted $\vec{v} \cdot \vec{u}$, by

$$\vec{v} \cdot \vec{u} = v_1^* u_1 + v_2^* u_2 + \ldots v_n^* u_n$$

where v_j^* denotes the complex conjugate of v_j. If B is an n x n matrix with complex number elements

$$B = \begin{bmatrix} B_{11} & \cdots & B_{1n} \\ \vdots & & \vdots \\ B_{n1} & \cdots & B_{nn} \end{bmatrix},$$

we define the adjoint of B, denoted B^+, as the n x n matrix

$$B^+ = \begin{bmatrix} B_{11}^+ & \cdots & B_{1n}^+ \\ \vdots & & \vdots \\ B_{n1}^+ & \cdots & B_{nn}^+ \end{bmatrix},$$

which satisfies the equation

$$\vec{u} \cdot (B\vec{v}) = (B^+\vec{u}) \cdot \vec{v}$$

for all ordered n-tuples of complex numbers, $\vec{v}$ and $\vec{u}$. In component form this reads

$$\sum_{j=1}^{n} \sum_{i=1}^{n} B_{ji} v_i u_j^* = \sum_{i=1}^{n} \sum_{j=1}^{n} v_i (B_{ij}^+ u_j)^*$$

which implies (adopting the summation convention) that

$$(B_{ji} - B_{ij}^{+*}) v_i u_j^* = 0.$$

Since the above equation is true for arbitrary $\vec{v} = (v_1,\ldots,v_n)$ and $\vec{u} = (u_1,\ldots,u_n)$ it follows that

$$B^+_{ij} = B^*_{ji}, \quad \text{for all } i,j = 1,2,\ldots,n,$$

so that $B^+ = (\tilde{B})^* = \tilde{B}^*$, the complex conjugate of the transpose of B.

We call

$$C = \begin{bmatrix} C_{11} & \cdots & C_{1n} \\ \vdots & & \vdots \\ C_{n1} & \cdots & C_{nn} \end{bmatrix}, \quad C_{ij} \text{ complex, } i,j = 1,2,\ldots,n,$$

a unitary matrix if $\tilde{C}^* = C^+ = C^{-1}$. Since $C^+C = I$

$$C^+_{ij}C_{j\ell} = C^*_{ji}C_{j\ell} = \delta_{i\ell};$$

also from $CC^+ = I$ one obtains the conjugate relation

$$C_{ij}C^+_{j\ell} = C_{ij}C^*_{\ell j} = \delta_{i\ell}.$$

Here $\delta_{i\ell}$ is the Kronecker delta function defined by

$$\delta_{i\ell} = \begin{cases} 1 & \text{if } i = \ell, \\ 0 & \text{if } i \neq \ell. \end{cases}$$

The determinant of a unitary matrix has unit modulus† since

$$\det(C\tilde{C}^*) = 1 \Rightarrow (\det C)(\det \tilde{C})^* = 1 \Rightarrow |\det C|^2 = 1.$$

Transformation by unitary matrices preserves inner products since, if

$$\vec{u}' = C\vec{u} \quad \text{and} \quad \vec{v}' = C\vec{v},$$

then, writing the sums out explicitly, we have

$$\vec{v}'\cdot\vec{u}' = \sum_{i=1}^{n} u'_i v'^*_i = \sum_{i=1}^{n} \left(\sum_{j=1}^{n} C_{ij}u_j\right)\left(\sum_{\ell=1}^{n} C^*_{i\ell}v^*_\ell\right)$$

†The modulus of a complex number Z is denoted $|Z|$ and defined by $|Z| = \sqrt{ZZ^*}$.

$$= \sum_{\ell=1}^{n} \sum_{j=1}^{n} \left(\sum_{i=1}^{n} C_{ij}C_{i\ell}^{*}\right) u_j v_\ell^{*}$$

$$= \sum_{j=1}^{n} \sum_{\ell=1}^{n} \delta_{j\ell} u_j v_\ell^{*} = \sum_{\ell=1}^{n} u_\ell v_\ell^{*} = \vec{v}\cdot\vec{u}.\dagger$$

In physical applications one is sometimes restricted to the use of real numbers only. The inner product between two ordered n-tuples with real components (i.e. the scalar product) $\vec{u} = (u_1,\ldots,u_n)$ and $\vec{v} = (v_1,\ldots,v_n)$ is a special case of the above definition which can be written (summation convention)

$$\vec{v}\cdot\vec{u} = v_i u_i$$

since the complex conjugate of a real number is the number itself. The adjoint of an n x n matrix B with real entries (i.e. B_{ij} real for $i,j = (1,2,\ldots,n)$) is again denoted B^+ and defined by $(B\vec{v})\cdot\vec{u} = \vec{v}\cdot(B^+\vec{u})$ for all ordered n-tuples of real numbers $\vec{v}$ and $\vec{u}$. Calculation reveals that $B^+ = \tilde{B}$, the transpose of B.

We call

$$C = \begin{bmatrix} C_{11} & \cdots & C_{1n} \\ \vdots & & \vdots \\ C_{n1} & \cdots & C_{nn} \end{bmatrix}, \qquad C_{ij} \text{ real, } i,j = 1,2,\ldots,n,$$

an orthogonal matrix if $\tilde{C} = C^+ = C^{-1}$. An orthogonal matrix is unitary but the converse is not necessarily true. From $CC^{-1} = I$ we have in a similar manner to the above

(1) $$C_{kj}C_{je}^{-1} = C_{kj}C_{ej} = \delta_{ke}$$

and, from $C^{-1}C = I$,

(2) $$C_{kj}^{-1}C_{je} = C_{jk}C_{je} = \delta_{ke}$$

---

†A shorter but somewhat less instructive proof is $\vec{u}'\cdot\vec{v}' = (C\vec{u})\cdot(C\vec{v}) = (C^+C\vec{u})\cdot\vec{v} = \vec{u}\cdot\vec{v}$ since $C^+C = I$.

where $C^{-1}_{\ell k}$ signifies the ℓ,kth matrix element of the inverse matrix C^{-1} and not $1/C_{\ell k}$. The determinant of an orthogonal matrix is ±1 since

$$C\tilde{C} = I \Rightarrow (\det C)(\det \tilde{C}) = 1 \Rightarrow (\det C)^2 = 1.$$

Consider now a linear transformation of the ordered n-tuples (with real components) $\vec{u} = (u_1,\ldots,u_n)$ and $\vec{v} = (v_1,\ldots,v_n)$. Put

$$\text{(3a)} \qquad u'_k = C_{kj}u_j \quad \text{and} \quad v'_k = C_{kj}v_j.$$

Transformation of $\vec{u}$ and $\vec{v}$ by the orthogonal matrix C leaves the inner product of $\vec{u}$ and $\vec{v}$ invariant (i.e. $\vec{u}\cdot\vec{v} = \vec{u}'\cdot\vec{v}'$). Since the norm or length of a vector and the angle between two vectors are given in terms of inner products, both are invariant under orthogonal transformation. (We have used the term vector for the ordered n-tuple of real numbers; see Appendix A.)

If we multiply both sides of the first equation of (3a) by C^{-1}_{ek} and sum over k, we obtain (using explicit sums)

$$\text{(3b)} \qquad \sum_{k=1}^{n} C^{-1}_{ek}\, u'_k = \sum_{j=1}^{n} \Big(\sum_{k=1}^{n} C^{-1}_{ek}\, C_{kj} \Big) u_j = u_e.$$

Hence, under orthogonal (or any invertible linear) transformation of vectors, the inverse matrix gives the inverse transformation with the same rules of combination (i.e. the running index goes over columns, the fixed index referring to a specific row).

PROBLEMS

1 Prove that the inner product of two n-tuples of complex numbers satisfies

(a) $\vec{v}\cdot(a_1\vec{u}_1 + a_2\vec{u}_2) = a_1(\vec{v}\cdot\vec{u}_1) = a_2(\vec{v}\cdot\vec{u}_2)$ for all $a_1, a_2 \in C$ and all $\vec{v}, \vec{u}_1, \vec{u}_2 \in C^n$;

(b) $\vec{v} \cdot \vec{u} = (\vec{u} \cdot \vec{v})^*$ for all $\vec{v}, \vec{u} \in C^n$;

(c) $\vec{v} \cdot \vec{v} \geq 0$ for all $\vec{v} \in C^n$, with equality if and only if $\vec{v} = 0$.

2 Prove that

(a) $(A_1 + A_2)^+ = A_1^+ + A_2^+$,

(b) $(aA)^+ = a^* A^+$,

(c) $(A^+)^+ = A$,

(d) $(A_1 A_2)^+ = A_2^+ A_1^+$,

where A_1, A_2, and A are n x n matrices with complex entries.

APPENDIX C
Tensors as equivalence classes

In Chapter 2 a Cartesian vector in n-dimensional space was defined as an ordered set of components $\{v_j\}$, $j = 1,\ldots,n$, which transform under rotations like

$$v_j' = a_{jk}v_k \quad \text{where } a_{jk} = \hat{e}_j'\cdot\hat{e}_k.$$

It is clear that the sets of components $\{v_j'\}$, $j = 1,\ldots,n$, and $\{v_j\}$, $j = 1,\ldots,n$, represent the same vector although they are not numerically equal (i.e., it is possible that $v_1' \neq v_1$, for example). It is for this reason that mathematicians sometimes prefer to use the following definition for a vector.

Let C denote an n-dimensional Cartesian coordinate system and $(\vec{v},C)$ be an ordered pair of an ordered n-tuple of real numbers with an n-dimensional Cartesian coordinate system:

$$(\vec{v},C) = ((v_1,\ldots,v_n),C).$$

An n-dimensional Cartesian vector is defined as an equivalence class of the set of all ordered pairs $(\vec{v},C)$ under the equivalence relation $\sim$ defined by $(\vec{v},C) \sim (\vec{v}',C')$ if and only if

$$((a_{1j}v_j,\ldots,a_{nj}v_j),C') = ((v_1',\ldots,v_n'),C')$$

where C is the Cartesian coordinate system with axes $\{\hat{e}_j\}$, $j = 1,2,\ldots,n$, C' is the Cartesian coordinate system with axes $\{\hat{e}_j'\}$, $j = 1,\ldots,n$, and $a_{ij} = \hat{e}_i'\cdot\hat{e}_j$.

Similar remarks hold for higher-ranked tensors.

APPENDIX D
Topological spaces

Set theory forms a basis for several branches of modern mathematics, such as the theory of functions, differential geometry, and topology. In this appendix we introduce the concept of a topology defined on a set S, which is then referred to as a topological space. For the most part, the topological spaces of interest in mathematical physics are well-behaved (i.e. Hausdorff spaces) and correspond with our intuitive ideas of one or more continuous, curved surfaces in some higher-dimensional Cartesian space. Nonetheless, important exceptions occur and even well-behaved spaces have subtle properties which are best understood in the language of set theory.

A topological space may be defined as an ordered pair (S,θ) where S is a set and θ a collection of subsets of S satisfying three axioms:

(a) The empty set $\emptyset$ and the complete set S are contained in θ.

(b) The union of sets from any subcollection of θ is in θ; for example, if J is some index set such that $X_\alpha \in \theta$ whenever $\alpha \in J$, then $\cup_{\alpha \in J} X_\alpha \in \theta$.

(c) If $X_i \in \theta$, $i = 1,2,\ldots,n$, then the intersection $\bigcap_{i=1}^{n} X_i \in \theta$.

Those subsets of S which are in θ are called open sets.

If (S,θ) is a topological space we say that θ defines a topology on S. A given set S may have several different topologies defined on it. For example, each of the following choices of θ makes (S,θ) a topological space for arbitrary S.

(i) θ contains all subsets of S;

---

†See, for example, J.R. Munkres, Topology: A First Course (Prentice-Hall, Englewood Cliffs, 1975).

(ii) θ contains the empty set $\emptyset$ and all subsets of S that contain all but a finite number of elements S;

(iii) $\theta = \{\emptyset, S\}$.

Suppose that θ and θ' are two topologies defined on S. We say that θ is weaker than θ' (or equivalently θ' stronger than θ) and write $\theta < \theta'$ if every subset of S contained in θ is also in θ'.

Let (S,θ) be a topological space. A neighbourhood of a point $p \in S$ is an open set containing p. The topological space (S,θ) is called Hausdorff if and only if for all p_1 and p_2 in S, $p_1 \neq p_2$, there are neighbourhoods O_1, O_2 such that $p_1 \in O_1$, $p_2 \in O_2$, and $O_1 \cap O_2 = \emptyset$. A sequence $p_i \in S$, $i = 1,2,\ldots,$ is said to converge to a limit $p \in S$ if and only if every neighbourhood of p contains all but a finite number of the points p_i. In general it is possible for a sequence to have several different limits; however, it is easily seen that in a Hausdorff space a sequence cannot have two distinct limits.

A set $C \subset S$ is called <u>closed</u> if and only if it is the complement of an open set. It is important to realize that in a topological space there exist subsets that are both open and closed, for example S and $\emptyset$. If a topological space has a non-trivial subset (i.e. a subset different from S and $\emptyset$) that is both open and closed, then S is the union of two disjoint non-empty closed sets and is called disconnected. A topological space that is not disconnected is called connected.

Let A be a subset of S. The closure of A, denoted by $\bar{A}$, is defined as the smallest closed set containing A and the interior of A, denoted by $A°$, is defined as the largest open set contained in A. The boundary of A is the difference set $A - A°$. Since the union of arbitrary open sets is an open set the intersection of arbitrary closed sets is a closed set. Hence for any subset A of S there exists a largest open set contained in A (namely $\cup O$ where the union is over all open sets O contained in A) and there exists a smallest closed set containing A (namely $\cap C$ where the intersection is over all closed sets C containing A).

If R' is the field of real numbers and $a,b \in R'$, the open interval (a,b) is defined as $\{x \in R' \mid a < x < b\}$. Let θ denote the collection of subsets $O \subset R'$ with the property that for each $x \in O$ there is an open interval A such that $x \in A \subset O$. Then θ defines a topology on R' called the natural topology. A family of subsets $\beta \subset \theta$ is called a base for a topology θ defined on a set S if and only if any $O \in \theta$ is of the form $O = \bigcup_\alpha \beta_\alpha$ for some family $\{\beta_\alpha\} \subset \beta$. For example, the set of all open intervals forms a base for the natural topology on R'.

Let (S',θ') and (S,θ) be two topological spaces. A function $f: S \to S'$ is called continuous if and only if $f^{-1}(X) \in \theta$ for every $X \in \theta'$ (i.e. the inverse image of any open set is open). A continuous bijection with continuous inverse is called a homeomorphism. Let β be a family of subsets of the Cartesian product $S \times S'$ defined by

$$\beta = \{O \times O' \mid O \in \theta,\ O' \in \theta'\}.$$

The set β forms a base for a topology on $S \times S'$ called the topological product space of the topological spaces (S,θ), (S',θ'). Similarly one can define the product space of any number of topological spaces. In this way we can define a topology on the space $R^n = R' \times R' \times \ldots \times R'$ (n factors) as the topological product space of the natural topology on R'. This is called the natural topology on R^n.

APPENDIX E
Measure and integration

A non-empty collection K of subsets of a set X is called a <u>Boolean algebra</u> if

(1a) for every $R, S \in K$, $R \cup S \in K$,

and

(1b) for every $R \in K$, $R' = (X - R) \in K$.

Since K is non-empty, there exists some $R \in K$. Consequently, $X = R \cup R'$ and $\emptyset = X'$ are elements of K. If in addition to axioms (1a) and (1b) K has the property that $\cup_{k=1}^{\infty} S_k \in K$ whenever $S_1, S_2, \ldots$ is a countable collection of sets from K, then K is called a <u>Boolean σ-algebra</u>. Given any family F of subsets of a set X there exists a unique smallest Boolean algebra containing F. It is called the Boolean algebra generated by F.

The Boolean algebra generated by the family I^n of all intervals, including the degenerate ones, of $R^n = R \times R \times \ldots \times R$, with R taken n times where $R = (-\infty, \infty)$, is of particular importance in wave mechanics. The family of all finite unions $I_1 \cup I_2 \ldots \cup I_k$ with $I_1, I_2, \ldots I_k \in I^n$ is identical with the Boolean algebra generated by I^n. Members of the Boolean σ-algebra generated by I^n are called Borel sets. The class of Borel sets is very large, and not every Borel set can be written as a countable union of sets from I^n. In fact every closed and every open set in R^n is a Borel set.

An ordered pair (A,X), where X is a set and A a Boolean σ-algebra of subsets of X, is called a <u>measurable space</u>. The sets S belonging to A are called measurable. A real-valued set function† $\mu(S)$ with domain of definition a Boolean σ-algebra A is called a measure if

(2a) $\mu(\phi) = 0$,

(2b) $\mu(S) \geq 0$ for every $S \in A$,

(2c) $\mu(\cup_{k=1}^{\infty} S_k) = \sum_{k=1}^{\infty} \mu(S_k)$ provided $S_i \cap S_j = \emptyset$

where $i \neq j$ and $S_k \in A$.

If $S \in A$, $\mu(S)$ is called the measure of the set S. When $\mu(S) < \infty$ we say S has finite measure. If S is a countable union of sets with finite measure, then the measure of S is called σ-finite. A measure $\mu(S)$ on a Boolean σ-algebra A is called finite (σ-finite) if the universal set X has σ-finite measure.

The Lebesgue measure is defined on the Borel sets of the real line (or more generally R^n) and assigns a measure $\mu([a,b]) = b - a$ to each interval [a,b] of the real line.†† The Lebesgue measure is not finite since $\mu(R) = \infty$.

A real-valued function f(x), $x \in X$, defined on (the elements of) a measurable subset Q of a measurable space is called measurable on Q if for every open interval $I \subset R$ the subset

(3) $f^{-1}(I) = \{x \mid x \in Q \text{ and } f(x) \in I\}$

of X is measurable.

A measure space is an ordered pair $((A,X),\mu)$ where (A,X) is a measurable space and μ is a measure defined on the Boolean σ-algebra A. Of particular importance in the development of integration theory are the simple functions f(x) which may be written in the form

---

†In fact a measure is an extended real-valued function since the measure of a set may be infinite.

††The Lebesgue measure of any Borel set may be determined from the measure of intervals. See, for example, A. Kolmogorov and S. Fomin, Functional Analysis (Graylock Press, Albany, N.Y., 1961).

$$f(x) = \sum_{k=1}^{n} a_k \chi_{S_k}(x),$$

where $S_i \cap S_j = \emptyset$ for $i \neq j$, where $a_1, \ldots, a_n$ are real numbers, and where $\chi_{S_k}(x)$ is the characteristic function of the measurable set S_k defined by $\chi_{S_k}(x) = 1$ if $x \in S_k$ and 0 if $x \notin S_k$. A simple function

(4) $$f(x) = \sum_{k=1}^{n} a_k \chi_{S_k}(x)$$

defined on a measurable set Q of the measure space $((A,X),\mu)$ is called integrable on Q if

(5) $$\mu(Q \cap \{x| \ f(x) \neq 0\}) < \infty$$

and its integral with respect to μ is given by

(6) $$\int_Q f(x)d\mu(x) = \sum_{k=1}^{n} a_k \mu(S_k \cap Q).$$

It is easy to extend the above definition of integration to a much wider class of functions. A real-valued, non-negative function $f(x)$ defined on a measurable set Q of a measure space $((A,X),\mu)$ is integrable on Q with respect to the measure μ provided that

(i) $f(x)$ is measurable on Q,

(ii) there exists a non-decreasing sequence

(7) $$f_1(x) \leq f_2(x) \leq \ldots$$

of simple functions, each integrable on Q, converging pointwise to $f(x)$ in such a manner that

(8) $$\lim_{n\to\infty} \int f_n(x)d\mu(x)$$

exists and is finite.

The integral of f(x) on Q with respect to μ is defined by

$$\int_Q f(x)\,d\mu(x) = \lim_{n\to\infty} \int f_n(x)\,d\mu(x). \tag{9}$$

One can prove† that this limit is independent of the particular choice of the sequence $\{f_n(x)\}$ so that the above definition makes sense.

The Riemann integral of a function f(x) is defined by dividing up its domain of definition into finer and finer pieces (Figure E.1),

$$\int_{[a,b]} f(x)\,dx \sim \sum_m f(x_m)[x_m - x_{m-1}], \tag{10}$$

whereas the definition of the integral given in equation (9) corresponds to dividing up the range of the function into finer and finer pieces (Figure E.2),

$$\int_{[a,b]} f(x)\,d\mu(x) \sim \lim_{n\to\infty} \sum_{m=1}^{\infty} \left(\frac{m}{n}\right) \mu\left(f^{-1}\left(\left[\frac{m-1}{n},\frac{m}{n}\right]\right)\right). \tag{11}$$

The integral of functions f(x) which attain both positive and negative values is defined in terms of their positive and negative parts, $f^+(x) = \max\{f(x),0\}$ and $f^-(x) = \max\{-f(x),0\}$, respectively, that is

$$\int_Q f(x)\,d\mu(x) = \int_Q f^+(x)\,d\mu(x) - \int_Q f^-(x)\,d\mu(x), \tag{12}$$

the function f(x) being integrable with respect to μ (or μ-integrable) on Q provided both its positive and negative parts are μ-integrable. If a bounded function is Riemann integrable on an interval [a,b], then its integral with respect to the Lebesgue measure (i.e. Lebesgue integral) exists and has the

---

†See, for example, E. Prugovecki, Quantum Mechanics in Hilbert Space (Academic Press, New York, 1971).

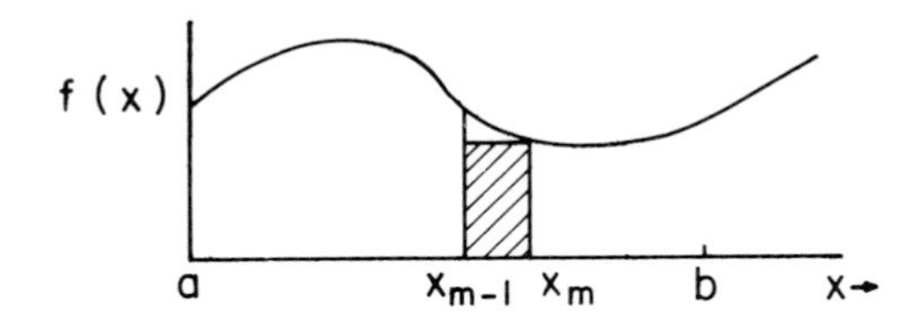

Figure E.1. Structure of the Riemann integral.

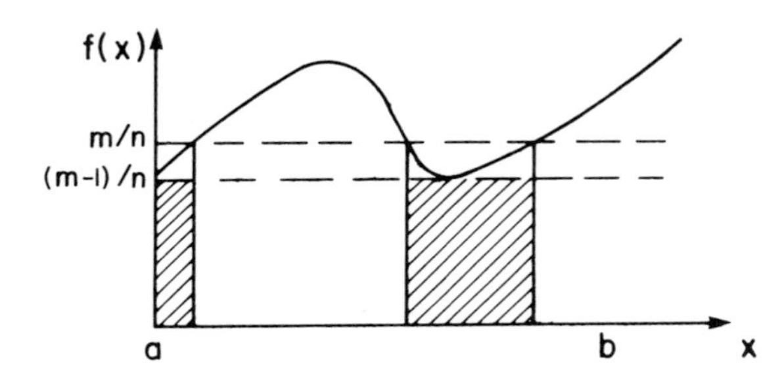

Figure E.2. Structure of the Lebesgue integral.

same value as the Riemann integral. However, there exist functions which are Lebesgue integrable but not Riemann integrable. For example, the function

$$f(x) = \begin{cases} 0 & \text{if } x \notin [0,1] \text{ or if } x = m/2^n,\ n,m = 0,1,2\ldots, \\ 1 & \text{otherwise,} \end{cases}$$

is not Riemann integrable on the interval [0,1] since if τ_1 and τ_2 are step functions which bound f from above and below respectively then $\tau_1 - \tau_2 \geq 1$ for all x in [0,1]. On the other hand the non-decreasing sequence of step functions (note that step functions are simple)

$$\sigma_n(x) = \begin{cases} 0 & \text{if } x \notin [0,1] \text{ or if } x = k \cdot 2^{-n},\ k = 0,1,\ldots,2^n - 1, \\ 1 & \text{otherwise,} \end{cases}$$

converges pointwise to f and

$$\int_{[0,1]} \sigma_n(x)\,d\mu(x) = 1,$$

implying that f is Lebesgue integrable.

Finally a complex-valued function $f(x) = u(x) + iv(x)$ is integrable provided its real and imaginary parts are integrable, where the integral is defined by

$$\int_Q f(x)\,d\mu(x) = \int_Q u(x)\,d\mu(x) + i\int_Q v(x)\,d\mu(x).$$

APPENDIX F
Hilbert spaces

Let V denote a vector space defined over the complex numbers. An inner product of two vectors f and g, denoted $\langle f|g \rangle$, is a mapping of V x V into the field of complex numbers $\mathcal{C}$ with the following properties:

(i) $\langle f|f \rangle \geq 0$ for every $f \in V$, with equality if and only if $f = 0$;

(ii) $\langle f|g \rangle = \langle g|f \rangle^*$ for every $f,g \in V$;

(iii) $\langle f|\alpha g \rangle = \alpha \langle f|g \rangle$ for every $f,g \in V$ and $\alpha \in \mathcal{C}$;

(iv) $\langle f|g + h \rangle = \langle f|g \rangle + \langle f|h \rangle$ for every $f,g,h \in \mathcal{C}$.

The norm of a vector f, denoted $||f||$, is defined by $\langle f|f \rangle^{\frac{1}{2}}$ and is a real number. A vector space with an inner product defined is called a Euclidean space.

In Chapter 7 a proof is given of the Cauchy-Schwartz inequality: for any $f,g \in V$,

$$|\langle f|g \rangle| \leq ||f||\cdot||g||,$$

with equality if and only if f is a multiple of g ($f = \alpha g$, $\alpha \in \mathcal{C}$). The triangular inequality

$$||f + g|| \leq ||f|| + ||g||$$

can easily be proved using the Cauchy-Schwartz inequality.

An infinite sequence of vectors $f_1, f_2, \ldots$ in a Euclidean space E is said to converge to a vector $f \in E$ if for every $e > 0$ there exists an $N(e) > 0$ such that for all $n > N(e)$, $||f - f_n|| < e$. The infinite sequence is called a Cauchy sequence if for every $\varepsilon > 0$ there exists $M(\varepsilon) > 0$ such that for all $m,n > M(\varepsilon)$, $||f_n - f_m|| < \varepsilon$. It can easily be shown that every convergent sequence is a Cauchy sequence.

A Euclidean space E is called a Hilbert space if every Cauchy sequence converges to an element of E.

If f is an element of a Euclidean space E, the set of all points (i.e. vectors) satisfying $||f - g|| < \varepsilon$ for some $\varepsilon > 0$ is called an ε neighbourhood of f. Let S be some subset of E. Then f is called an accumulation point of S if every neighbourhood of f contains an element of S. The set $\bar{S}$ consisting of S plus all its accumulation points is called the closure of S. Clearly S is either contained in $\bar{S}$ or is identical with it. If $S = \bar{S}$ the set S is called closed. The complement of a closed set is called an open set. The ordered pair (E,θ) where θ is the collection of open sets on E defines a topology on E (see Appendix D). A subset S of E is called dense in E if $\bar{S} = E$.

Let S be a set of orthonormal vectors such that $S \subset H$ where H is a Hilbert space. If the closure of the linear vector space spanned by S equals H the set S is said to form a basis for H. A Hilbert space is separable if and only if it has an orthonormal basis with a countable number of elements. The necessary and sufficient condition for the countable set $T = \{e_1, e_2, \ldots\}$ of orthonormal vectors to be a basis for H can be stated in any one of the following four ways:

(i) If $\langle e_k | f \rangle = 0$ for all k, then f is the zero vector.

(ii) $\lim_{n\to\infty} \left|\left| f - \sum_{k=1}^{n} \langle e_k | f \rangle e_k \right|\right| = 0$ for every $f \in H$, or symbolically

$$f = \sum_{k=1}^{\infty} \langle e_k | f \rangle e_k .$$

(iii) $\langle f | g \rangle = \sum_{k=1}^{\infty} \langle f | e_k \rangle \langle e_k | g \rangle$ for all $f,g \in H$. Thus

$\sum_{k=1}^{\infty} |e_k\rangle\langle e_k|$ is the 'unit operator.'

(iv) $||f||^2 = \sum_{k=1}^{\infty} |\langle e_k|f\rangle|^2$ for every $f \in \mathcal{H}$.

A complex-valued function $f(x)$, $x \in X$, defined almost everywhere[†] on the measure space $((A,X),\mu)$ is called square-integrable on X if

$$\int_X |f(x)|^2 d\mu(x) < \infty.$$

If $f(x)$ is not defined everywhere, we can extend its definition by defining $f(x) = 0$ at all points x where it was not originally defined. The set $L_2(X,\mu)$ of all complex-valued functions which are square-integrable forms a linear vector space over the field of complex numbers. This set can be made into a separable Hilbert space with an inner product defined by

$$\langle f|g\rangle = \int_X f^*(x)g(x)d\mu(x) \quad \text{for every } f,g \in L_2.$$

It is clear that $L_2(X,\mu)$ is the Hilbert space of interest in quantum wave mechanics.

A linear operator A in a Hilbert space $\mathcal{H}$ is a linear transformation of a subspace $\mathcal{D}_A$ of $\mathcal{H}$ into $\mathcal{H}$. $\mathcal{D}_A$ is called the <u>domain of definition</u> of A and $\mathcal{R}_A = \{Af|\ f \in \mathcal{D}_A\}$, where Af is the vector mapping of f by A, is the <u>range</u> of A. If $||Af|| \leq a||f||$ for every $f \in \mathcal{D}_A$, where a is a finite number, A is said to be <u>bounded</u> on $\mathcal{D}_A$. Let $\mathcal{D}_A^+$ be the set of all vectors $g \in \mathcal{H}$ each of which can be associated with a unique vector $g^* \in \mathcal{H}$ such that

$$\langle g^*|f\rangle = \langle g|Af\rangle \quad \text{for every } f \in \mathcal{D}_A.$$

Call the linear mapping $A^+: g \to g^* = A^+g$ the <u>adjoint</u> of A. It can be shown[††] that the adjoint exists if and only if $\bar{\mathcal{D}}_A = \mathcal{H}$,

---

[†]I.e. defined everywhere on a set $S \subset X$ whose complement $S' = X - S$ has measure $\mu(X - S) = 0$.
[††]See E. Prugovecki, Quantum Mechanics in Hilbert Space (Academic Press, New York, 1971).

i.e. if $\mathcal{D}_A$ is dense on $\mathcal{H}$.

The linear operator A acting on a Hilbert space is called symmetric if its adjoint A^+ exists and contains A or is identical with A; in the latter case $A = A^+$ and the operator is said to be self-adjoint.

It can be shown that for every closed subspace M' of $\mathcal{H}$ the orthogonal subspace M" is also closed and each vector $f \in \mathcal{H}$ can be uniquely decomposed in the form $f = f' + f''$ where $f' \in M'$ and $f'' \in M''$. The linear mapping $f \rightarrow f' = E_{M'}(f)$ is defined on the entire Hilbert space and is called the projection operator onto M'. It can easily be shown that the projection operator $E_{M'}$ is self-adjoint and satisfies $E_{M'}^2 = E_{M'}$. Projection operators are bounded since $||E_{M'}(f)|| \leq ||f||$ for all $f \in \mathcal{H}$.

The product of two projection operators is also a projection operator if and only if the two projection operators commute; the direct sum is also a projection operator if and only if the corresponding subspaces are mutually orthogonal. Let A(t) be a one-parameter family of operators defined on the entire Hilbert space. If for every $f \in \mathcal{H}$

$$\lim_{t \to t_o} ||(A(t) - A)f|| = 0$$

for some operator A, then A is called the strong limit of A(t) and is written $A = \text{s-lim}A(t)$. This definition can be used to define the sum of an infinite number of projection operators.

A spectral measure E(S) on a measurable space (A,X) is a function that assigns a projection operator E(S) to each subset S of A and satisfies:

(i) $E(X) = 1$,

(ii) $E(\bigcup_{k=1}^{\infty} S_k) = \text{s-lim}_{n \to \infty} \sum_{k=1}^{n} E(S_k)$

where $S_1, S_2, \ldots$ is any sequence of disjoint measurable sets. Clearly one has $E(\phi) = 0$ and $E(R)E(S) = 0$ whenever R and S are disjoint. It can also be shown, for any $f,g \in \mathcal{H}$, that

$$\mu_{f,g}(S) = \langle f|E(S)g\rangle = \sum_{\alpha=1}^{n} C_\alpha \mu_\alpha(S)$$

where the $\mu_\alpha(S)$ are finite measures on (A,X) and the C_α are arbitrary complex coefficients. The integral of a measurable complex function f(x), x ε X (measurable with respect to each measure μ_α), with respect to the complex measure $\mu_{f,g}$ is defined by

$$\int_Q f(x)\,d\mu_{f,g}(x) = \sum_{\alpha=1}^{n} C_\alpha \int_Q f(x)\,d\mu_\alpha(x).$$

Clearly f is integrable on Q ε A if and only if it is integrable with respect to each of the measures μ_α, $\alpha = 1,2,\ldots,n$.

We can now state the fundamental spectral theorem:

To each self-adjoint operator A in a Hilbert space $\mathcal{H}$ there corresponds a unique spectral measure $E^A(S)$ on the Borel sets of the real line such that

$$\langle f|Ag\rangle = \int_{\mathcal{R}} \lambda\,d\mu_{f,g}(\lambda), \quad \text{for every } f \,\varepsilon\, \mathcal{H},\ g \,\varepsilon\, \mathcal{D}_A,$$

where

$$\mathcal{D}_A = \{g|\int_{\mathcal{R}} \lambda^2 d\mu_{g,g}(\lambda) < \infty\}$$

and

$$\mu_{f,g}(S) = \langle f|E^A(S)g\rangle.$$

A point $\lambda \,\varepsilon\, \mathcal{R}$ belongs to the spectrum of the self-adjoint operator A if $E^A(I)$ is non-zero for every open interval I containing λ. One can easily show that points not belonging to the spectrum form an open set, so that spectral values form a closed set. A point in the spectrum of A is an element of the point spectrum of A if and only if $E^A(\{\lambda\}) \neq 0$. Those elements of the spectrum which do not belong to the point spectrum constitute the continuous spectrum of A. It can be shown that

$E^A(S^A) = 1$ where S^A is the spectrum of A.

In wave mechanics a measurable quantity is associated with a self-adjoint operator $A = A^+$. A normalized vector Ψ in the Hilbert space is associated with a state of the system at time t, and the quantity

$$P^A_\Psi(S) = ||E^A(S)\Psi||^2$$

is the probability that a measurement of observable A on a state Ψ gives a result in the set S.

Let μ and ν be two measures defined on the same measurable space. ν is said to be absolutely continuous with respect to μ if and only if $\mu(Q) = 0$ implies that $\nu(Q) = 0$. The Radon-Nikodym theorem states that ν is absolutely continuous with respect to μ if and only if there is a measurable function $d\nu(x)/d\mu$ such that

$$\nu(Q) = \int_Q \frac{d\nu(x)}{d\mu}\, d\mu(x)$$

for any measurable set Q, in which case

$$\int_Q g(x)\,d\nu(x) = \int_Q g(x)\,\frac{d\nu(x)}{d\mu}\, d\mu(x)$$

for any g integrable over Q with respect to ν.

Consider the one-dimensional Hilbert space $L_2(\mathcal{R})$ and the complex measure

$$\mu_{f,g}(S) = \int_S f^*(x)g(x)\,dx$$

which is absolutely continuous with respect to Lebesgue measure. Let X be the position operator and take $f \in \mathcal{H}$ and $g \in \mathcal{D}_X$. Since

$$\begin{aligned}\langle f|Xg\rangle &= \int f^*(x)\,xg(x)\,dx \\ &= \int x\,\frac{d\mu_{f,g}}{dx}\,dx = \int x\,d\mu_{f,g}(x)\end{aligned}$$

the spectral measure is obtained for X by observing that

$$\mu_{f,g}(S) = \langle f|E^X(S)g\rangle$$

where $E^X(S)$ is the projection operator defined by

$$(E^X(S)\psi)(x) = \psi_S(x)\psi(x) \quad \text{for every } \psi \in \mathcal{H}.$$

Finally, we note that the momentum operator P defined by

$$(P\psi)(x) = -i\hbar d\psi(x)/dx$$

where

$$\mathcal{D}_P = \{\psi \in L_2(\mathcal{R}) \mid d\psi/dx \text{ exists for all } x \in \mathcal{R}\}$$

is symmetric but not self-adjoint. The domain of definition of P, however, may be expanded to define a self-adjoint extension of P. Moreover, in many problems in quantum mechanics the potential energy function has a smoothing effect so that P in the Hamiltonian may be treated as if it were self-adjoint.

Index